AF803146

HIGH-VALUE PLANTS

Novel Insights and Biotechnological Advances

HIGH-VALUE PLANTS

Novel Insights and Biotechnological Advances

Edited by

Thangjam Anand Singh, PhD

Prakash Kumar Sarangi, PhD

Ch. Brajakishore Singh, PhD

AAP | APPLE ACADEMIC PRESS

First edition published 2025

Apple Academic Press Inc.
1265 Goldenrod Circle, NE,
Palm Bay, FL 32905 USA

760 Laurentian Drive, Unit 19,
Burlington, ON L7N 0A4, CANADA

CRC Press
2385 NW Executive Center Drive,
Suite 320, Boca Raton FL 33431

4 Park Square, Milton Park,
Abingdon, Oxon, OX14 4RN UK

© 2025 by Apple Academic Press, Inc.

Apple Academic Press exclusively co-publishes with CRC Press, an imprint of Taylor & Francis Group, LLC

Reasonable efforts have been made to publish reliable data and information, but the authors, editors, and publisher cannot assume responsibility for the validity of all materials or the consequences of their use. The authors are solely responsible for all the chapter content, figures, tables, data etc. provided by them. The authors, editors, and publishers have attempted to trace the copyright holders of all material reproduced in this publication and apologize to copyright holders if permission to publish in this form has not been obtained. If any copyright material has not been acknowledged, please write and let us know so we may rectify in any future reprint.

Except as permitted under U.S. Copyright Law, no part of this book may be reprinted, reproduced, transmitted, or utilized in any form by any electronic, mechanical, or other means, now known or hereafter invented, including photocopying, microfilming, and recording, or in any information storage or retrieval system, without written permission from the publishers.

For permission to photocopy or use material electronically from this work, access www.copyright.com or contact the Copyright Clearance Center, Inc. (CCC), 222 Rosewood Drive, Danvers, MA 01923, 978-750-8400. For works that are not available on CCC please contact mpkbookspermissions@tandf.co.uk

Trademark notice: Product or corporate names may be trademarks or registered trademarks and are used only for identification and explanation without intent to infringe.

Library and Archives Canada Cataloguing in Publication

Title: High-value plants : novel insights and biotechnological advances / edited by Thangjam Anand Singh, PhD, Prakash Kumar Sarangi, PhD, Ch. Brajakishore Singh, PhD.
Names: Singh, Thangjam Anand, editor. | Sarangi, Prakash Kumar, editor. | Singh, Ch. Brajakishore (Chingakham Brajakishor), editor.
Description: First edition. | Includes bibliographical references and index.
Identifiers: Canadiana (print) 20240400690 | Canadiana (ebook) 20240400755 | ISBN 9781774914748 (hardcover) | ISBN 9781774914755 (softcover) | ISBN 9781003457572 (ebook)
Subjects: LCSH: Specialty crops—Economic aspects—India. | LCSH: Cash crops—India. | LCSH: Agricultural innovations—India. | LCSH: Food industry and trade—India.
Classification: LCC SB87.I42 H54 2025 | DDC 630—dc23

Library of Congress Cataloging-in-Publication Data

CIP data on file with US Library of Congress

ISBN: 978-1-77491-474-8 (hbk)
ISBN: 978-1-77491-475-5 (pbk)
ISBN: 978-1-00345-757-2 (ebk)

About the Editors

Thangjam Anand Singh, PhD
Scientist, Directorate of Research at Central Agricultural University, Imphal, India

Thangjam Anand Singh, PhD, is a Scientist working with the Directorate of Research at Central Agricultural University in Imphal, India. He earned his PhD degree in the broad area of therapeutic enzymes associated with fermented foods of Northeast India from the Biotechnology Department, Gauhati University, Guwahati, India, although the major bulk of his PhD research was carried out at the Institute of Bioresources and Sustainable Development (IBSD), Imphal, India. During his research career, he has worked with many renowned institutes in the Northeastern Hill Region (NEH) of India, including the North Eastern Regional Institute of Science and Technology, Nirjuli, Arunachal Pradesh; Indian Council of Agricultural Research, NEH region of India at Barapani, Meghalaya; and IBSD. Dr. Singh has firsthand knowledge of the region from his many field trips, reaching nooks and corners of this remote area. Dr. Singh has been working on traditional ethnic foods of the region. His research is focused on ITK, fermented foods, traditional food processing, ethnobiology, food microbiology, nutraceutical and functional aspects of ethnic food, and postharvest engineering and technology. He has more than 15 years of teaching and research experience in medicinal and aromatic plant, microbial biotechnology, downstream processing, food biochemistry, food quality, and molecular biology. He also published more than 35 research articles in peer-reviewed international and national journals, authored several book chapters, and reviewed articles for many international journals. He is a member of the Association of Microbiologists of India and Indian Science Congress Association.

Prakash Kumar Sarangi, PhD
Scientist in Microbiology, Central Agricultural University, Imphal, India

Prakash Kumar Sarangi, PhD, is a scientist with specialization in microbiology at the Central Agricultural University in Imphal, India. He has a PhD degree in Microbial Biotechnology from Ravenshaw University, India, and a MTech degree from the Department of Agriculture and Food Engineering, Indian Institute of Technology Kharagpur, India. Dr. Sarangi's research is focused on bioprocess engineering, biochemicals, biomaterials, and sustainable development. His expertise is in bioconversion of biomass into biofuels, biochemicals, and nutraceuticals, as well as environmental waste remediation. He has taken leading roles as the principal investigator in different R&D projects on biomass conversion and microbial biodegradation into value-added products. He has more than 12 years of teaching and research experience in biochemical engineering, microbial biotechnology, downstream processing, food microbiology, and molecular biology. He is serving as editorial board member for many international journals, including *PLOS One* and *Biofuels, Bioproducts and Biorefining*. He is serving as a guest editor for several journals. He has published more than 60 research articles in peer-reviewed journals and authored more than 25 book chapters. He has presented at many national and international conferences. He is associated with many scientific societies as a fellow member (Society for Applied Biotechnology) and a life member (Biotech Research Society of India; Society for Biotechnologists of India; Association of Microbiologists of India; Indian Science Congress Association; Forum of Scientists, Engineers & Technologists; and International Association of Academicians and Researchers). He has published many books on biomass conversion and value addition.

Ch. Brajakishore Singh, PhD
Scientist, Institute of Bioresources and Sustainable Development, Imphal, India

Ch. Brajakishor Singh, PhD, is currently working as a Scientist at the Institute of Bioresources and Sustainable Development, Imphal, India. He has supervised several PhD and postdoctoral students. He has published more than 90 articles in peer-reviewed journals and filed one patent. His research areas of interest are medicinal plants, natural product chemistry, and synthesis of small heterocyclic molecules. He worked as a Research Associate at the Manipur University Chemistry Department and the Chemistry Department of the Indian Institute of Technology, Assam, India. He has worked at the National Center for Natural Research at the University of Mississippi, USA, as a visiting fellow. He also visited Tokushima Bunri University, Japan, for two months. Dr. Singh received his PhD and master's degree from Manipur University, India.

Contents

Contributors .... *xi*

Abbreviations .... *xiii*

Preface .... *xv*

1. ***Kaempferia parviflora*: A Black Ginger with Promising Ornamental and Medicinal Value** **1**
Lhaineichong Khongsai and Chingakham Brajakishor Singh

2. ***Elaeagnus latifolia* L. (Elaeagnaceae): A Promising and Emerging High-Value Bioresource with Nutraceutical and Pharmaceutical Prospects** **23**
Evanylla Kharlyngdoh, Longing Basuk Langstieh, Jitendra Kumar Shukla, Pardeep Kumar Bhardwaj, and Pulok Kumar Mukherjee

3. ***Swertia chirayita* A Critically Endangered High-Value Medicinal Plant of the Himalayan Region** **51**
Jitendra Kumar Shukla, Bharti Kashyap, Kishor Basor, Preeti Dhakal, Evanylla Kharlyngdoh, Pardeep Kumar Bhardwaj, and Pulok Kumar Mukherjee

4. **Genus *Garcinia* in Northeast India: Folklore Applications, Scope, and Opportunities** **69**
J. C. Bora and D. D. Joshi

5. ***Parkia timoriana* (DC.) Merr.: An Important Economic Plant** **99**
Maibam Beebina Chanu and Ch. Brajakishor Singh

6. **King Chili: A Future Potential Source of Pharmacological Applications** **117**
Wahengbam Kabita Chanu and Chingakham Brajakishor Singh

7. **Chenopodium: Future Smart Crop from Indian Himalaya** **139**
Rajkumari Jashmi Devi, Nikhil Kumar Chrungoo, and Biseshwori Thongam

8. **Unearthing of Sea Buckthorn (*Hippophae* spp) in India** **167**
Thangjam Anand Singh, Prakash K Sarangi, and Ng. Joykumar Singh

9. **Perspectives and Potential of Indian Organic Basmati Rice** **183**
Ankur Saxena and Ashutosh Singh

10. High-Value Plant Resources in Riparian Areas 205
Poppy Rajbongshi and Tapati Das

11. Processing and Potential of Broom Grass (*Thysanolaena latifolia*)........229
Achintya Kumar Sinha, Mihir Kumar Reang, and Sourabh Deb

12. Lichens in the Himalayan Region: Present and Future Perspectives 243
Awadhesh Kumar, Nurpen Meitei Thangjam, TBC Laldingliani, and
Ch Brajakishor Singh

Index..263

Contributors

Kishor Basor
Institute of Bioresources and Sustainable Development, Sub-centre: Meghalaya, Upper Shillong, Meghalaya, India

Pardeep Kumar Bhardwaj
Institute of Bioresources and Sustainable Development, Meghalaya Node, Upper Shillong, Meghalaya, India
Institute of Bioresources and Sustainable Development, Takyelpat, Imphal, Manipur, India

J. C. Bora
Institute of Advanced Studies in Science and Technology, Guwahati, Assam, India

Maibam Beebina Chanu
Plant Bioresource Division, Institute of Bioresources and Sustainable Development, Takyelpat, Imphal West, Manipur, India

Wahengbam Kabita Chanu
Institute of Bioresources and Sustainable Development, Takyelpat, Imphal West, Manipur, India

Nikhil Kumar Chrungoo
Department of Botany, North-Eastern Hill University, Shillong, Meghalaya, India

Tapati Das
Department of Ecology and Environmental Science, Assam University, Silchar, India

Sourabh Deb
Department of Forestry and Biodiversity, Tripura University, Agartala, India

Rajkumari Jashmi Devi
Plant Systematics and Conservation Laboratory, Institute of Bioresources and Sustainable Development, Takyelpat, Imphal, Manipur, India

Preeti Dhakal
Agricultural Technology Management Agency (ATMA), Mangan, North Sikkim, Sikkim, India

D. D. Joshi
Amity Herbal Consortium, Amity University Uttar Pradesh, Noida, Uttar Pradesh, India

Bharti Kashyap
Institute of Bioresources and Sustainable Development, Sub-centre: Meghalaya, Upper Shillong, Meghalaya, India

Evanylla Kharlyngdoh
Institute of Bioresources and Sustainable Development, Meghalaya Node, Upper Shillong, Meghalaya, India

Lhaineichong Khongsai
Institute of Bioresources and Sustainable Development-Imphal Centre, Takyelpat, Imphal West, Manipur, India

Awadhesh Kumar
Department of Horticulture Aromatic and Medicinal Plants, School of Earth Sciences and Natural Resource Management, Mizoram University, Aizawl, Mizoram, India

TBC Laldingliani
Department of Horticulture Aromatic and Medicinal Plants, School of Earth Sciences and Natural Resource Management, Mizoram University, Aizawl, Mizoram, India

Longing Basuk Langstieh
Institute of Bioresources and Sustainable Development, Meghalaya Node, Upper Shillong, East Khasi Hills District, Meghalaya, India

Pulok Kumar Mukherjee
Institute of Bioresources and Sustainable Development, Takyelpat, Imphal, Manipur, India

Poppy Rajbongshi
Department of Ecology and Environmental Science, Assam University, Silchar, India

Mihir Kumar Reang
Department of Forestry and Biodiversity, Tripura University, Agartala, India

Prakash K. Sarangi
College of Agriculture, Iroisemba, Central Agricultural University, Imphal, Manipur, India

Ankur Saxena
College of Agribusiness Management, G. B. Pant University of Agriculture and Technology, Pantnagar, Uttarakhand, India

Jitendra Kumar Shukla
Institute of Bioresources and Sustainable Development, Meghalaya Node, Upper Shillong, East Khasi Hills District, Meghalaya, India

Ashutosh Singh
College of Agribusiness Management, G. B. Pant University of Agriculture and Technology, Pantnagar, Uttarakhand, India

Chingakham Brajakishor Singh
Plant Bioresource Division, Institute of Bioresources and Sustainable Development-Imphal Centre, Takyelpat, Imphal West, Manipur, India

Ng. Joykumar Singh
College of Agriculture, Iroisemba, Central Agricultural University, Imphal, Manipur, India

Thangjam Anand Singh
College of Agriculture, Iroisemba, Central Agricultural University, Imphal, Manipur, India

Achintya Kumar Sinha
Gorkha Basti, Kunjaban, Agartala, Tripura, India

Nurpen Meitei Thangjam
Department of Horticulture Aromatic and Medicinal Plants, School of Earth Sciences and Natural Resource Management, Mizoram University, Aizawl, Mizoram, India

Biseshwori Thongam
Plant Systematics and Conservation Laboratory, Institute of Bioresources and Sustainable Development, Takyelpat, Imphal, Manipur, India

Abbreviations

AD	Alzheimer's disease
AFLP	amplified fragment-length polymorphism
BAT	brown adipose tissue
BMC	biodiversity management committee
CART	classification and regression trees
CGIAR	Consultative Group on International Agriculture
cGMP	cyclic guanosine monophosphate
CGRP	calcitonin gene-related peptide
DAMDA	directed amplification of minisatellite DNA
DASP	diversified agriculture support project
DBM	diamondback moth
DRC	domestic resource cost
EC	executive committee
EMR	effective multiplex ratio
ERS	endoplasmic reticulum stress
FDA	Forest Development Agency
FSH	follicle-stimulating hormone
GB	general body
GI	geographical indication
HCA	hydroxycitric acid
ICD	inland container depot
ICM	integrated crop management
iNOS	inducible nitric oxide synthase
IPGRI	International Plant Genetic Resources Institute
IPNI	International Plant Names Index
ITS	internal transcribed spacer region
JFMC	Joint Forest Management Committees
LAMPS	large-sized multipurpose society
LDL	low-density lipoprotein
LH	luteinizing hormone
LSC	large single copy
MIA	monoiodoacetic acid
MMPs	matrix metalloproteinases
MOMA	Manipur Organic Mission Agency
MTP	mitochondrial transmembrane potential
NCDF	NTFP conservation and development fees
NO	nitric oxide

NPC	nominal protection coefficient
NTFP	non-timber forest produce
OA	osteoarthritis
OBEP	Organic Basmati Export Program
PACS	primary agricultural credit society
PDR	pleiotropic drug resistance
PGS	participatory guarantee system
PIC	polymorphic information content
PMF	polymethoxy flavonoid-rich fraction
PPAR	peroxisome proliferator-activated receptor
PPF	polymethoxy flavonoid-poor fraction
QA	quinolinic acid
RAPD	random amplified polymorphic DNA
RCA	revealed comparative advantage
RSCA	revealed symmetric comparative advantage
SB	sea buckthorn
SD	Sprague-Dawley
SHG	self-help groups
SHU	Scoville heat unit
SSC	small single copy
SYK	spleen tyrosine kinase
TCM	traditional system of Chinese medicine
TEM	transmission electron microscope
TRPC	Tripura Rehabilitation Plantation Corporation Limited
TRPV1	transient receptor potential vanilloid subfamily member 1
TSM	traditional system of medicines
TSOD	Tsumura Suzuki obese diabetes
UOCB	Uttarakhand Organic Commodity Board
UPGMA	unweighted pair group method using arithmetic averages
VPA	valproic acid
VR1	vanilloid receptor 1
XRD	X-ray diffractometer

Preface

High-value crops are non-staple plants that produce much higher return per unit area. Many staple agricultural products do not give enough or encouraging returns to the farmers that make traditional agriculture an unattractive option to the youth. However, on the other hand, the economy of our country depends on the agricultural sector. Identification and investing on high-value plants could enhance the income of prospective farmers making it more attractive as a livelihood or profession.

The demand for high-value food and other products is on the increase, day by day. The agriculture sector is tremendously dominated by small farmers, and it is very important to transform farm economy to diversify into high-value crops for enhancing farmers' income in spite of its own set of risks involved. Many high-value crops are either cultivated or naturally harvested from the natural forest. With the declining share of traditional agricultural commodities in production, consumption and trade, horticultural and other nontraditional high-value agricultural crops present themselves as a major scope for higher income in farming communities.

The high-value agriculture strategy also provides ample opportunities for small scale agricultural farmers to commercialize their product in the global market. Despite its enormous potential, the trade of high-value agricultural products is very small, though increasing steadily. This book aims to explore biotechnological applications for the unique high-value crops toward impending high-value agriculture revolution which will make agriculture an attractive and challenging venture to the youth.

This book discusses the traditional knowledge, nutrition, and photochemical, value addition, and postharvest management of some select unique high-value crops. The editors are grateful to all the authors for contributing their valuable chapters to this book. The editors are also grateful to Ashish Kumar and Sandy Jones Sickels from Apple Academic Press, Inc., for their kind support and coordination during the entire project.

Dr. Th. Anand Singh
Scientist, Central Agricultural University, Imphal, India

Dr. Prakash Kumar Sarangi
Scientist, Central Agricultural University, Imphal, India

Dr. Ch. Brajakishore Singh
Scientist, IBSD, Imphal, Manipur, India

CHAPTER 1

Kaempferia parviflora: A Black Ginger with Promising Ornamental and Medicinal Value

LHAINEICHONG KHONGSAI and CHINGAKHAM BRAJAKISHOR SINGH

Institute of Bioresources and Sustainable Development-Imphal Centre, Takyelpat, Imphal West, Manipur, India

ABSTRACT

Kaempferia parviflora Wall. Ex Baker is a rhizomatous, herbaceous plant belonging to the family *Zingiberaceae* and is native to Thailand. The rhizomes of *K. parviflora*, known as black ginger, is popular as health-promoting herbs. Traditionally, it is used as a folk medicine for managing multiplex diseases including colic disorder, inflammation, ulcers, abscesses, allergy, gout, and osteoarthritis. Pharmacological researches on black ginger have also claimed for its potential benefits on treating numerous diseases. Besides medicinal value, *K. parviflora* makes a suitable potted ornamental plant, indoor plant, and also a perfect addition to herbal or edible gardens. It can also be used as ground covers in urban landscaping. It has a promising potential to be reintroduced as an ornamental plant with medicinal value and developed into a variety of commercial products. We have currently several studies to improve availability of the plant and affirm its medicinal value. Various pharmacological benefits includes antimicrobial, anti-inflammatory, antioxidative activity, antispasmodic, hepatoprotective, cardiogenic, vasorelaxation, improvement of blood fluidity, anti-gastric ulcer activity, antioxidant, antixanthine, anti-allergic, cellular metabolism-regulating activity, anticancer activity, enhance sexual activity, neuroprotective activity, antiosteoarthritis activity, and inducing transdermal permeable activity.

High-Value Plants: Novel Insights and Biotechnological Advances. Thangjam Anand Singh, Prakash Kumar Sarangi, and Ch. Brajakishore Singh (Eds.)
© 2025 Apple Academic Press, Inc. Co-published with CRC Press (Taylor & Francis)

However, there are limited clinical trials on the aforementioned activities. So, researchers and scientists are on continuous investigation on it. Therefore, future studies with regards to both safety and efficacy outcomes are needed on black ginger.

1.1 INTRODUCTION

The *Zingiberaceae* family is the largest of the Zingiberales order's families. The *Zingiberaceae* family can be found all across the tropics, especially in Southeast Asia. In the South-Eastern region, a number of Zinger species are utilized as spices, traditional remedies, flavoring agents, and as a source of various colors (Tewtrakul and Subhadhiasakul, 2007). *Kaempferia parviflora*, often known as black ginger, is a *Zingiberaceae* family medicinal plant. Kaempferia species can be found all over the world, from India to Southeast Asia, with Thailand having the most biodiversity, with over 20 native species (Sirirugsa, 1992; Larsen, 2006). In Thailand, the rhizomes of this plant are used as seasoning agent in local food and as a traditional medicine for treating a variety of ailments (Sutthanut et al., 2007; Konkumnerd et al., 2010). Other traditional applications of *K. parviflora* include the treatment of colic, peptic, and duodenal ulcers, and lowering blood pressure, improve blood circulation, increase vitality, and also as an aphrodisiac (Yenjai et al., 2004; Akase et al., 2011). *K. parviflora* possesses a considerable potential for numerous health benefits, which sparked a slew of pharmacological studies on these benefits (Chivapat et al., 2010), such as studies on its anti-inflammatory (Tewtrakul et al., 2008), antipyretic, antimycobacterial, antifungal, antiplasmodial (Yenjai et al., 2004), antimalarial, anti-HIV-1 protease activity (Sookkongwaree et al., 2006), cytotoxic activity, artery dilation effects, and smooth muscle relaxation properties (Wattanapitayakul et al., 2008). Secondary metabolites of flavonoids, such as 5-hydroxy-7-methoxyflavonone, 5,7-dimethoxyflavone, and 3,5,7-trimethoxyflavone, have been found as responsible for a wide spectrum of biological functions (Jaipetch et al., 1983; Sutthanut et al., 2007; Leardkamolkarn et al., 2009). Recent research has also found that the primary bioactive flavonoids of *K. parviflora*—5,7 dimethoxyflavone, 3,5,7,3′,4′-pentamethoxyflavone and 5,7,4′-trimethoxyflavone—have chemopreventive activities in a range of human cell lines (Yenjai and Wanich, 2010; Leardkamolkarn et al., 2009). The rhizomes contain volatile oil, chalcones, and phenolic compounds (Herunsalee, 1987: Wongsinkongman et al., 2003; Azuma et al., 2008).

Several serious issues in medicinal plant production such as quality control of active ingredients, safety and environmental conservation, environmental factors, biotic and abiotic contaminants, and insects can all affect the concentration of secondary metabolites in field-grown plants, which potentially fluctuates the concentration of active constituents thereby leading toward risk for consumers (Mosaleeyanon et al., 2005). The accumulation level of a special metabolite in plant tissues and chemical profile of plants can be influenced by a variety of environmental conditions such as temperature, quality of light, and light intensity. In this regard, determining the optimal light intensity for chemical accumulation as well as plant growth and development is a key topic in getting larger phytochemical concentrations.

Studies on human cell lines have shown that the primary bioactive flavonoids of this plant have anti-cancer properties (Yenjai and Wanich, 2010; Leardkamolkarn et al., 2009).

1.2 TAXONOMIC STUDY

Taxonomic classification:

Kingdom: Plantae
Phyllum: Tracheophyta
Class: Lilopsida
Order: Zingiberales
Family: *Zingibereceae*
Genus: Kaempferia L.
Species: *Kaempferia parviflora* Wall. Ex Baker

Synonyms

Kaempferia rubromarginata (S. Q. Tong) R. J. Searle, *Stahlianthus rubromarginatus* S. Q. Tong [The plant List, 2013]

Vernacular names:

English:	Black ginger, black galingale, black ginseng (Asiatische et al., 2008)
India:	Kalahood (Quattrocchi, 2012)
Thailand:	Krachai dam (Asiatische et al., 2008)
Myanmar:	Gamin-ni (Kress et al., 2003)
Manipur:	Yaimu

1.3 BOTANICAL DESCRIPTION

The *Zingiberaceae* family includes *K. parviflora*. It's a rhizomatous perennial herb with no stems that can grow up to 20 cm tall. The rhizomes are tiny, heavily branching, and dark purple to black in color, with slender root fibres. The leaves are 8–16 cm long, thin, and plain green at the base. The petiole is channeled and short. The flowers are few in a sessile central tuft, white and purple tinged with darker in the middle of the lip; has eight flowers, blooms only one flower flowered per day; scent is sweet; lanceolate bracts, greenish slightly, 1.4–1.7 × 3–3.5 cm; calyx is white with 2.4 × 0.7 cm, 1.5 cm white linear corolla lobes; linear lateral staminodes, 1.1 × 0.3 cm; lateral staminodes linear, 1.1. April to September are the best months for flowering and fruiting [Hooker JD].

FIGURE 1.1 *K. parviflora's plant*

FIGURE 1.2 Flowering *K. parviflora's*

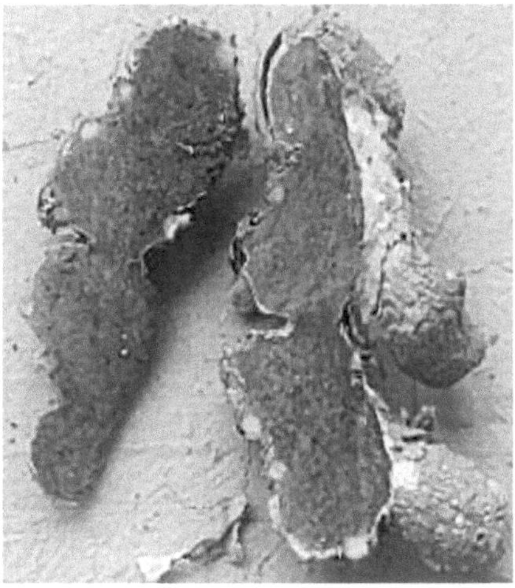

FIGURE 1.3 Cut Rhizome with rhizome plant

1.4 GEOGRAPHICAL DISTRIBUTION

The ginger family, *Zingiberaceae*, includes 52 genera and around 1300 plant species. Kaempferia is found in 60 different species throughout Southeast Asia, including Indonesia, India, Malaysia, Myanmar, Cambodia, and China, as well as Thailand (Sirirugsa, 1992; Larsen and Saksuwan Larsen, 2006).

1.5 PHYTOCHEMICAL CONSTITUENTS

K. parviflora has been reported to contain 1α,3α,4β-trihydroxy-9-cadinen-8-one, 4α-acetoxycadina-2,9-diene-1,8-dione; 5,7-dimethoxyflavone, 5,7,4′-trimethoxyflavone, 5,7,3′,4′-tetramethoxyflavone, 3,5,7,4′-tetramethoxy-flavone, 3,5,7,3′,4′-pentamethoxyflavone, 5-hydroxy-7-methoxyflavone, 5-hydroxy-3,7-dimethoxyflavone, rel-(5aS,10bS)-5a,10b-dihydro-1,3,5a,9-tetrahydroxy-8-methoxy-6H-benz[b]indeno[1,2-d]furan-6-one 5a-O-[α-L-rhamnopyranosyl-(1-->6)-β-d-glucopyranoside], rel-5aS,10bR isomer; (2R,3S,4S)-3-O-[α-L-rhamnopyranosyl-(1-->6)-β-d-glucopyranosyl]-3′-O-methyl-ent-epicatechin-(2α-->O-->3,4α-->4)-(5aS,10bS)-5a,10b-dihydro-1,3,5a,9-tetrahydroxy-8-methoxy-6H-benz[b]indeno[1,2-d]furan-6-one 5a-O-[α-L-rhamnopyranosyl-(1-->6)-β-D-glucopyranoside], flav-3-en-3-ol glycosides, and kaempferiaosides A-F (Yenjai et al., 2004; Sutthanut et al., 2007; Azuma et al., 2008; Sadhu et al., 2009; Mekjaruskul, 2013; Sookkong-waree et al., 2006; Chaipech et al., 2011, 2012).

1.6 TRADITIONAL USES

Several gingers from the *Zingiberaceae* family, particularly the Kaempferia genus, are used in Malaysian folk medicine to treat stomach problems, vomiting, cough, sore throat, bruises, epilepsy, nausea, rheumatism, wounds, eyewash, childbirth, liver problems, muscular pains, ringworm, malignancies, asthma, fever, swelling, and a variety of other ailments (Vimala et al., 1998). Black ginger has traditionally been used as a male tonic and aphrodisiac to promote male sexual performance. Around Southeast Asia, it's known as "Thai Viagra." Other traditional uses of *K. parviflora* in Thailand and Southeast Asia include—gastric ulcer treatment; increasing vitality and improving physical performance; anti-malarial medication; diabetes; and weight loss The dried rhizome is commonly used to make tea bags, while the fresh rhizome is used to make brew wine. Wine preparation is becoming more popular in Thailand as a tonic and aphrodisiac (Saokaew et al., 2017; Toda et al., 2016). As nutritional supplements, several preparations such as medicinal liquor or liquor plus honey, pills (powdered rhizome with honey), capsules, and tablets have been developed. For ages, the rhizome has been widely utilized as a traditional medicine in Thailand's north and northeast (Mekjaruskul et al., 2012). Traditional remedies in Laos and Thailand used the plant rhizomes to treat inflammation, hypertension, erectile dysfunction, abdominal disorders (Wutythamawech, 1997; Yenjai et al., 2004), and to promote vitality and blood flow (Wutythamawech, 1997; Yenjai et al., 2004; Akase et al., 2011). Folkloric medicine in Japan has proven a favorable effect of this plant extract as a food supplement and for the treatment of metabolic diseases. Gout, aphthous ulcer, peptic ulcer, allergies, osteoarthritis, and abscesses have all been treated with the rhizome of black ginger, which is popular as a health-promoting plant (Sae-Wong et al., 2009). In Thailand, the rhizome had been promoted for health and treatment of digestive disorders and gastric ulcer (Kobayashi et al., 2015).

1.7 PHARMACOLOGICAL ACTIVITY

1.7.1 ANTIPLASMODIAL ACTIVITY

The antiplasmodial activity of *K. Parviflora* has been documented by Yenjai C et al. There were nine flavonoids discovered. The flavonoids 5,7,49-trimethoxyflavone(8) and 5,7,49-trimethoxyflavone(8) had antiplasmodial, antifungal, and antimycobacterial activity. These extracted flavonoids had no cytotoxicity when tested against the KB, BC, and NCI-H187 cell lines, implying that the rhizomes of *K. Parviflora* may be safe to use as a traditional medicine ingredient (Yenjai et al., 2004).

K. parviflora rhizome flavonoids (5,7,4′-trimethoxyflavone and 5,7,3′,4′-tetramethoxyflavone) had antiplasmodial activity against Plasmodium falciparum giving a significant IC50 values of 3.70 and 4.06 g/mL, respectively. The flavones 3,5,7,4′-tetramethoxyflavone and 5,7,4′-trimrthoxyflavone are also active against Candida albicans and Mycobacterium sp.. Antiplasmodial activity was assessed using the microdilution radioisotope technique (Trager and Jensen, 1976; Desjardins et al., 1979), whereas antifungal activity was assessed using the modified formazan assay (Trager and Jensen, 1976; Desjardins et al., 1979; Scudiero et al., 1988). Micro Alamar Blue assay technique was used to test antimycobacterial activity (Collins and Franzblau, 1997). The cytotoxicity of the KB (oral human epidermoid carcinoma), BC (breast cancer), and NCI-H187 (human, small cell lung cancer) cell lines was investigated using a colorimetric technique (Skehan et al., 1990).

1.7.2 ANTIVIRAL ACTIVITY

According to Sookkongwaree et al. (2006), flavonoids of *K. parviflora* were found to have antiviral activity. With IC50 values of 19 M, two flavones (5-hydroxy-7-methoxyflavone and 5,7-dimethoxyflavone) showed antiviral efficacy against HIV-1 protease. HCV protease, and HCMV proteases were both inhibited by 5-hydroxy-3,7-dimethoxyflavone, which had IC50s of 190 and 250 M, respectively. These findings demonstrate black ginger's antiviral activity.

1.7.3 ANTI-INFLAMMATORY ACTIVITY

The crude ethanol and chloroform rhizome extracts generated a number of anti-inflammatory methoxyflavonoid compounds (5,5-hydroxy-3,7,3′,4′-tetramethoxyflavone; 5,7-dimethoxyflavone; trimethylapigenin; and tetramethylluteolin). According to the findings, ethanolic and 5,5-hydroxy-3,7,3′,4′-tetramethoxyflavone lowered inducible nitric oxide synthase (iNOS) mRNA expression in a dose-dependent manner, but only minimally altered cyclooxygenase-2 (COX-2) mRNA expression. In RAW264.7 cells activated with lipopolysaccharide, some of the active substances inhibited the generation of nitric oxide (NO) (LPS). They mildly decreased TNF-alpha generation and strongly suppressed iNOS mRNA, and protein expression in a dose-dependent manner, but they had no effect on p-ERK or p-JNK protein expression. Trimethylapigenin had no effect on inhibitor of kB kinases or mitogen-activated protein kinases enzyme activity; however, it did reduce the activity of spleen tyrosine kinase (SYK).

According to the findings, the anti-inflammatory effect of *K. parviflora* rhizome is attributable to inhibition of iNOS mRNA expression, mRNA COX-2 expression, and inhibition of SYK, all of which may be important in suppressing LPS-induced signaling in macrophages (Sae-wong et al., 2009, 2011).

1.7.4 HEPATOPROTECTIVE ACTIVITY

The hepatoprotective activity of components extracted from *K. parviflora* rhizome was investigated by Chaipech et al. (2012). The impact was measured using primary-cultured mouse hepatocytes and the 3-(4,5-dimethylthiazol-2-yl)-2,5-diphenyltetrazolium bromide (MTT) colorimetric test (Yoshikawa et al., 2006; Ninomiya et al., 2007, 2008; Zhang et al., 2007; Nakamura et al., 2007, 2008; Morikawa, 2010). When evaluated on D-GalN-induced cytotoxicity in primary-cultured mouse hepatocytes, 5,3′-dihydroxy-3,7,4′-trimethoxyflavone demonstrated strong hepatoprotective action compared to the commercial drug silybin, with IC50 values of 18.4 and 38.8 M, respectively. The findings demonstrate that components extracted from the rhizome of *K. parviflora* have potent hepatoprotective properties (Chaipech et al., 2012).

1.7.5 CYTOCHROME P450 ENZYMES ACTIVITY

In vitro tests on the extract of *K. parviflora* impacted the activities of hepatic CYP450 metabolic enzymes such as CYP1A1, CYP1A2, CYP2B, and CYP2E1 by non-competitive, mixed-competitive, competitive, and uncompetitive processes, respectively, using non-induced mice hepatic microsomes. Furthermore, when fed orally to mice for 7, 14, and 21 days, *K. parviflora* (250 mg/kg) significantly increased CYP1A1, CYP1A2, CYP2B, and CYP2E1 enzyme activity in distinct ways. The extract had no effect on the activity of the CYP3A enzyme. As a result, combining the extract with other plants or medications revealed possible drug–herb interactions (Mekjaruskul et al., 2012).

1.7.6 CARDIOGENIC ACTIVITY

The report on cardiogenic activity of *K. parviflora's* rhizome extract was given by Weerateerangkul et al. (2012). The study was conducted using a rat model. The findings demonstrated a beneficial effect on the nitric oxide (NO) signaling pathway, resulting in decreased defibrillation efficacy. In the used rat models, the extract increased cyclic guanosine monophosphate (cGMP) levels and decreased cardiac function and calcium (Ca) transient in ventricular myocytes, which altered the physiology of the normal heart by attenuating cardiac function and lowering Ca transient. In another swine heart investigation, Weerateerangkul et al. (2013) found that a high *K. parviflora* extract concentration reduced defibrillation efficiency and increased cardiac sensitivity to arrhythmia in normal swine heart.

1.7.7 VASORELAXATION ACTIVITY

Vasorelaxation activity of *K. parviflora* rhizome extracts has been reported in several studies. When evaluated, an ethanolic extract of the rhizome demonstrated the capacity to encourage vasorelaxant effects, according to preliminary research. The elimination of endothelium, N(G)-nitro-l-arginine, and 1H-[1,2,4] oxadiazolol [4,3-a]quinocaline-1-one were reported to limit the vasorelaxant action. At the same time, the extract was discovered to suppress exogenous Ca2+-induced contractile response (Malakul et al., 2011).

In another study, two substances isolated from the *K. parviflora* rhizome—5,7-dimethoxyflavone (DMF) and 3, 5, 7, 3′, 4′-pentamethoxyflavone (PMF)—were found to have vasorelaxant effect when evaluated on isolated rat aortic rings and isolated human cavernosum, respectively. DMF produced rat aortic ring relaxation in a dose-dependent manner. The process is thought to be driven in part by the NO-cGMP and cycloxygenase pathways, and is primarily caused by an increase in K+ efflux and a decrease in Ca2+ influx from the extracellular space. In the meantime, PMF had only a minor effect on NO release and had no effect as a K(ATP)- or K(Ca) channel opener, phosphodiesterase inhibitor, stored-operated Ca2+ channel blocker, or Rho-kinase inhibitor. PMFs vasorelaxant effect in the human cavernosum is thought to be mediated through voltage-dependent Ca2+ channels and other calcium mobilization pathways (Tep-Areenan et al., 2010; Jansakul et al.)

1.7.8 ANTISPASMODIC ACTIVITY

According to a study, when precontracted with acetylcholine, Ethanol extract of *K. parviflora* 's rhizome has the potential to relax the phenylephrine-precontracted aortic ring and have antispasmodic effect in the ileum in a dose-dependent manner (Wattanapitayakul et al., 2007).

1.7.9 IMPROVEMENT OF BLOOD FLUIDITY

By reducing blood flow time via a micro slit by using a disseminated intravascular coagulation model, the active principle of *K. parviflora* rhizome extract (methoxyflavones) was found to be responsible for enhancing blood fluidity (Murata et al., 2012).

1.7.10 ANTIGASTRIC ULCER ACTIVITY

Ethanol extract of *K. parviflora* rhizome administered orally to indomethacin, HCl/EtOH, and water immersion-induced gastric ulcer formation showed gastroprotective activity partly due to accumulation of gastric mucus secretion and not related to hindrance of secretion of gastric acid (Rujjanawate et al., 2005).

1.7.11 ANTICHOLINESTERASE ACTIVITY

The extract of the *K. parviflora* rhizome exhibited significant acetycholinesterase and butrylcholinesterase inhibitory activity. Two methyl-flavones, namely 5,7,4′-trimethoxyflavone and 5,7-dimethocyflavone, were identified to have the highest potential inhibitory activity specific to each enzyme with the inhibitory activity percentage varying over 42–86%. They are specifically inhibited acetylcholinesterase and butrylcholinesterase, respectively (Sawasdee et al., 2009).

1.7.12 ANTIXANTHINE OXIDASE ACTIVITY

The extract of *K. parviflora* rhizomes demonstrated significant antixanthine oxidase activity, inhibiting xanthine oxidase by 38% at 500 g/mL (XOD).

3,5,7,4′,5′-pentamethoxyflavone and 3′,4′,5,7-tetramethoxyflavone, both obtained from *K. parviflora* rhizome extracts, demonstrated inhibitory action with IC50 values of 0.85 and >4 mM, respectively. These two compounds had a competitive/non-competitive mixed type of action (Nakao et al., 2011).

1.7.13 ANTIOXIDANT ACTIVITY

The extract of *K. parviflora* rhizomes demonstrated significant antixanthine oxidase activity, inhibiting xanthine oxidase by 38% at 500 g/mL (XOD). 3,5,7,4′,5′-pentamethoxyflavone and 3′,4′,5,7-tetramethoxyflavone, both obtained from *K. parviflora* rhizome extracts, demonstrated inhibitory action with IC50 values of 0.9 and >4 mM, respectively. These two compounds had a competitive/non-competitive mixed type of action (Nakao et al., 2011).

1.7.14 SEXUAL BEHAVIORAL ACTIVITY

In rats and mice, *K. parviflora* extract has been shown to boost desire and sexual performance. The size and weight of testes are unaffected by these changes. The effectiveness of the extract to improve circulation to the testes was attributed primarily to the large increase in serum testosterone seen in one trial. Because *K. parviflora* had no influence on the amount of follicle stimulating hormone (FSH), luteinizing hormone (LH), progesterone, or corticosterone in the blood of rats treated with the extracts, the testosterone-like effect does not disrupt the hypothalamic-pituitary-testicular axis. The 7-methoxyflavone group of chemicals derived from *K. parviflora* demonstrated inhibitory effect on phosphodiesterase type 6 (PDE6) and phosphodiesterase type 5 (PDE5) and implying its potential use in erectile dysfunction treatment. The most potent PDE5 inhibitor was found to be 5,7-dimethoxyflavone, with an IC50 value of 10.64 2.09 M, and the most potent PDE6 inhibitor was found to be 3,5,73′,4′-pentamethoxyflavone, with an IC50 value of 7.83 M, while their inhibitory selectivity (PDE6/PDE5) was found to be 3.71 and 0.16 M, respectively (Sudwan et al., 2006; Trisomboon et al., 2007; Chaturapanich et al., 2008, 2011; Temkitthawon, 2007).

1.7.15 ANTIALLERGIC ACTIVITY

In a test conducted in the Zingiberacea family, ethanol extract of *K. parviflora* showed the most powerful anti-allergic action against antigen-induced hexosaminidase release as a marker of degranulation in RBL-2H3 cells, with an IC50 value of 10.7 g/mL (Tewtrakul and Subhadhirasakul, 2007). Several methoxyflavones have been reported to have anti-allergic activity in additional investigations, with 5-hydroxy-3,7,3′,4′-tetramethoxyflavone being the most active, followed by 5-hydroxy-7-methoxyflavone and 5-hydroxy-7,4′-dimethoxyflavone. The activity is related to the listed chemicals' suppression of cell degranulation, which involves $Ca2+$ influx into the cells (Tewtrakul et al., 2008).

1.7.16 ANTITUMOR ACTIVITY

Various studies have reported the anti-tumor activity of K. parviflora rhizome. It have shown treatment potential for HL-60 cells, human promonocytic leukaemia U937, KB cell line, human cholangiocarcinoma (HUCCA-1 and RMCCA-1), and NCI-H187 cancer cell lines. The general features of the effects are decrease in cell growth and viability. Microscopic studies showed deformed cell morphology with thick nuclei and apoptotic bodies. The mitochondrial transmembrane potential (MTP) was reduced. There is also activation of the caspase-3 in these cells. It was found that methoxyflavones were responsible for these actions, and the most active being 5,7,4-trimethoxyflavone (Banjerdpongchai et al., 2008, 2009; Wanich et al., 2009; Leardkamolkarn, 2009; Yenjai et al., 2010).

1.7.17 ANTIMULTIDRUG RESISTANCE ACTIVITY

In LLC-GA5-COL150 and A549 cells, an ethanol extract of the *K. parviflora* rhizome increased the accumulation of rhodamine 123, daunorubicin, calcein, and P-gp substrates. Dose and timing are both factors that influence the outcome. The presence of methoxyflavones in the *K. parviflora* utilised is linked to the action. The most effective inhibitor of MRP (multiple-drug resistance proteins) function was 5,7-methoxyflavone. Furthermore, 3,5,7,3′,4′-pentamethoxyflavone and 5,7-dimethoxyflavone reduced doxorubicin resistance in A549 cells (Patanasethanont et al., 2007a, 2007b).

1.7.18 NEUROPROTECTIVE ACTIVITY

In vivo and in vitro studies have looked into the effects of black ginger extracts on neuropharmacological and neuroprotective actions. The ethanol extract of *K. parviflora* downregulates 14 proteins and upregulates 37 proteins expression in the hippocampus of Sprague-Dawley (SD) rats, according to two-dimensional gel electrophoresis (2Dgel), with the expressions of GFAP and DPYSL2 being associated with antioxidative and microtubule-forming activities, respectively. For validation, a Western-blot test was performed, and the results revealed that the gene was considerably elevated. Surprisingly, however, phosphorylated DPYSL2 expression was shown to be downregulated. Furthermore, the extract boosted norepinephrine (NE), serotonin (5-HT), and dopamine (DA) synthesis in the rat hippocampus (Plaingam et al., 2017). The extract of *K. parviflora* has been used to treat cognitive enhancement and psychiatric disorders. The precise mechanisms, on the other hand, are still being investigated. Beta-site amyloid precursor protein cleaving enzyme 1 (BACE1), a rate-limiting enzyme linked to aberrant expression of amyloid peptides, has been linked to the pathological development of Alzheimer's disease (AD). Some compounds have been reported to reduce BACE1 activity significantly without any impact on secretase or other serine proteases in a noncompetitive way (Youn et al., 2016). Using a forced swimming test, the ethanol extract of black ginger rhizome was shown to have antidepressant-like effects in rats. It can effectively guard against the loss of proliferating cells and valproic acid (VPA)-induced cognitive impairment. The extract upregulates the expression of doublecortin (DCX) in the hippocampus. These findings suggest that black ginger extract may restore spatial memory and cell proliferation in the subgranular zone, which has been affected by VPA (Welbat et al., 2016). *K. parviflora* extract consistently exhibits neuroprotective activity against learning and memory impairment caused by persistent stress restraint, as evidenced by the increased overall neuron density of the hippocampus (Wattanathorn et al., 2013).

1.7.19 ANTIOSTEOARTHRITIS ACTIVITY

Osteoarthritis (OA) is a degenerative disease that affects the joints. Nonsteroidal anti-inflammatories (NSAIDs) have been shown to be effective in improving joint function. In rat OA models caused by monoiodoacetic acid (MIA), black ginger extract has been demonstrated to increase the cartilage lesions severity. The extract of black ginger and its separate components were reported to reduce the matrix metalloproteinases (MMPs) expression, which breakdown articular cartilage's collagen, in IL-1-treated human chondrocytes (Kobayashi et al., 2018). The levels of IL-1, TNF, and PGE2 in rat OA knee synovial fluid were reported to be reduced by an isolated compound, which is also one of the primary methoxyflavones discovered in Murraya exotica.

Compound-3 suppresses the EP/cAMP/PKA signaling pathway and the catenin signaling pathway in cultured rat chondrocytes, protecting the articular cartilage and chondrocytes (Wu et al., 2014). Furthermore, as evidenced by downregulation of IRE1-JNK signaling, ATF6 signaling, PERK-CHOP signaling, and GSK-3 upregulation and expression of XBP1 and GRP78 expression, a compound has chondroprotective effects on attenuation of endoplasmic reticulum stress (ERS) induced by proinflammatory cytokines PGE2 (Yang et al., 2015) GSK-3 expression is increased by ERS, which causes ERS to become activated. The deletion or overexpression of GSK-3 in transfected chondrocytes is established to illustrate the primary mechanism of ERS in triggering chondrocyte death. GSK-3 can dramatically exacerbate chondrocyte apoptosis mediated by ERS (Yuan et al., 2017). IRE1 deficiency causes apoptosis in ERS-induced chondrocytes in the early stages of the disease. A compound has been found to upregulate the expressions of Bcl-2, XBP1s, and IRE1α, and reversely and downregulate the expressions of CHOP, p-JNK, and caspase-3 (Wu et al., 2018).

1.7.20 TRANSDERMAL PERMEABLE ACTIVITY

Most methoxyfavones from black ginger have physicochemical properties that comply with requirements of transdermal delivery, such as low logP, low molecular weight, and low melting point. An appropriate monolithic drug-in-adhesive patch containing major constituents from the ginger has been examined to avoid first pass metabolism and boost the limited bioavailability of black ginger. The unsaturated fatty acid oleic acid has the largest effect on methoxyfavone permeability. In rats, a formulation containing 15% KP and 3% each of oleic acid and menthol was found to increase the penetration activity of methoxyfavones. The Cmax of total methoxyfavones in vivo is 217.08 ng/mL, with a T_{max} of 8 h, according to a pharmacokinetic research (Tuntiyasawasdikul et al., 2015). Similarly, a solid lipid nanoparticle containing a dichloromethane extract of black ginger enhanced transdermal permeability (Sutthanut et al., 2009).

The ginger extract was observed to reduce the subcutaneous fat layer thickness and adipocyte infiltration into the dermis in TSOD mice. Furthermore, it was discovered that *K. parviflora* protects the skin from UVB-induced denaturation of collagenous fibres (Hidaka et al., 2017).

1.7.21 *ANTI-OBESITY ACTIVITY*

An obesity-induced dermatopathy system was used as an obesity model, with Tsumura Suzuki obese diabetes (TSOD) mice serving as the obesity model. Ethanolic extract, a polymethoxyflavonoid-rich fraction (PMF), and a polymethoxyflavonoid-poor fraction (PPF) from *K. parviflora* were examined (Hidaka, Horikawa, Akase, Makihara, Ogami, Tomozawa, Tsubata, Ibuki, and Matsumoto). The results showed that ethanolic extract reduced mouse body weight and subcutaneous fat layer thickness more than PMF fraction, which is used as a dietary supplement to treat obesity-related skin disorders (Hidaka et al., 2017).

1.8 CLINICAL STUDIES

After 1 and 2 months of treatment, individuals receiving 90 mg/day of *K. parviflora* extract showed a substantial improvement in all parameters with respect to response latency time to visual erotic stimuli, length, and size of penis both in flaccid and erectile stages compared to placebo. It stated that nitric oxide was involved in the consequences. *K. Parviflora* extract has been shown to boost protein expression and endothelial nitric oxide synthase in umbilical vein endothelial cells of human in experiments studied (Guohua et al., 2009). Abundance of endothelial nitric oxide synthase in sinusoidal endothelium and penile vascular endothelium within the corpora carvernosa may thus improve penile erection (Tajuddin Ahmad et al., 2003, 2004, 2005).

In knee osteoarthritis, a comparison of analgesic cream with *K parviflora* cream was discovered in a study for pain markers (Chalee, 2010). Although there was a considerable reduction in pain in all parameters after four weeks when compared to baseline, there was no meaningful difference when compared to analgesic cream.

Furthermore, the effect of *K. parviflora* on energy expenditure was linked to brown adipose tissue activity, a location of nonshivering thermogenesis (Yoshino et al., 2014).

The constituents of black ginger extract stimulate hormone-sensitive lipase in adipocytes, according to a prior study (Okabe et al., 2014). A study found that *K. parviflora* extract could possibly increase the expenditure of whole-body energy by activating brown adipose tissue, suggesting that it could be useful as an anti-obesity treatment (Matsushita et al., 2015).

1.9 ADVERSE EFFECTS OF K. PARVIFLORA

In terms of safety, all investigations on *K. parviflora* have found no negative consequences. An animal investigation on long-term toxicity of its extract was undertaken. Five groups of 120 Wistar rats were grouped, each with 24 rats (12 males and 12 females). The extract were given orally to three treatment groups at doses of 5, 50, and 500 mg/kg/day for six months, which were comparable to 1, 10, and 100 times that of human use, respectively, while two control groups were given 1.0% tragacanth and distilled water respectively. The histological examination of visceral organs indicated no notable abnormalities due to *K. parviflora* extract toxicity (Chivapat et al., 2010).

1.10 CONCLUSIONS

K. parviflora has been used in traditional medicine since time immemorable around the world. Numerous therapeutic drug leads have arisen from the extracted chemicals and components, as well as data on its biological activity. These plants will continue to be a promising source for therapeutic natural goods and food materials and industrial products, based on proven chemical and biological results. More clinical experiments on the sample, on the other hand, are needed to demonstrate its safety in use.

ACKNOWLEDGEMENTS

The authors are thankful to the Institute of Bioresources and Sustainable Development for the providence of every necessary support in writing this book chapter.

KEYWORDS

- **Kaempferia**
- **parviflora**
- **pharmacological**

REFERENCES

Akase, T.; Shimada, T.; Terabayashi, S.; Ikeya, Y.; Sanada, H.; Aburada, M. Antiobesity Effects of *Kaempferia parviflora* in Spontaneously Obese Type II Diabetic Mice. *Japanese So. Pharmacogn. Springer* **2010,** *65*, 73–80.

Asiatische Heilkräuter und Heilpflanzen. *Kaempferia parviflora*

Azuma, T.; Tanaka, Y.; Kikuzaki, H. Phenolic Glycosides from *Kaempferia parviflora. J. Phytochem.* **2008,** 69, 2743–2748.

Banjerdpongchai, R.; Chanwikruy, Y.; Rattanapanone, V.; Sripanidkulchai, B. Induction of Apoptosis in the Human Leukemic U937 Cell Line by *Kaempferia parviflora* Wall. ex. Baker Extract and Effects of Paclitaxel and Camptothecin. *Asian Pac. J. Cancer Prev.* **2009,** *10* (6), 1137–1140.

Banjerdpongchai, R.; Suwannachot, K.; Rattanapanone, V.; Sripanidkulchai, B. Ethanolic Rhizome Extract from *Kaempferia parviflora* Wall. ex. Baker Induces Apoptosis in HL-60 Cells. *Asian Pac. J. Cancer Prev.* **2008,** *9* (4), 595–600.

Chaipech, S.; Morikawa, T.; Ninomiya, K.; Yoshikawa, M.; Pongpiriyadacha, Y.; Hayakawa T, Muraoka O. Structures of Two New Phenolic Glycosides, Kaempferiaosides A and B, and Hepatoprotective Constituents from the Rhizomes of *Kaempferia parviflora. Chem. Pharm. Bull. Tokyo* **2012,** *60* (1), 62–69.

Chaipech, S.; Morikawa, T; Ninomiya, K.; Yoshikawa, M.; Pongpiriyadacha, Y.; Hayakawa, T.; Muraoka, O. New flav-3-en-3-ol Glycosides, Kaempferiaosides C and D, and Acetophenone Glycosides, Kaempferiaosides E and F, from the Rhizomes of *Kaempferia parviflora. J. Nat. Med.* **2012,** *66* (3), 486–492. DOI: 10.1007/s11418–011–0611–1. Epub 2011 Nov 20.

Chalee, S. *Study of Efficacy and Anti-Inflammatory Effect of K. parviflora Extract Cream in Osteoarthritis of Knee*; Khon Kaen University: Khon Kaen, Thailand, 2010.

Chaturapanich, G.; Chaiyakul, S.; Verawatnapakul, V.; Pholpramool, C. Effects of *Kaempferia parviflora* Extracts on Reproductive Parameters and Spermatic Blood Flow in Male Rats. *Reproduction* **2008,** *136* (4), 515–522.

Chaturapanich, G.; Chaiyakul, S.; Verawatnapakul, V.; Yimlamai, T.; Pholpramool C. Enhancement of Aphrodisiac Activity in Male Rats by Ethanol Extract of *Kaempferia parviflora* and Exercise Training. *Andrologia* **2012,** *44* (Suppl 1), 323–328.

Chivapat, S.; Chavalittumrong, P.; Attawish, A.; Rungsipipat, A. Chronic Toxicity Study of *Kaempferia parviflora* Wall ex. Extract. *Thai J. Vet. Med.* **2010,** *40*, 377–383.

Collins, L. A.; Frazblau, S. G. *Antimicrob. Agents Chemother.* **1997,** *41*, 1004.

Desjardins, R. E.; Canfield, C. J.; Haynes, J. D.; Chulay, J. D. *Antimicrob. Agents Chemother.* **1979,** *16*, 710.

Guohua, H.; Yanhua, L.; Rengang, M.; Dongzhi, W.; Zhengzhi, M.; Hua, Z. Aphrodisiac Properties of *Allium tuberosum* Seeds Extract. *J. Ethnopharmacol.* **2009,** *122*, 579–582.

Herunsalee, A.; Pancharoen, O.; Tuntiwachwuttikul, P. Further Studies of Flavonoids of the Black Rhizomes *Boesenbergia pandurata. J. Sci. Soc. Thai* **1987,** *13*, 119–122.

Hidaka, M.; Horikawa, K.; Akase, T. et al. Efcacy of *Kaempferia parviflora* in a Mouse Model of Obesity-Induced Dermatopathy. *J. Nat. Med.* **2017,** *71* (1), 59–67.

Hooker, J. D. *Flora of British India*, Vol. 6; L. Reeve & Co. Ltd.: Ashford, 1894; p 22.

Jaipetch, T.; Reutrakul, V.; Tuntiwachwuttikul, P.; Santisuk, T. Flavonoids in the Black Rhizomes of *Boesenbergia pandurata. J. Phytochem.* **1983,** *22*, 625–626.

Jansakul, C.; Tachanaparuksa, K.; Mulvany, M. J.; Sukpondma, Y. Relaxant Mechanisms of 3, 5, 7, 3′, 4′-Pentamethoxyflavone on Isolated Human Cavernosum. *Eur. J. Pharmacol.* **2012,** *691* (1–3), 235–244.

Kobayashi, H.; Suzuki, R.; Sato, K. et al. Efect of *Kaempferia parviflora* Extract on Knee Osteoarthritis. *J. Nat. Med.* **2018,** *72* (1), 136–144.

Kobayashi, S.; Kato, T.; Azuma, T.; Kikuzaki, H.; Abe, K. Antiallergenic Activity of Polymethoxyflavones from *Kaempferia parviflora. J. Funct. Foods* **2015,** *13*, 100–107.

Konkumnerd, J.; Karnchanatat, A.; Sangvanich, P. A Thermostable Lectin from the Rhizomes of *Kaempferia parviflora. J. Sci. Food Agric.* **2010,** *90*, 1920–1925.

Kress, W. J.; DeFilipps, R. A.; Farr, E.; Kyi, D. Y. Y. A Checklist of the Trees, Shrubs, Herbs, and Climbers of Myanmar (Revised from the Original Works by Lace, J. H., Rodger, R., Hundley, H. G., Ko, U. C. K. on the "List of Trees, Shrubs, Herbs and Principal Climbers, etc. Recorded from Burma), Vol. 45; Department of Systematic Biology-Botany, National Museum of Natural History: Washington, DC, 2003; p 123.

Kusirisin, W.; Srichairatanakool, S.; Lerttrakarnnon, P.; Lailerd, N.; Suttajit, M.; Jaikang, C.; Chaiyasut, C. Antioxidative Activity, Polyphenolic Content and Anti-Glycation Effect of Some Thai Medicinal Plants Traditionally Used in Diabetic Patients. *Med Chem.* **2009,** *5* (2), 139–147.

Larsen, K.; Larsen S. S. *Gingers of Thailand*; Queen Sirikit Botanic Garden: Chiang Mai, 2006; pp 1–130.

Leardkamolkarn, V.; Tiamyuyen, S.; Sripanidkulchai, B. O. Pharmacological Activity of *Kaempferia parviflora* Extract Against Human Bile Duct Cancer Cell Lines. *Asian. Pac. J. Cancer. Prev.* **2009,** *10*, 695–698.

Malakul, W.; Ingkaninan, K.; Sawasdee, P.; Woodman, O. L. The Ethanolic Extract of *Kaempferia parviflora* Reduces Ischaemic Injury in Rat Isolated Hearts. *J. Ethnopharmacol.* **2011,** *137* (1), 184–191.

Malakul, W.; Thirawarapan, S.; Ingkaninan, K.; Sawasdee, P. Effects of *Kaempferia parviflora* Wall. ex. Baker on Endothelial Dysfunction in Streptozotocin-Induced Diabetic Rats. *J. Ethnopharmacol.* **2011,** *133* (2), 371–377.

Matsushita, M.; Yoneshiro, T.; Aita, S. *Kaempferia parviflora* Extract Increases Whole-Body Energy Expenditure in Humans: Roles of Brown Adipose Tissue. *J. Nutr. Sci. Vitaminol. (Tokyo)*. **2015,** *61*, 79–83.

Mekjaruskul, C.; Jay, M.; Sripanidkulchai, B. Modulatory Effects of *Kaempferia parviflora* Extract on Mouse Hepatic Cytochrome P450 Enzymes. *J. Ethnopharmacol.* **2012,** *141* (3), 831–839.

Mekjaruskul, C.; Yang, Y. T.; Leed, M. G.; Sadgrove, M. P.; Jay, M.; Sripanidkulchai, B. Novel Formulation Strategies for Enhancing Oral Delivery of Methoxyflavones in *Kaempferia parviflora* by SMEDDS or Complexation with 2-Hydroxypropyl-β-Cyclodextrin. *Int. J. Pharm.* **2013,** *445* (1–2), 1–11.

Morikawa, T.; Pan, Y.; Ninomiya, K.; Imura, K.; Matsuda, H.; Yoshikawa, M.; Yuan, D.; Muraoka, O. *Bioorg. Med. Chem.* **2010,** *18*, 1882–1890.

Murata, K.; Deguchi, T.; Fujita, T; Matsuda, H. Improvement in Blood Fluidity by *Kaempferia parviflora* Rhizome. *J. Nat. Med.* **2012.**

Nakamura, S.; Li, X.; Matsuda, H.; Ninomiya, K.; Morikawa, T.; Yamaguti, K.; Yoshikawa, M. *Chem. Pharm. Bull.* **2007,** *55*, 1505–1511.

Nakamura, S.; Okazaki, Y.; Ninomiya, K.; Morikawa, T.; Matsuda, H.; Yoshikawa, M. *Chem. Pharm. Bull.* **2008,** *56*, 1704–1709.

Nakao, K.; Murata, K.; Deguchi, T.; Itoh, K.; Fujita, T.; Higashino, M.; Yoshioka, Y.; Matsumura, S.; Tanaka, R.; Shinada, T.; Ohfune, Y.; Matsuda, H. Xanthine Oxidase Inhibitory Activities and Crystal Structures of Methoxyflavones from *Kaempferia parviflora* Rhizome. *Biol. Pharm. Bull.* **2011,** *34* (7), 1143–1146.

Ninomiya, K.; Morikawa, T.; Xie, H.; Matsuda, H.; Yoshikawa, M. *Heterocycles* **2008,** *75*, 1983–1995.

Ninomiya, K.; Morikawa, T.; Zhang, Y.; Nakamura, S.; Matsuda, H.; Muraoka, O.; Yoshikawa, M. *Chem. Pharm. Bull.* **2007,** *55*, 1185–1191.

Okabe, Y.; Shimada, T.; Horikawa, T. Suppression of Adipocyte Hypertrophy by Polymethoxyflavonoids Isolated from *Kaempferia parviflora*. *Phytomedicine* **2014,** *21*, 800–806.

Patanasethanont, D.; Nagai, J.; Matsuura, C.; Fukui, K.; Sutthanut, K.; Sripanidkulchai, BO.; Yumoto, R.; Takano, M. Modulation of Function of Multidrug Resistance Associated-Proteins by *Kaempferia parviflora* Extracts and Their Components. *Eur J Pharmacol.* **2007,** *566* (1–3), 67–74. Epub 2007 Apr 14.

Patanasethanont, D.; Nagai, J.; Yumoto, R.; Murakami, T.; Sutthanut, K.; Sripanidkulchai, B. O.; Yenjai, C.; Takano, M. Effects of *Kaempferia parviflora* Extracts and Their Flavone Constituents on P-glycoprotein Function. *J. Pharm. Sci.* **2007,** *96* (1), 223–233.

Plaingam, W.; Sangsuthum, S.; Angkhasirisap, W.; Tencomnao, T. *Kaempferia parviflora* Rhizome Extract and *Myristica fragrans* Volatile Oil Increase the Levels of Monoamine Neurotransmitters and Impact the Proteomic Profles in the Rat Hippocampus: Mechanistic Insights into Their Neuroprotective Effects. *J. Tradit. Complement. Med.* **2017,** *7* (4), 538–552.

Quattrocchi, U. *CRC World Dictionary of Medicinal and Poisonous Plants: Common Names, Scientific Names, Eponyms, Synonyms, and Etymology*, Vol. III E-L.; CRC Press: Boca Raton, 2012; p 653.

Rujjanawate, C.; Kanjanapothi, D.; Amornlerdpison, D.; Pojanagaroon, S. Anti-Gastric Ulcer Effect of *Kaempferia parviflora*. *J. Ethnopharmacol.* **2005,** *102* (1), 120–122.

Sadhu, S. K.; Tamaki, M.; Ohtsuki, T.; Toume.; Koyano, T.; Kowithayakorn, T.; Ishibashi, M.; Cadinane Sesquiterpenes from *Curcuma parviflora*. *J. Nat. Prod.* **2009,** *72* (4), 782–783.

Sae-wong, C.; Tansakul, P.; Tewtrakul, S. Anti-Inflammatory Mechanism of *Kaempferia parviflora* in Murine Macrophage Cells (RAW 264.7) and in Experimental Animals. *J. Ethnopharmacol.* **2009,** *124* (3), 576–580.

Sae-Wong, C.;Matsuda, H.; Tewtrakul, S.; Tansakul, P.; Nakamura, S.; Nomura, Y.; Yoshikawa, M. Suppressive Effects of Methoxyflavonoids Isolated from *Kaempferia parviflora* on Inducible Nitric Oxide Synthase (iNOS) Expression in RAW 264.7 Cells. *J. Ethnopharmacol.* **2011,** *136* (3), 488–495.

Saokaew, S.; Wilairat, P.; Raktanyakan, P. et al. Clinical Effects of Krachaidum (*Kaempferia parviflora*): A Systematic Review. *Evid.-Based Complem. Altern. Med.* **2017,** *22* (3), 413–428. View at: Publisher Site | Google Scholar.

Sawasdee, P.; Sabphon, C.; Sitthiwongwanit, D.; Kokpol, U. Anticholinesterase Activity of 7-Methoxyflavones Isolated from *Kaempferia parviflora*. *Phytother Res.* **2009,** *23* (12), 1792–1794.

Scudiero, D. A.; Shoemaker, R. H.; Paull, K. D.; Monks, A.; Tierney, S.; Nofziger, T. H. et al. *Cancer Res.* **1988,** *48*, 4827.

Sirirugsa, P. Taxonomy of the Genus Kaempferia (Zingiberaceae) in Thailand. *Thai. For. Bull.* **1992,** *19*, 1–15.

Skehan, P.; Storreng, R.; Scudiero, D.; Monks, A.; McMahon, J.; Vistica, D. et al. *J. Natl. Cancer Inst.* **1990,** *82*, 1107.

Sookkongwaree, K.; Geitmann, M.; Roengsumran, S.; Petsom, A.; Danielson, U. H. Inhibition of Viral Proteases by Zingiberaceae Extracts and Flavonoids Isolated from *Kaempferia parviflora*. *Pharmazie* **2006,** *61*, 717–721.

Sudwan, P.; Saenphet, K.; Saenphet, S.; Suwansirikul, S. Effect of *Kaempferia parviflora* Wall. ex. Baker on Sexual Activity of Male Rats and Its Toxicity. *Southeast Asian J. Trop. Med. Public Health*. **2006,** *37* (Suppl 3), 210–215.

Sutthanut, K.; Lu, X.; Jay, M.; Sripanidkulchai, B. Solid Lipid Nanoparticles for Topical Administration of *Kaempferia parviflora* Extracts. *J. Biomed. Nanotechnol.* **2009,** *5* (2), 224–232.

Sutthanut, K.; Sripanidkulchai, B.; Yenjai, C.; Jay, M. Simultaneous Identification and Quantitation of 11 Flavonoid Constituents in *Kaempferia parviflora* by Gas Chromatography. *J. Chromatogr A* **2007,** *1143*, 227–233.

Tajuddin Ahmad, S.; Latif, A.; Qasmi, I. A. Aphrodisiac Activity of 50% Ethanolic Extracts of *Myristica fragrans* Houtt. (nutmeg) and *Syzygium aromaticum* (L) Merr. & Perry. (Clove) in Male Mice: A Comparative Study. *BMC Complement Altern Med.* **2003,** *3*, 6.

Tajuddin Ahmad, S.; Latif, A.; Qasmi, I. A. Effect of 50% Ethanolic Extract of *Syzygium aromaticum* (L.) Merr. & Perry. (Clove) on Sexual Behaviour of Normal Male Rats. *BMC Complement. Altern. Med.* **2004,** *4*, 17.

Tajuddin Ahmad, S.; Latif, A.; Qasmi, I. A.; Amin, K. M. An Experimental Study of Sexual Function Improving Effect of *Myristica fragrans* Houtt. (Nutmeg). *BMC Complement. Altern. Med.* **2005,** *5*, 16.

Temkitthawon, P.; Hinds, TR.; Beavo, JA.; Viyoch, J.; Suwanborirux, K.; Pongamornkul, W.; Sawasdee, P.; Ingkaninan, K. *Kaempferia parviflora*, a Plant Used in Traditional Medicine to Enhance Sexual Performance Contains Large Amounts of Low Affinity PDE5 Inhibitors. *J. Ethnopharmacol.* **2011,** *137* (3), 1437–1441.

Tep-Areenan, P.; Sawasdee, P.; Randall, M. Possible Mechanisms of Vasorelaxation for 5,7-Dimethoxyflavone from *Kaempferia parviflora* in the Rat Aorta. *Phytother. Res*. **2010,** *24* (10), 1520.

Tewtrakul, S.; Subhadhirasakul, S. Anti-Allergic Activity of Some Selected Plants in the Zingiberaceae Family. *J. Ethnopharmacol.* **2007,** *109* (3), 535–538. Epub 2006 Aug 15.

Tewtrakul, S.; Subhadhirasakul, S.; Kummee, S. Anti-Allergic Activity of Compounds from *Kaempferia parviflora*. *J. Ethnopharmacol.* **2008,** *116*, 191–193.

Tewtrakul, S.; Subhadhirasakul, S.; Kummee, S. Anti-Allergic Activity of Compounds from *Kaempferia parviflora*. *J. Ethnopharmacol.* **2008,** *116* (1), 191–193. Epub 2007 Nov 6.

Toda, K.; Hitoe, S.; Takeda, S.; Shimoda, H. Black Ginger Extract Increases Physical Fitness Performance and Muscular Endurance by Improving Inflammation and Energy Metabolism. *Heliyon* **2016,** *2* (5).

Trager, W.; Jensen, J. B. *Science* **1976,** *193*, 673.

Trisomboon, H.; Tohei, A.; Malaivijitnond, S.; Watanabe, G.; Taya, K. Oral Administration of *Kaempferia parviflora* Did Not Disturb Male Reproduction in Rats. *J. Reprod. Dev*. **2008,** *54* (5), 375–380. Epub 2008 Jul 2.

Trisomboon, H; Watanabe, G.; Wetchasit, P.; Taya, K. Effect of Daily Treatment with Thai Herb, *Kaempferia parviflora*, in Hershberger Assay Using Castrated Immature Rats. *J. Reprod. Dev*. **2007,** *53* (2), 351–356. Epub 2006 Dec 20.

Tuntiyasawasdikul, S.; Limpongsa, E.; Jaipakdee, N.; Sripanidkulchai, B. A Monolithic Drug-in-Adhesive Patch of Methoxyfavones from *Kaempferia parviflora*: In Vitro and In Vivo Evaluation. *Int. J. Pharma.* **2015,** *478* (2), 486–495.

Vimala, S.; Norhanom, A.; Yadav, M. Anti-Tumour Promoter Activity in Malaysian Ginger Rhizobia Used in Traditional Medicine. *Br. J. Cancer.* **1999,** *80*, 110.

Wanich, S.; Yenjai, C.; Amino and nitro derivatives of 5,7-Dimethoxyflavone from *Kaempferia parviflora* and Cytotoxicity Against KB Cell Line. *Arch. Pharm. Res.* **2009,** *32* (9), 1185–1189. DOI: 10.1007/s12272–009–1901-y. Epub 2009 Sep 26.

Wannanon, P.; Wattanathorn, J.; Tong-Un, T.; et al. Efficacy assessment of *Kaempferia parviflora* for the Management of Erectile Dysfunction. *Online J. Biol. Sci.* **2012,** *12*, 149–155.

Wattanapitayakul, S. K.; Chularojmontri, L.; Herunsalee, A.; Charuchongkolwongse, S.; Chansuvanich, N.; Vasorelaxation and Antispasmodic Effects of *Kaempferia parviflora* Ethanolic Extract in Isolated Rat Organ Studies. *Fitoterapia* **2008,** *79* (3), 214–216.

Wattanapitayakul, S. K.; Suwatronnakorn, M.; Chularojmontri, L.; Herunsalee, A.; Niumsakul, S.; Charuchongkolwongse, S.; Chansuvanich, N. *Kaempferia parviflora* Ethanolic Extract Promoted Nitric Oxide Production in Human Umbilical Vein Endothelial Cells. *J. Ethnopharmacol.* **2007,** *110* (3), 559–562.

Wattanathorn, J.; Tong-Un, T.; Muchimapura, S.; Wannanon, P.; Sripanidkulchai, B.; Phachonpai, W. Anti-Stress Efects of *Kaempferia parviflora* in immobilization subjected rats," American Journal of Pharmacology and Toxicology, vol. 8, no. 1, pp. 31–38, 2013.

Weerateerangkul, P.; Palee, S.; Chinda, K.; Chattipakorn, S. C.; Chattipakorn N. Effects of *Kaempferia parviflora* Wall. ex. Baker and Sildenafil Citrate on cGMP Level, Cardiac Function, and Intracellular Ca2+ Regulation in Rat Hearts. *J. Cardiovasc. Pharmacol.* **2012,** *60* (3), 299–309.

Weerateerangkul, P.; Surinkaew, S.; Chattipakorn, S. C.; Chattipakorn, N. Effects of *Kaempferia parviflora* Wall. Ex. Baker on Electrophysiology of the Swine Hearts. *Indian J. Med. Res.* **2013,** *137* (1), 156–163.

Welbat, J. U.; Chaisawang, P.; Chaijaroonkhanarak, W. et al. *Kaempferia parviflora* Extract Ameliorates the Cognitive Impairments and the Reduction in Cell Proliferation Induced by Valproic Acid D Treatment in Rats. *Ann. Anatomy* **2016,** *206*, 7–13.

Wongsinkongman, P.; Mongkolchaipak, N.; Chansuvanich, N.; Techadumrongsin, Y.; Boonruad, T. Quality Evaluation of Crude Drugs and Volatile Oil of Krachai-Dam Rhizomes. *Bull. Dept. Med. Sci* **2003,** *45* (1), 1–16.

Wu, L.; Liu, H.; Li, L. et al. 5,7,3′, 4′ -Tetramethoxyfavone Exhibits Chondroprotective Activity by Targeting β-Catenin Signaling In Vivo and In Vitro. *Biochem. Biophys. Res. Commun.* **2014,** *452* (3), 682–688.

Wu, L.; Liu, H.; Li, L. et al. 5,7,3′,4′-Tetramethoxyfavone Protects Chondrocytes from ER Stress-Induced Apoptosis Through Regulation of the IRE1α Pathway. *Connective Tissue Res.* **2018,** *59* (2), 157–166.

Wutythamawech, W. *Encyclopedia of Thai Herbs*; OS Printing; Bangkok, Thailand, 1997; p 365.

Yang, J.; Liu, H.; Li, L.; Liu, H. et al. Te Chondroprotective Role of TMF in PGE2-Induced Apoptosis Associating with Endoplasmic Reticulum Stress. *Evid.-Based Complem. Altern. Med.* **2015,** *2015*, Article ID 297423, 7.

Yenjai, C.; Prasanphen, K.; Daodee, S.; Wongpanich, V.; Kittakoop, P. Bioactive Flavonoids from *Kaempferia parviflora*. *Fitoterapia* **2004,** *75*, 89–92.

Yenjai, C.; Prasanphen, K.; Daodee, S.; Wongpanich, V.; Kittakoop, P. Bioactive Flavonoids from *Kaempferia parviflora*. *Fitoterapia* **2004,** *75* (1), 89–92.

Yenjai, C.; Prasanphen, K.; Daodee, S.; Wongpanich, V.; Kittakoop, P. Bioactive Flavonoids from *Kaempferia parviflora*. *Fitoterapia* **2004,** *75*, 89–92.

Yenjai, C.; Wanich, S. Cytotoxicity Against KB and NCI-H187 Cell Lines of Modified Flavonoids from *Kaempferia parviflora*. *Bioorg. Med. Chem. Lett.* **2010,** *20*, 2821–2823.

Yenjai, C.; Wanich, S. Cytotoxicity Against KB and NCI-H187 Cell Lines of Modified Flavonoids from *Kaempferia parviflora*. *Bioorg. Med. Chem. Lett.* **2010,** *20* (9), 2821–2813. DOI: 10.1016/j.bmcl.2010.03.054. Epub 2010 Mar 15.

Yoshikawa, M.; Nishida, N.; Ninomiya, K.; Ohgushi, T.; Kubo, M.; Morikawa, T.; Matsuda, H. *Bioorg. Med. Chem.* **2006,** *14*, 456–463.

Yoshino, S.; Kim, M.; Awa, R.; Kuwahara, H.; Kano, Y.; Kawada, T. *Kaempferia parviflora* Extract Increases Energy Consumption Through Activation of BAT in Mice. *Food Sci. Nutr.* **2014,** *2*, 634–637.

Youn, K.; Lee, J.; Ho, C-T.; Jun, M. Discovery of Polymethoxyfavones from Black Ginger (*Kaempferia parviflora*) as Potential α-Secretase (BACE1) Inhibitors. *J. Funct. Foods* **2016,** *20*, 567–574.

Yuan, X.; Li, L.; Shi, W. TMF Protects Chondrocytes from ER Stress-Induced Apoptosis by Down-Regulating. *Biomed. Pharmacother. = Biomedecine pharmacotherapie* **2017,** *89*, 1262–1268.

Zhang, Y.; Morikawa, T.; Nakamura, S.; Ninomiya, K.; Matsuda, H.; Muraoka, O.; Yoshikawa, M. *Heterocycles* **2007,** *71*, 1565–1576.

CHAPTER 2

Elaeagnus latifolia L. (Elaeagnaceae): A Promising and Emerging High-Value Bioresource with Nutraceutical and Pharmaceutical Prospects

EVANYLLA KHARLYNGDOH[1], LONGING BASUK LANGSTIEH[1] JITENDRA KUMAR SHUKLA[1], PARDEEP KUMAR BHARDWAJ[1], and PULOK KUMAR MUKHERJEE[2]

[1]*Institute of Bioresources and Sustainable Development, Meghalaya Node, Upper Shillong, East Khasi Hills District, Meghalaya, India*

[2]*Institute of Bioresources and Sustainable Development, Takyelpat, Imphal, Manipur, India*

ABSTRACT

Elaeagnus latifolia Linn. native to Asia (commonly known as Bastard Oleaster or Silver berry), is an underutilized edible plant traditionally used for nutrition and as a natural remedy for varied ailments and diseases. This chapter focuses on the botanical description, ethnobotany, nutritional, phytochemistry, pharmacological, and therapeutic properties of *E. latifolia* using a literature survey, search engines, and online databases. *E. latifolia* possessed a broad spectrum of nutritional, pharmacological, and therapeutic properties including antioxidant, antimicrobial, insecticidal, antiarthritic, cardioprotective, antinociceptive, anti-inflammatory, and gastroprotective activity. Fruits are eaten raw or as pickles, juice, chutney, or jam. Traditional healers use it for the treatment of ailments, namely, blood alcohol removal, pain alleviation, wound–healing, muscle relaxant, rheumatism, removal

High-Value Plants: Novel Insights and Biotechnological Advances. Thangjam Anand Singh, Prakash Kumar Sarangi, and Ch. Brajakishore Singh (Eds.)
© 2025 Apple Academic Press, Inc. Co-published with CRC Press (Taylor & Francis)

of retained placenta, and threatened abortion. Phytochemical constituents, including carbohydrates, phenols, flavonoids, fatty acids, phytosterols, ascorbic acid, tannins, phytosterols, glycosides, saponins, amino acids, and nutritional components, namely, amino acids, vitamins, minerals, and fatty acids have been identified from *E. latifolia*. This review underlines the current status and future prospects of *E. latifolia* with ecological, nutritional, phytochemical, and therapeutic properties, with further experimental and clinical studies on nutraceutical and pharmaceutical aspects including effective agro & breeding techniques for sustainable utilization of the crop.

2.1 INTRODUCTION

Elaeagnaceae family also known as the oleaster family has about 70–80 species found worldwide under three genera: *Hippophae*, *Shepherdia*, and *Elaeagnus* (Nazir et al., 2020). The Northeastern region of India has a rich diversity of these species and most of them are being exploited for their medicinal and nutritional properties which in turn are being used in modern-day medicine, pharmaceutical industries, and breeding programs. *Elaeagnus latifolia* Linn. commonly known as Bastard Oleaster or Silver berry is one such high-value plant species with immense nutritional and ethnomedicinal importance. This review paper aims to study the current status of *E. latifolia* in terms of its nutritional, ethnomedicinal, phytochemical constituents, and therapeutic values. It also aims in providing brief information about the species ecology, its economic importance, and research gaps relating to the present knowledge, thereby providing more data for further research opportunities, especially in the areas of nutraceutical and pharmaceutical industries.

2.2 MATERIAL AND METHODS

Available literature survey of *E. latifolia* using search engines such as Google, Google Scholar, Web of Science, PubMed, PubChem, Science Direct, Springer, Elsevier, Scopus, Taylor and Francis, Wiley Online Library, and online databases such as GBIF, The Plant List, etc. were screened to collect relevant data. The scientific names and synonyms were verified using "The Plant List," "International Plant Name Index," and Kew Botanical Garden Plant name databases.

2.3 RESULTS

Taxonomy: *E. latifolia* Linn. belongs to the Elaeagnaceae family with synonyms: *Elaeagnus kologa* subsp. *ceylanica* Servett., *E. kologa* subsp. *grisebachii* Servett., *E. kologa* subsp. *macrophylla* Servett., *E. kologa* subsp. *wightii* Servett., *Elaeagnus punctata* Schltdl., *E. rotundifolia* Servett., and *Elaeagnus thwaitesii* Schltdl (http://www.theplantlist.org). The vernacular name of this fruit is known as *Sohshang* (Khasi), *Slangi* (Jaintia), *Chhokhutt* (Garo), *Heiyai* (Manipuri), *Sarzuk* (Mizo), *Musleri*, *Mallero* (Sikkimese), *Mirika–tenga* (Assamese), and in Karnataka, as *Hulige hannina gida* (Kannada).

Geographical distribution: *E. latifolia* plant is native to Asia and is mainly distributed in China, Myanmar, Thailand, Vietnam, Bangladesh, Srilanka, Indonesia, Philippines, Nepal, India, and subtropical and temperate Himalayas. It is also found in Southern Europe, North America, and Australia (GBIF.org; Lachowicz et al., 2019). In India, it is distributed in Karnataka, Odisha, Kerala, Tamil Nadu, Assam, Manipur, Meghalaya, Mizoram, Nagaland, and Sikkim (Schwalbe et al., 2007; Patel et al., 2008; Yingthongchai et al., 2014a). In the Northeast region, the species is found in restricted habitats in semi-wild form and also planted in backyard gardens.

Botanical description: *E. latifolia* is a perennial, semi-deciduous, spreading, multi-stem woody shrub that can grow up to 30-m tall (Fig. 2.1). The stems are hardy and often thorny. Inflorescence is a raceme, flowers are light yellow colored, hermaphrodite, actinomorphic, and give a strong aroma. The ovary is inferior surrounded by hypanthium (Rymbai et al., 2017). Flowering occurs during September–December and pollination is carried out by bees (Patel et al., 2008). Fruiting starts in February and continues till April. Fruits are oblong in shape with dark pinkish–red color when ripen and are harvested during March–April in three to four pickings. Each fruit contains a seed that is also edible. Fruits are highly perishable with a shelf life of 3–5 days at room temperature (23°C). Unripe fruits are astringent but they are often sweet with a tinge of sourness when fully ripe.

Ethnomedicinal uses: *E. latifolia* fruits and seeds are edible and fruits are mainly eaten raw or locally prepared as pickles, jams, juices, and fruit leather. In traditional folk medicine, the fruits, leaves, and roots of *E. latifolia* are being used for blood alcohol removal, pain alleviation, wound healing,

FIGURE 2.1 *E. latifolia*: Fruiting plant (a), fruits (b), seeds (c), seeds without seed coat (d).

muscle relaxant, antimicrobial properties, and also as anti-inflammatory, antipyretic, analgesic, astringent (Saboonchian, 2014; Patel, 2015), anticancer (Lachowicz et al., 2019) and cardiac problem treatment (Sarwar, 2015). Ripe fruits are used as painkillers and for rheumatism (Murty et al., 2015). The flowers and fruit are astringent (Choudhury and Patel, 2016). In Mizoram, the root juice is prescribed orally (05–10 mL) twice a day for retained placenta removal. A root decoction is taken orally (10–15 mL) thrice a day for the treatment of miscarriage (Sharma et al., 2001) and two spoonfuls of fruit juice are taken twice daily for the treatment of anemic patients (Hazarika et al., 2016). In Karnataka, *E. kologa* Schldl (Syn: *E. latifolia* L.), is used for treatment against cutaneous infections, dropsy, anasarca, venereal sores, deafness, and dysentery (Gowda, 2004). The non-edible parts of the plant are used for fuel and fodder. Similar ethnomedicinal properties have also been reported in other *Elaeagnus* species. The flowers, fruits, and leaves of *Elaeagnus angustifolia* are being used in traditional medicine for treatment of sore throat, cough, flu, cold, fever, nausea, vomiting, jaundice, asthma, diarrhea (Mirhydar, 1998; Qureshi et al., 2006; Natanzi et al., 2012), tetanus (Hosseinzadeh et al., 2003), rheumatoid arthritis (Ramezani et al., 2001), headaches, and rickets (Farahbakhsh et al., 2011). Leaves and fruits are used for ulcer treatment (Gurbuz et al., 2003), muscle relaxant (Hosseinzadeh et al., 2003), antipyretic (Farahbakhsh et al., 2011), antinociceptive, anti-inflammatory (Ahmadiani et al., 2000; Ramezani et al., 2001) and wound healing (Natanzi et al., 2012). The popular herbal tea known as "Zhourat" in Lebanon and Syria contains a mixture of herbs and flowers, *E. angustifolia* acting as a digestive and soothing agent (Obón et al., 2014).

Propagation: *E. latifolia* can be propagated through seeds or mature stem cuttings. Seeds of *E. latifolia* were reported to show early germination owing to its thin seed coat however, with low seed viability (Sundriyal and Sundriyal, 2003). Mature stem cuttings (10–12 cm length) with 3 nodes intact, are planted during the months from June–July. After 15–20 days the cuttings start to sprout and within 3–4 months after sprouting, the plants are transplanted in the field. Disease and insect–pest attack on *E. latifolia* plant are reported to be low; however, to avoid infestation of fruit borer, Endosulphan @2 mL/L is sprayed during fruit development (Patel et al., 2008). The plant starts bearing fruit after 2 years with a yield of 10–15 kg fruit/tree. According to Sundriyal and Sundriyal (2003), *E. latifolia* tree density was reported at 15.0 trees/ha and the average fruit yield ranged from 9 to 155 kg/tree.

Species ecology: *E. latifolia* is mostly found in semi-wild habitats between 1500 and 2600 m a.s.l (Gupta, 1945; Gamble, 1972) in the Himalayas. In Northeastern region, this species is also planted as a home–garden fruit plant or as an ornamental and fencing hedge since it causes less disease and insect/pest infestation (Krupinsky and Frank, 1986; Patel, 2015). The species grows well in forest openings with well-drained light sandy, medium loamy, and heavy clay soils; however, it can also grow in nutritionally poor soil and can withstand drought conditions, and high wind velocity (Yingthongchai et al., 2014a; Rymbai et al., 2016, 2017) and can tolerate acidic to alkaline soil conditions (Tuksan and Laughlin, 1991). The plant can tolerate a minimum temperature of –7°C and can grow up to 2.7–4.0 m in height (Panja et al., 2014). Like other *Elaeagnus* species (*Elaeagnus umbellata*), *E. latifolia* is also valued for its ability to prevent erosion and to attract wildlife (Zarger, 1980; Shaw, 1988).

Trees of *Elaeagnus* species have a symbiotic relationship with *Actinomycetes* that are responsible for producing root nodules and fixing atmospheric nitrogen (Follstad et al., 2010). The nitrogen-fixing ability of *Elaeagnus* species was also reported earlier by Baker et al. (1979) and Bargali (2011). Trees of *E. latifolia* are actinorhizal and form symbiotic relationships with nitrogen-fixing bacteria *Actinomycetes* and *Frankia* (Handique and Parkash,

2014). Nodules in *E. latifolia* have many globular nodule lobes arranged in clusters of two or three without bearing nodule roots. Further, screening of rhizospheric microbial association from *E. latifolia* plant showed 90%, hyphal, 25% vesicular, and 70% arbuscular root infection/colonization, whereas dominant rhizosphere endomycorrhizae belongs to *Glomus, Acaulospora*, and *Gigaspora* genera (Handique and Parkash, 2014). Because of its ability to fix atmospheric nitrogen, a nearly 10% increase in fruit production on intercropping with plums and nuts was reported (Rymbai et al., 2017). Other *Elaeagnus* species, namely, *E. umbellata* can also fix atmospheric nitrogen (Gardner, 1958; Kim et al., 1993). It performs very well on less fertile soils because of its root nodules that have nitrogen–fixing actinomycetes (Sternberg, 1982) and yield 236 kg of mineral nitrogen/ha/year within 20 cm of soil (Paschke et al., 1989). Planting of *E. umbellata* with black walnut controls mycosphaerella leaf spots and walnut anthracnose (Kessler, 1990).

Research on genetic diversity among *E. latifolia* populations has been carried out only in a few countries. Yingthongchai et al. (2014a, 2014b) performed genetic diversity and relationships study among 88 accessions *E. latifolia* populations collected from eight provinces distributed in the upper North of Thailand, using 14 primers and ISSR markers. The study showed that fourteen selected primers produced 278 discernible bands, with 264 (94.96%) being polymorphic and 86 out of 88 accessions could be distinguished while two of them showed great similarity, indicating high genetic diversity in *E. latifolia* populations.

Economic importance: Fruits of *E. latifolia* are prized for their unique taste and adaptability in adverse edaphic and climatic conditions. They are easily available and are nutritionally adequate for consumption in both urban and rural areas. Processing of fruits into pickles, jams, jelly, juice, herbal tea, beverage, vinegar, ketchup, dessert, ice cream, and fruit leather adds to the economic value of this fruit. However, people in the Northeast region are unaware of the value addition that can be obtained from processing this fruit and so a huge quantity of it usually gets damaged and spoiled during transport and handling which in turn leads to huge losses to the farmers. In Meghalaya and other Northeast states, *E. latifolia* fruit is sold at various local markets for a period of 1.5 months in 3–4 pickings starting from the last week of February to March and rarely till April second week. The average sale of raw fruit in the Northeast region is ₹80/kg.

Physical properties: Physical properties of *E. latifolia* showed that the fruit weight ranged between 11.63–22.00 g; fruit length, 3.20–4.46 cm; fruit diameter, 2.26–2.84; and fruit volume, 12.58–16.72 mL., whereas seed weight, length, and breadth ranged s1.72–4.51 g, 2.83–3.56 cm, and 1.20–1.47 cm, respectively (Patel et al., 2008; Paulsamy et al., 2010; Nath et al., 2011; Seal, 2011; Rymbai et al., 2014, 2020; Kumar and Said, 2018; Pandey et al., 2018). The fruits contain high moisture content (62–89.1%) which is comparable to other *Elaeagnus* species namely, *E. pyriformis* (82.0%) (Khomdram et al., 2014) and commercial fruits such as apple (84%), litchi (84%), mango ripe (81%), papaya ripe (91%) (Seal, 2011). Pulp and juice content varies between 56.24–84.68% and 25.11–28.72%, respectively, and pulp:seed ratio ranged between 2.6 and 4.82. The total mineral content of the fruit ranges from 0.61 to 3.60% which is slightly higher than other commercial fruits such as apple (1.2%), litchi (1.0%), Mango ripe (1.1%), Papaya ripe (1.3%) (Seal, 2011).

Nutritional properties: Studies on the nutritional composition of *E. latifolia* have been carried out by various workers including comparative studies with other local and commercial fruits. Fruits are considered a good source of vitamins (A, B, C, and E), secondary metabolites, carbohydrates, organic acids, minerals, bioactive compounds (phenolics, flavonoids, lipids, carotenoids, and ascorbic acid), and essential fatty acids (Patel et al., 2008; Nath et al., 2011; Seal, 2011; Wu et al., 2011; Panja et al., 2014; Rymbai et al., 2016). The nutritive value of *E. latifolia* fruit was reported at 332.10–370.27 kcal/100 g which is higher as compared to apple (58.00 kcal/100 g), litchi (61.00 kcal/100 g), mango ripe (74.00 kcal/100 g), papaya ripe (32.00 kcal/100 g), and wood apple (134.00 kcal/100 g) (Seal, 2011).

Total soluble solids in *E. latifolia* fruits range from 6.02 to 14.3%, which were higher than that of orange (7.56), banana (12.04), apple (1.00), guava (10.00), and grape (9.5). The acids available in *E. latifolia* fruits (2.04–4.03%) are higher than that present in *E. kologa* (1.28%) and other commercial fruits, namely, banana (1.07%), apple (0.19%), and guava (0.75%), orange (2.14%) and grape (1.50%) (Patel et al., 2008; Paulsamy et al., 2010; Nath et al., 2011; Seal, 2011; Rymbai et al., 2014, 2020; Kumar and Said, 2018; Pandey et al., 2018).

Available carbohydrates in *E. latifolia* fruits were reported at 74.06% (Seal, 2011), which was higher than the common commercial fruits and nuts such as almond (10.50%) apple (13.70%), cashewnut (22.30%), litchi

(13.60%), mango ripe (16.90%), papaya ripe (7.20%), walnut (11.00%), and wood apple (18.10%) (Seal, 2011).

Reducing sugars, non–reducing sugars, and total sugars in *E. latifolia* fruits ranged between 1.0–4.02%, 0.27–1.22%, and 2.4–6.10%, respectively. The percentages of reducing sugars are higher than *E. pyriformis* (0.0178%) (Khomdram et al., 2014) but similar to *E. umbellata* (6.8–8.4%) (Sabir and Riaz, 2005), *E. kologa* (8.02%) (Paulsamy et al., 2010) and other commercial fruits, namely, guava (4.20%) and Orange (3.55%) but lower than apple (56.0%), banana (24.60%), and grape (32.35%) (Nath et al., 2011; Patel et al., 2013; Rymbai et al., 2014, 2016; Pandey et al., 2018). The non-reducing sugar reserves are comparable to that of *E. kologa* (0.27%) (Paulsamy et al., 2010), *E. umbellata* (1.4–2.2%) (Sabir and Riaz, 2005), *E. pyriformis* (0.016%) (Khomdram et al., 2014) and other commercial fruits such as apple (0.45%), grape (0.68%), and orange (0.01%) (Nath et al., 2011; Patel et a., 2013; Rymbai et al., 2014, 2016; Pandey et al., 2018). Total sugars are significantly higher than the reported 0.0339% in *E. pyriformis* (Khomdram et al., 2014), however relatively lower than apple (70.0%), banana (23.0%), grape (14.87%), guava (8.0%), and orange (10.20%) (Nath et al., 2011; Patel et a., 2013; Rymbai et al., 2014, 2016; Pandey et al., 2018).

The concentration of crude fiber (5.54–9.30%), crude fat (0.52%), and protein (4.8–7.80%) in *E. latifolia* fruit was relatively higher compared to other commercial fruits (crude fiber, crude fat, and protein percentages of apple: 3.40, 0.31 and 0.76%; litchi: 0.50, 0.20 and 1.10%; mango ripe: 0.70, 0.40 and 0.60%; and papaya ripe: 0.80, 0.10 and 0.60%, respectively) (Paulsamy et al., 2010; Seal, 2011). The protein percentage in *E. umbellata* was reported at 2.5–5.1% (Sabir and Riaz, 2005). Seeds of the *Elaeagnus* genus are known to have good fat and protein content (Patel, 2015).

Carotenoids (vitamin A): The most widespread group of phytonutrients in vegetables and fruits are carotenoids, and lycopene is the most predominant pigment mainly responsible for the deep-red color of ripe vegetables/fruits such as tomato (*Lycopersicon esculentum*), berries, watermelon, guava and red bell pepper (Suwanaruang, 2016). Lycopene is widely believed to protect cells against myocardial infection (Clinton, 1998), including prostate cancer (Giovannucci et al., 1995). The link between lycopene and the prevention of prostate cancer in humans showed a positive effect (Minorsky, 2002) including prostate cancer cell behavior (Soares et al., 2017). It has been reported that higher consumption of lycopene intake reduces the prostate cancer risk in numerous human cohort studies (Wan et al., 2014).

Efficient absorption of lycopene in humans has been reported to be higher in processed/cooked tomato products compared to fresh tomatoes (Gartner et al., 1997; Kirsh et al., 2006), due to the formation of lycopene *cis* isomers during food processing and storage thereby enhancing its biological activity (Minorsky, 2002). Besides preventing prostate cancer, studies have shown that lycopene enhances low-density lipoprotein degradation and inhibits cholesterol synthesis (Arab and Steck, 2000).

So far, tomato and watermelon are the two main food sources with abundant lycopene content (Pinto et al., 2011) and more than 80% of the dietary lycopene intake comes from raw and processed tomato products consumption (USDA, 2003). As per USDA.NCC reports (1998), the consumption of 20 mg/day of lycopene-rich watermelon or canned tomato juice, resulted in a 100–200% increase in human plasma lycopene. As the demand for lycopene-rich food products are growing, research into finding alternative lycopene-rich food sources is also increasing (Minorsky, 2002).

Studies have shown that *Elaeagnus* species are highly rich in lycopene content. A study by Fordham et al. (2001) showed that lycopene of the red-pigmented autumn olive (*E. umbellata*) accounted for 72–82% of the total carotenoids and lycopene content ranged 15–54-mg/100 g fresh fruit which was 12 times higher compared to fresh tomato fruit (0.88–4.20 mg 100 g^{-1}) (Ahmad et al., 2005). This new source of lycopene from *E. umbellata* may provide a substitution for tomato as a dietary source of lycopene (Kohlmeier et al., 1997, Fordham et al., 2001). Lycopene content in *E. latifolia* fruit pulp were reported at 0.115–0.215 mg/100 g (Nath et al., 2011; Seal, 2011) which was comparatively lower than that of *E. umbellata* (15–54 mg 100 g^{-1}). Besides lycopene, *E. latifolia* fruit also has high α–carotene content ranging 3.13–3.58 mg/100 g (Nath et al., 2011) as compared to other *Elaeagnus* species (*E. umbellata*: 0.031–0.068 mg/100 g) (Fordham et al., 2001), while β-carotene was 1.12 μg/100 g (Rymbai et al., 2014, 2016). Anthocyanin content of *E. latifolia* ranges from 0.112 to 16.33 mg/100 g (Nath et al., 2011; Rymbai et al., 2014).

Ascorbic acid (Vitamin C) is the most essential nutrient found in fruits and vegetables and has various antioxidant properties (Carr and Maggini, 2017). The fruits of *E. latifolia* showed the presence of a good source of vitamin C (7.17–37.31 mg/100 gm) which is significantly higher than *E. umbellata* (7.04%) (Gamba et al., 2020), *E. kologa* (13.5%) (Paulsamy et al., 2010), and other commercial fruits such as apple (24%), banana (5.12%), grape (3.44%), guava (34%), and orange (8.72%) (Paulsamy et al., 2010; Nath et al., 2011; Rymbai et al., 2014; Seal et al., 2017). *E. latifolia* fruits are fairly abundant in antioxidants (0.038–10.17 μmol/trolax/g) (Nath et al., 2011; Panja et al., 2014; Rymbai et al., 2014).

There are various species of *Elaeagnus* which has effective cytotoxic activities against cancer cells like *E. umbellata* (Wang et al., 2007), *Elaeagnus glabra* (Li et al., 2009; Roomi et al., 2013), *E. multiflora* (Lee et al., 2010), *E. angustifolia* (Okmen and Turkcan, 2014), *Elaeagnus oldhamii* (Liao et al., 2014) and research on anticancer effect and their underlying mechanism have already been carried out with positive results. However, the potential of *E. latifolia* fruits in cancer prevention has not been investigated.

Thiamine (Vitamin B1) is a colorless, water-soluble vitamin found in plants and microorganisms and plays an essential role in maintaining cellular and organ functions (Pourcel et al., 2013). Studies on vitamin B1 in varieties of foods have shown that thiamine content ranged from 0.001 to 0.672 mg/100 g, while the recommended daily intake for infants and adults ranges from 0.2 to 1.4 mg per day (Institute of Medicine, 1998). Thiamine content in *E. latifolia* was 1.20 mg/100 g (Seal et al., 2017) which is significantly higher than other commercial vegetables such as carrot, brinjal, okra, spinach (0.02, 0.06, 0.14, and 0.08 mg/100 g) and fruits such as Banana and Guava (0.18 and 0.06 mg/100 g), respectively (Ismail et al., 2013).

Riboflavin (Vitamin B2) is a water-soluble vitamin that is required for body function and development and is used as dietary supplement (Suwannasom et al., 2020). Riboflavin (B2) content in *E. latifolia* fruit was reported at 0.05 mg/100 g (Seal et al., 2017).

Niacin (Vitamin B3): Niacin (B3) another crucial vitamin for growth and synthesis of hormones has also been reported in *E. latifolia* at 0.29 mg/100 g (Seal et al., 2017) which was comparable with the B3 content of other commercial vegetables such as cabbage, cucumber, cauliflowers, spinach and tomatoes (0.19–0.97 mg/100 g) (Ismail et al., 2013).

Pantothenic acid (Vitamin B5) is another essential vitamin involved in coenzyme-A biosynthesis and fatty acid, carbohydrates, and protein metabolism (Tahiliani and Beinlich, 1991). Among all B vitamins in *E. latifolia* fruit, pantothenic acid (vitamin B5) was detected highest at 1.76 mg/100 g (Seal et al., 2017).

Pyridoxine (Vitamin B6) is a water-soluble vitamin that is available in many foods, such as meat, fish, nuts, grains, fruits, and vegetable that are necessary for the metabolism of red blood cell, the nervous and immune system (Brown et al., 2021). Vitamin B6 was reported higher (1.72 mg/100 g) (Seal et al., 2017) in *E. latifolia* as compared to common vegetables and fruits such as banana (0.37 mg/100 g), guava (0.16 mg/100 g) avocados (0.29 mg/100 g), spinach (0.26 mg/100 g), broccoli (0.134 mg/100 g), cauliflower (0.115 mg/100 g), and cucumber (0.2 mg/100 g) (Ismail et al., 2013).

Folic acid (Vitamin B9) is a water-soluble B vitamin required for several body functions and development (Krebs et al., 2009). Folate which is mostly required during infancy and pregnancy plays an important role in cell division and growth (Mahmood, 2014). A deficiency of folate can cause anemia in adults and slow growth in infants. Rich sources of folate include vegetables, soybeans, whole grains, wheat, orange juice, avocado, and milk (Mahmood, 2014). The daily recommendations for dietary folic acid range between 65 μg/day for infants to 400 μg/day for adults and 500–600 μg/day for pregnant and breastfeeding women (Institute of Medicine, 1998). A fair amount of folic acid was also detected in *E. latifolia* fruits (0.013 mg/100 g) (Seal et al., 2017).

Minerals: The mineral composition of *E. latifolia* fruit showed rich source of macronutrients, namely, Ca (52–1470 mg/100 g), N (1022 mg/100 g), Na (51–1022 mg/100 g), K (41.6–910 mg/100 g), Fe (16.82–180.00 mg/100 g), S (712.5 mg/100 g), P (57–96 mg/100 g), and Mg (40 mg/100 g) followed by micronutrients, namely, Zn (1.186–4.94 mg/100 g), Mn (2.10–4.52 mg/100 g), and Cu (0.046–1.91 mg/100 g) (Patel et al., 2008; Seal, 2011, 2012; Rymbai et al., 2014). These mineral contents are comparatively higher than the mineral composition of *E. conferta* (Ca: 280.1; N: 0.57; Na: 184.3; Fe: 21.33; P: 1.29; Mn: 3.80; and Cu: 0.94 mg/100 g) (Gill and Gupta, 2018), *E. kologa* (Ca: 52.0; P: 57.0; K: 416.0; Mg: 40; and Fe: 8.0 mg/100 g) (Paulsamy et al., 2010), and commercial fruits such as apple (Na: 28; K: 75; Ca: 10; Mn: 0.14; Cu: 0.10; Fe: 0.66; and Zn: 0.060 mg/100 g), litchi (Na: 24.9; K: 159; Ca: 10; Cu: 0.30; and Fe: 0.70 mg/100 g), mango ripe (Na: 26; K: 205; Ca: 14; Mn: 0.13; Cu: 0.11; Fe: 1.30; and Zn: 0.270 mg/100 g), and papaya (Na: 6.0; K: 69; Ca: 17; Cu: 0.20; and Fe: 0.50 mg/100 g) (Seal, 2011, 2012, Seal et al., 2014).

Phytochemical constituents: Phytochemical assessments have revealed that *Elaeagnus* genus is known to contain various bioactive components, namely, carbohydrates (fructose and glucose), polysaccharides, phenolic acids (chlorogenic, *p*–formic acids, 4–hydroxybenzoic acid, and caffeic acid), flavonoids (rutin, quercetin, and elaeagnosides), coumarins, lipids, carotenoids, tannins, saponins, and vitamins (Ayaz and Bertoft, 2001; Fordham et al., 2001; Abizov et al., 2009; Wu et al., 2011; Saboonchian et al., 2014; Patel, 2015). The seeds have been evaluated to be rich in unsaturated fatty acids dominated by linoleic acid, oleic acid, and stearic acid and proteins with globulin and albumin being the most prevalent (Kadir and Kuerban–jiang, 2011). Leaves contain several monoterpenes and sesquiterpenes, namely, betulin, lupeol, kaempferol, oleanolic, ursolic, and syringic acid (Liao et al., 2014). Flavonol glycosides known as elaeagnosides were isolated from the flowers of *E. angustifolia* (Bendaikha et al., 2014). The floral volatiles of *E. umbellata* contain palmitic, C_{14}–C_{20} fatty acid, methyl esters, eugenol, 4-methyl phenol, phenylacetate aldehyde, and (E)-2-nonenol (Ahmad et al., 2005).

Phytochemical evaluation in *E. latifolia* showed the presence of mono- and disaccharides in leaves; fructose and L-ascorbic acid, phenols, flavonoids, and carotenoids in fruits; fatty acids and phytosterols in seed, leaves, and stems (Sakamura and Suga, 1987; Yingthongchai et al., 2014a); and phytosterols, glycoside and saponins in flower (Choudhury and Patel, 2016). Panja et al. (2014) evaluated the phytochemical constituents of *E. latifolia* fruit and found that the fruit had highest carbohydrates (10.96 mg glucose Eq) followed by total phenolics (7.04-mg gallic acid Eq), total flavonoids (5.44-mg quercetin Eq), total ascorbic acid (0.23-mg ascorbic acid Eq) and tannins (0.20-mg catechin Eq). Tannic acid, catechin, purpurin, quercetin, reserpine, and rutin were also reported (Panja et al., 2014). *E. latifolia* leaves contain a good source of phenols [61.15 μg/mL (GAE)] and flavonoids (15.12 μg/mL of quercetin) (Dutta et al., 2018). Phytochemical estimation of *E. kologa* leaf using various solvent extracts reported high phenol (23.11 mg/g), flavonoids (21.19 mg/g), and protein content (0.98 mg/g) in petroleum ether extract, whereas high carbohydrate content (0.94 mg/g) were extracted from water extract as compared to other solvents (Merculieff et al., 2014).

The presence of *γ*–aminobutyric acid (GABA), a non–proteinogenic amino acid, was also reported in *E. latifolia* fruit extract, with high recovery (Meeploy and Niwaspragrit, 2019). GABA content in *E. latifolia* fruit was reported to be highest in orangish red stage (0.1640 mg/g fresh weight) and lowest during the dark red stage (0.0913 mg/g fresh weight) (Meeploy and

Niwaspragrit, 2019). GABA content in tomatoes and cherry tomatoes was highest before maturation and further declined during the ripening stage (Nonaka et al., 2017). Studies have also shown that GABA lowers blood pressure of hypertensive patients except that of normotensive ones (Elliott et al., 1959; Takahashi et al., 1961, 1995).

E. latifolia fruit contains three biochemical compounds such as ascorbic acid, carotenoids, and phenolic compounds that can inhibit tyrosinase production, a compound that synthesis melanin in animals and browning in plants. The tyrosinase inhibitory activity of *E. latifolia* (Ma–Lod flesh) extracts examined by Meeploy and Niwaspragrit (2019) showed that *E. latifolia* could inhibit the tyrosinase activity owing to the presence of ascorbic acid which acts as a melanogenesis inhibitor, presence of some phenolic compounds, specific tyrosinase inactivators, and inhibitors thus slowing down its catalytic activity (Meeploy and Niwaspragrit, 2019).

PHARMACEUTICAL PROPERTIES

Antioxidant: *E. latifolia* fruit is reported to have natural antioxidant components such as vitamins, Phenolics, flavonoids, and carotenoids. The fruits were also reported to be capable of reducing the growth of cancer cells (Matthews, 1994). Kedtien and Joomwong (2018) studied the effect of antioxidant and total phenolic content under three maturity stages (early ripe, moderately ripe, and lately ripe) of *E. latifolia* fruit. The results showed that both antioxidant activity and total phenolic content decreased with the fruit maturity stage. Various in vitro assays evaluated by Panja et al. (2014) also indicated similar effects which might be helpful in preventing the progress of oxidative stress in the body system. Fruit extracts of *E. kologa* also exhibited effective reductive capacity, strong ferric ion chelation, and lipid peroxidation reduction. Evaluation of total phenolic and flavonoid contents and antioxidant capacity of *E. kologa* Schldl. fruits showed positive correlations between polyphenolic and antioxidant capacities (Sasikumar et al., 2012). Methanol extracts of *E. kologa* leaf were not only rich in proteins and carbohydrates but also exhibited high free radical scavenging activity indicating significant antioxidant properties (Vinayaka et al., 2009; Merculieff et al., 2014).

Antimicrobial: It is well documented that plants of the *Elaeagnus* genus are capable of arresting pathogenic microbes. Marshillong (2017) observed antimicrobial properties against pathogens *Escherichia coli*, *Bacillus subtilis*, *Shigella dysentriae*, and *Aspergillus niger* from methanolic extract of *E. latifolia* leaf fruit and seeds. In–vitro antimicrobial activities of leaf and fruit extracts were examined against five pathogenic bacteria and the result showed significant antioxidant and antimicrobial activity (Dutta et al., 2018). Kekuda et al. (2009) studied the efficacy of *E. kologa* Schldl (Syn: *E. latifolia*) fruit using various solvent extracts for antibacterial activity. The susceptibility toward these solvent extracts was tested against *Staphylococcus aureus*, *Bacillus cereus*, *E. coli*, and *Klebsiella pneumonia.* The results indicated that the crude solvent extracts showed antibacterial activities toward the Gram-positive bacteria to more extent than Gram-negative bacteria. Among extracts tested, methanol and chloroform extracts caused more inhibition of *S. aureus* while ethyl acetate and acetone extracts were found to inhibit *B. cereus* to more extent. Among Gram-negative bacteria, *E. coli* exhibited more sensitivity to most of the extracts than *K. pneumoniae.* The methanolic extract of *E. kologa* leaf demonstrated significant antibacterial action toward *Bacillus subtilis* (Merculieff et al., 2014). Antimicrobial properties of other *Elaeagnus* species such as *E. umbellata* and *E. angustifolia* have also been assessed (Sabir et al., 2007; Si et al., 2010; Okmen and Turkcan, 2014; Khan et al., 2016).

Insecticidal: Plant metabolites such as phenols, phenolic acids, flavonoids, catechins, etc. have a crucial role in insecticidal and antimicrobial activity (Savadi et al., 2020). Studies have shown that *E. latifolia* has insecticidal properties which make the plant resistant and less affected by disease and insect pests (Pumnuan et al., 2019a, 2019b). Ethanol extract from leaves showed insecticidal properties against diamondback moth (*Plutella xylostella*) (Pumnuan et al., 2019a) and that of fruits and leaves in killing mealybug nymph (Pumnuan et al., 2019b). Pumnuan et al. (2019a) evaluated the effects of *E. latifolia* leaves and fruit against larvae of diamond black moth (DBM). The result showed that ethanolic extracts from leaves had insecticidal properties against DBM with LC50 at 24 h (4.94–5.47%) and 48 h (3.42–3.63%). Both the extracts also showed more than 45.0% efficacy within 24 h in antifeedant while the most-effective inhibiting DBM growth was at 5% concentration causing 67–87% adult mortality. Vinayaka et al. (2009) studied the insecticidal efficacy of *E. kologa* leaf extract against second instar larvae of mosquito, *Aedes aegypti* using various solvents and results showed that larval mortality was recorded as 100% at 15 and 20 mg/mL of methanol, ethyl acetate and acetone extracts.

Anthelmintic: Helminth infections are among the most prevalent infectious agents in humans and animals particularly in developing countries and many of which still use anthelmintic plants for the treatment of such diseases (Nunomura et al., 2006). *Elaeagnus* species is also known for its antihelminthic property. Kekuda et al. (2009) studied the efficacy of *E. kologa* Schldl (Syn: *E. latifolia*) fruit using various solvents extracts for anthelmintic activity in worms with results showing petroleum ether extract as good anthelmintic activity followed by chloroform extract.

Anti-inflammation: Inflammation is a dynamic biological process for controlling and balancing the body's homeostasis (Fürst and Zündorf, 2014) in response to mechanical injuries, microbial infections, burns, and other hazardous stimuli. Pain and allergic conditions such as asthma, bronchitis, and atopic dermatitis are also indications of inflammation (Patel, 2014). The application of plant-derived bioactives for controlling and suppressing inflammatory crises has been in practice for ages and their efficacy is being validated at present with the empirical findings. To date, the anti-inflammation potential of *E. latifolia* fruits has not been investigated or validated. However, the efficacy of other *Elaeagnus* species in reducing pain and stiffness has been reported in *E. oldhamii* Maxim (Liao et al., 2013) and *E. angustifolia* (Esmaeili and Niknam, 2013; Ebrahimi et al., 2014; Nikniaz et al., 2014), and evaluation of respiratory tract disorders indicated positive results in *E. henryi*, *E. lanceolata*, and *Elaeagnus pungens* leaf concoctions (Ge et al., 2009, 2015).

Application in Nanoscience: Nanoscience deals with the alteration and technologies that are employed for synthesizing nanoparticles of specific size and shape (Jahn, 1999; Nalwa, 2000; Murphy, 2008). Nanotechnology has numerous applications including its uses in the pharmaceutical field to improve drug solubility, bioavailability, and delivery and to develop new and improved therapeutic devices (Chapman, 2005). The major drawback of producing nanoparticles is that they are very expensive, toxic, and hazardous to the environment and humans (Sastry et al., 2004). Alternatives for synthesizing ecofriendly nanoparticles using plants (Ganesan et al., 2013a), fungi (Saifuddin et al., 2009), seaweed (Ganesan et al., 2013b), bacteria (Sachin et al., 2012), and enzymes (Willner et al., 2007) have been employed since it offers many advantages such as time reducing, cost-effective, ecofriendly, and non-toxic. Besides, they are much cheaper than gold nanoparticles and are used as antibacterial or antifungal agents (Rauwel et al., 2015). Works on the synthesis of silver nanoparticles from plants have been reported in *Aloe vera* (Gardea et al., 2003, Chandran et al., 2006), *Emblica officinalis* (Ankamwar, et al., 2005a), *Tamarindus indica* (Ankamwar, et al., 2005b), *Jatropha curcas* (Harekrishna et al., 2009), *Justicia genderussa* (Chinna and Hema 2012), *Pisonia grandis* (Jannathul et al., 2012), leaves of *Merremia tridentata* (Astalakshmi et al., 2013), barks of *Eucalyptus globulus* (Ganesan et al., 2013a), seaweed *Kappaphycus alvarezii* (Ganesan et al., 2013b), pods of *Peltophorum pterocarpum* (Nima et al., 2013), *Ficus bengalensis* (Ganesan et al., 2015), and *Indigofera linifolia* (Nima and Ganesan, 2015).

In the case of *Elaeagnus* species, limited studies are available on nanotechnology research with only few reports on silver nanoparticles synthesis. Phanjom et al., 2012 extracted silver nanoparticles from *E. latifolia* leaf extracts using ultraviolet (UV)-visible spectrophotometer, Transmission electron microscope (TEM), and X-ray diffractometer (XRD). The result reported highly stable almost spherical-shaped *E. latifolia* silver nanoparticles in the average size range from 30 to 50 nm indicating a good source of silver nanoparticles synthesis. Moreover, the biosynthesis method does not require high pressure, energy, temperature, and toxic chemicals as the chemical and physical methods require, and these nanoparticles have a great application in the field of pharmacological and electrical industries. Silver nanoparticles were also extracted from other *Elaeagnus* species, namely, *E. indica* to evaluate toxicity in human pathogens and its therapeutic potential (Natarajan et al., 2013). The effect of incorporating *E. angustifolia* extract for bone tissue engineering (Hokmabad et al., 2019), as well as *in vivo* studies on polyamide nanocapsules containing *E. angustifolia* L., have also been carried out (Esmaeili and Niknam, 2014).

Value Addition: To date, consumption of *E. latifolia* fruits is limited and its application as a functional ingredient in the food or pharmaceutical industry has not yet been reached. In the Northeastern parts of India, fruits are consumed in raw form or occasionally processed as pickles or juice at the household level. Research on the processing and value addition of this fruit was carried out by Nath et al. (2011) and products like jam, fruit leather, and powder have been prepared. In European countries, in-depth research on the development of value-added products from *E. angustifolia* has been carried out. For example, flour produced from *E. angustifolia* dried fruit pulp has been used as a main ingredient in the production of food products to enhance their physical, chemical, sensory value, color properties, and antioxidant potential of the product (Sahan et al., 2013, 2015; Çakmakçi et al., 2015).

2.4 RESEARCH GAPS AND FUTURE PROSPECTS

Nutritional and phytochemical studies have shown that *E. latifolia* has a good source of vitamins, phenols, and flavonoids which corresponds to high antioxidant, antimicrobial, and insecticidal properties. However, only a fraction of its nutritional and phytochemical richness has been studied so far. Therefore, research gaps that have not been carried out earlier in these fields including primary and secondary metabolites quantification, value addition, propagation techniques, germplasm conservation, and commercialization are some important steps toward enhancing the utility of this fruit to its full potential.

Detailed experimental and clinical studies of various biological activities including antimicrobial, antioxidant, antiproliferative, anthelmintic, anticancer, and anti-inflammation from *E. latifolia* need to be carried out in the near future. Research on its active constituents related to nanoparticle synthesis needs to be explored further. The production of silver nanoparticles from *E. latifolia* through eco-friendly green synthesis process would offer numerous benefits for pharmaceutical and other bio-medical applications.

Studies have shown that fruits of *Elaeagnus* species are rich in lycopene content with *E. umbellata* fruit as high (17 times greater) as that of tomato (Kohlmeier, 1997; Fordham et al., 2001). Fruits rich in lycopene, an antioxidant carotenoid, have high potential in fighting against myocardial infarction and various forms of cancer (Giovannucci et al., 1995; Clinton, 1998; Fordham et al., 2001; Farzaei et al., 2015). Thus, if lycopene alone, or in synergy with other carotenoids, is responsible for cancer prevention, identifying *Elaeagnus* species with rich lycopene source and related carotenoids may prove valuable in drug designing through modern science and tools to combat cancer and other related diseases. The lycopene-rich *Elaeagnus* fruit may also act as a substitution for tomato as a lycopene dietary source (Minorsky, 2002).

As of now, *Elaeagnus* fruits including *E. latifolia* are largely underutilized. The fruits are highly perishable with a shelf life of 3–5 days (23°C). The availability of this fruit can be extended by value addition so that products will be available throughout the year. Like *E. angustifolia*, using various processing techniques, fruits of *E. latifolia* can also be value added into various products like jam, jelly, pickle, powder, leather, sauce, flour, herbal tea, wine, candy, ice cream topping, etc., and thereby can minimize the post-harvest losses during peak season. The non–edible parts of the plant can be used for fuel, fodder, medicine, construction, or other value-added products.

E. latifolia is regarded as a useful perspective crop owing to its drought, fire, disease tolerance level, good adaptability in different habitats and soil conditions, early bearing habits, erosion prevention, resistance to high wind velocity, high yields, and varied uses by the local people (Patel et al., 2008; Yingthongchai et al., 2014a, 2014b; Patel, 2015). These characteristics make this species worthy of rehabilitation and restoration of degraded and arid landscapes where only such plants can thrive. This will further prevent soil erosion, and provide windbreaks besides bearing fruit for consumption thereby improving socioeconomy and livelihood with their plantation. Owing to its nitrogen-fixing abilities, *E. latifolia* has also been recognized as an actinorhizal plant that could be beneficial as a host in rehabilitation and ecological succession. In addition, future research on the diversity of microendosmybionts and the host relationship may help unravel their role and function in bioactive compound synthesis (Handique and Parkash, 2014).

E. latifolia, therefore, deserves effective methods of cultivation and management for sustainable production. In Northeast India, this plant is found naturally and sparsely growing in wild habitats and is also planted by villagers in their home–gardens. Therefore, for commercial cultivation and utilization of this fruit, standardization of the propagation techniques is required. Research on genetic diversity linkage among *E. latifolia* populations for the development of improved cultivars with high yield should be taken into consideration, and proper agro & breeding techniques should be developed for higher productivity. Although research on the propagation of this fruit through seeds and cuttings has been carried out, the application of this technique at a commercial scale has not been done. Further, studies on pest and fruit borer attacks during *E. latifolia* fruiting phase also need to be looked upon as this would help to obtain healthy and disease-free fruit.

2.5 CONCLUSIONS

The above review discussed the importance of *E. latifolia* with reference to its traditional, ecological, nutritional, and medicinal uses, its phytochemical composition, and its potential for pharmaceutical and nutraceutical applications. It is evident that *E. latifolia* fruits have rich contents of total soluble solids, acids, ash, protein, iron, and good sources of vitamin C, sugars, essential fatty acids, and minerals such as calcium, phosphorus, potassium, and magnesium. The fruit also has low fat, higher nutritive value, and dietary fiber content compared to other commercial fruits such as apple, litchi, mango, and papaya. Plants with a high nutritive value are considered as a good dietary source and therefore, *E. latifolia* may be included as a part of dietary food for better health care. Besides food, the *E. latifolia* plant has been in use for many generations by tribal people and traditional healers in traditional folk medicine for curing various ailments and diseases such as blood alcohol removal, pain alleviation, wound healing, muscle relaxant, rheumatism, removal or retained placenta and treatment of threatened abortion. The plant is also used as an astringent and for treatment against cutaneous infections, dropsy, anasarca, venereal sores, deafness, and dysentery. Current evidence showed *E. latifolia* as a good source of phytochemicals which however needs further experimental and clinical studies to reveal the beneficial nutritional and therapeutic properties of this plant and for extending its applications in the nutraceutical and pharmaceutical industries. In addition, proper agro and breeding techniques for sustainable utilization of *E. latifolia* should be explored and implemented as it is a high-value plant that can also contribute toward nutrition and food security.

ACKNOWLEDGMENTS

The authors are thankful to the Director of the Institute of Bioresources and Sustainable Development (IBSD), Imphal, Manipur for his support and encouragement in publishing the work. The above work was funded by the Department of Biotechnology, Government of India under the IBSD Core Fund.

KEYWORDS

- ***Elaeagnus latifolia***
- **phytochemical**
- **pharmacological**
- **nutraceutical**

REFERENCES

Abizov, E. A.; Tolkachev, O. N.; Mal'tsev, S. D.; Abizova, E. V. Composition of Biologically Active Substances Isolated from the Fruits of Russian Olive (*Elaeagnus angustifolia*) Introduced in the European part of Russia. *Pharm. Chem. J.* **2009,** *42*, 696–698.

Ahmad, S. D.; Sabir, M. S.; Juma, M.; Asad, H. S. Morphological and Biochemical Variations in *Elaeagnus umbellata* Thunb. from Mountains of Pakistan. *Acta Bot. Croat.* **2005,** *64*, 121–128.

Ahmadiani, A.; Hosseiny, J.; Semnanian, S.; Javan, M.; Saeedi, F.; Kamalinejad, M.; Saremi, S. Antinociceptive and Antiinflammatory Effects of *Elaeagnus angustifolia* Fruit Extract. *J. Ethnopharmacol.* **2000,** *72*, 287–292.

Ankamwar, B.; Chaudhary, M.; Sastry, M. Gold Nanotriangles Biologically Synthesized Using Tamarind Leaf Extract and Potential Application in Vapor Sensing. *Synth. React. Inorg. Metal Organ. Nanotechnol. Metal Chem.* **2005b,** *35*, 19–26.

Ankamwar, B.; Damle, C.; Ahmad, A.; Sastry, M. Biosynthesis of Gold and Silver Nanoparticles Using *Emblica officinalis* Fruit Extract, Their Phase Transfer and Transmetallation in an Organic Solution. *J. Nanosci. Nanotechnol.* **2005a,** *5*, 1665–1671. DOI: 10.1166/jnn.2005.184. PMID: 16245525.

Arab, L.; Steck, S. Lycopene and Cardiovascular Disease. *Am. J. Clin. Nutr.* **2000,** *71*, 1691S–1695S.

Astalakshmi, A.; Nima, P.; Arunkumar, C.; Ganesan, V. Synthesis and Characterization of Silver Nanoparticles Using *Merremia tridentata* (L.) Hall.f., *IJCS*. **2013,** *6*, 87–93.

Ayaz, F. A.; Bertoft, E. Sugar and Phenolic Acid Composition of Stored Commercial Oleaster Fruits. *J. Food Compos. Anal.* **2001,** *14*, 505–511.

Baker, D.; Torrey, J. G.; Kidd, G. H. Isolation of Sucrose Density Fraction and Cultivation of In Vitro of Actinomycetes from Nitrogen–Fixing Root Nodules. *Nature* **1979,** *281*, 76–78.

Bargali, K. Actinorhizal Plants of Kumaun Himalaya and Their Ecological Significance. *Afr. J. Plant Sci.* **2011,** *5*, 401–406.

Bendaikha, S.; Gadaut, M.; Harakat, D.; Magid, A. Acylated flavonol Glycosides from the flower of *Elaeagnus angustifolia* L. *Phytochemistry* **2014,** *103*, 129–136.

Brown, M. J.; Ameer, M. A.; Beier, K. Vitamin B6 Deficiency. In: *StatPearls* [Internet]. StatPearls Publishing: Treasure Island, 2021. PMID: 29261855.

Çakmakçi, S.; Topda¸s, E. F.; Kalin, P.; Han, H.; ¸Sekerci, P.; Köse, L. P.; Gülçin, I. Antioxidant Capacity and Functionality of Oleaster (*Elaeagnus angustifolia* L.) Flour and Crust in a New Kind of Fruity Ice Cream. *Int. J. Food Sci. Technol.* **2015,** *50*, 472–481. DOI: 10.1111/ijfs.12637.

Carr, A. C.; Maggini, S. Vitamin C and Immune Function. *Nutrients* **2017,** *9*, 1211. DOI: 10.3390/nu9111211.

Chandran, S.P; Chaudhary, M.; Pasricha, R.; Ahmad, A.; Sastry, M. Synthesis of Gold Nanotriangles and Silver Nanoparticles Using *Aloe vera* Plant Extract. *Biotechnology* **2006,** *13*, 1357.

Chapman, P. Nanotechnology in the Pharmaceutical Industry. *Expert Opin. Therap. Patents*, **2005,** *15*, 249–251. DOI: 10.1517/13543776.15.3.249.

Chatterjea, M. N.; Shinde, R. *Textbook of Medical Biochemistry*, 8th ed.; JP Medical Ltd., 2011; pp 163–196.

Chinna, M.; Hema, P. Green Synthesis of Highly Stable Silver Nanoparticles Using *Justicia gendenussa*. *Int. J. Nanotechnol. App.* **2012,** 39–57.

Choudhury, A.; Patel, N. A. HPTLC Fingerprint Analysis of Bastard Oleaster (*Elaeagnus latifolia* Linn): An Edible Plant Used by the Tribal Community of Northeast India. *Int. J. Pharmacog. Phytochem. Res.* **2016,** *8*, 1920–1925.

Clinton, S. K. Lycopene: Chemistry, Biology, and Implications for Human Health and Disease. *Nutr. Rev.* **1998,** *56*, 35–51.

Dutta, I.; Gogoi, B.; Sharma, R.; Sharma, H. K. In Vitro Evaluation of Leaves and Fruits of *Elaeagnus latifolia* L. for Antioxidant and Antimicrobial Activities. *Asian J. Chem.* **2018,** *30*, 2433–2436.

Ebrahimi, A. A.; Nikniaz, Z.; Ostadrahimi, A.; Mahdavi, R.; Nikniaz, L. The Effect of *Elaeagnus angustifolia* L. Whole Fruit and Medulla Powder on Women with Osteoarthritis of the Knee: A Randomized Controlled Clinical Trial. *Eur. J. Integr. Med.* **2014,** *6*, 672–679.

Elliott, K. A. C.; Hobbiger, F. Gamma Aminobutyric Acid: Circulatory and Respiratory Effects in Different Species; Re–investigation of the Anti–strychnine Action in Mice. *J. Physiol.* **1959,** *146*, 70–84.

Esmaeili, A.; Niknam, S. Characterization of Nanocapsules Containing *Elaeagnus angustifolia* L. Extract Prepared Using an Emulsion–Diffusion Process, Flavour. *Fragr. J.* **2013,** *28*, 309–315.

Esmaeili, A.; Niknam, S. Preparation of Polyamide Nanocapsules of *Elaeagnus angustifolia* L. Delivery with In Vivo Studies. *Ind. Crops Prod.* **2014,** *55*, 49–55.

Farahbakhsh, S.; Arbabian, S.; Emami, F.; Moghadam, B. R.; Ghoshooni. H.; Noroozzadeh, A.; Sahraei, H.; Golmanesh, L.; Jalili, C.; Zrdooz, H. Inhibition of Cyclooxygenase Type 1 and 2 Enzyme by Aqueous Extract of *Elaeagnus angustifolia* in Mice. *Auton Neurosci–Basic.* **2011,** *2*, 31–37.

Farzaei, M. H.; Bahramsoltanic, R.; Abbasabadia, Z.; Rahimib, R. A Comprehensive Review on Phytochemical and Pharmacological Aspects of *Elaeagnus angustifolia* L. *J. Pharm. Pharmacol.* **2015,** *67*, 1467–1480.

Follstad, S. J. J.; Harner, M. J.; Tibbets, T. M. *Elaeagnus angustifolia* Elevates Soil Inorganic Nitrogen Pools in Riparian Ecosystems. *Ecosystems*, **2010,** *13*, 46–61.

Fordham, I. M.; Clevidence, B. A.; Wiley, E. R.; Zimmerman, R. H. Fruit of Autumn Olive: A Rich Source of Lycopene. *HortScience* **2001,** *36*, 1136–1137.

Fürst, R.; Zündorf, I. Plant–Derived Anti–Inflammatory Compounds: Hopes and Disappointments regarding the Translation of Preclinical Knowledge into Clinical Progress. *Hindawi Publishing Corporation Mediators of Inflammation*, 2014; pp 1–9. http://dx.doi.org/10.1155/2014/146832.

Gamba, G.; Donno, D.; Mellano, M. G.; Riondato, I.; Biaggi, M. D.; Randriamampionona, D.; Beccaro, G. L. Phytochemical Characterization and Bioactivity Evaluation of Autumn Olive (*Elaeagnus umbellata* Thunb.) Pseudodrupes as Potential Sources of Health–Promoting Compounds. *Appl. Sci.* **2020,** *10*, 4354. DOI: 10.3390/app10124354.

Gamble, J. S. *A Manual of Indian Timbers: An Account of the Growth, Distribution, and Uses of the Trees and Shrubs of India and Ceylon with Descriptions of Their Wood–Structure*; S. Low, Marston & Co. Ltd.: London, 1972.

Ganesan, V.; Astalakshmi, A.; Malathi, R. Possible Biochemical Mechanism in Aerial Roots of *Ficus bengalensis* L. Mediated Synthesis of Silver Nanoparticles. *J. Nanotech.* **2015,** *104*, 130–137.

Ganesan, V.; Astalaksmi, A.; Nima, P. A Green Approach in the Synthesis of Silver Nanoparticles Using Bark of *Eucalyptus globulus*, Labill. *Int. J. Pharm. Sci. Rev. Res.* **2013a,** *23*, 47–52.

Ganesan, V.; Devi, J. A.; Astalakshmi, A.; Nima, P.; Amirthalingam, T. Eco–Friendly Synthesis of Silver Nanoparticles Using a Sea Weed, *Kappaphycus alavarezii*. *Int. J. Eng. Adv. Rese*. **2013b,** *2*, 559–563.

Gardea, T. J. L.; Gomez, E.; Peralta, V. J.; Parsons, J. G.; Troiani, H. E.; Jose, Y. Synthesis of Gold Nanotriangles and Silver Nanoparticles Using *Aloe vera* Plant Extract. *Langmuir* **2003,** *13*, 1357.

Gardner. Nitrogen Fixation in *Elaeagnus* Roots Nodules. *Nature* **1958,** *181*, 717–718.

Gartner, C.; Stahl, W.; Sies, H. Lycopene Is More Bioavailable from Tomato Paste than from Fresh Tomatoes. *Am. J. Clin. Nutr.* **1997,** *66*, 116–122.

GBIF: The Global Biodiversity Information Facility. https://www.gbif.org/occurrence/search?q=Elaeagnus%20latifolia

Ge, Y.; Liu, J.; Su, D. In Vivo Evaluation of the Anti–Asthmatic, Antitussive and Expectorant Activities of Extract and Fractions from *Elaeagnus pungens* Leaf. *J. Ethnopharmacol.* **2009,** *126*, 538–42. DOI: 10.1016/j.jep.2009.08.042.

Ge, Y.; Zhang, F.; Qin, Q.; Shang, Y.; Wan, D. In Vivo Evaluation of the Antiasthmatic, Antitussive, and Expectorant Activities and Chemical Components of Three Elaeagnus Leaves. *Evid. Based Complement Alternat. Med.* **2015,** *2015*, 1–7. DOI: 10.1155/2015/428208.

Gill, N. S.; Gupta, M. *Elaeagnus conferta*: A Comprehensive Review. *Res. J. Pharm. Tech.* **2018,** *11*, 2667–2671.

Giovannucci, E.; Ascherio, A.; Rimm, E. B.; Stampfer, M. J.; Colditz, G. A.; Willett, W. C. Intake of Carotenoids and Retino in Relation to Risk of Prostate Cancer, *JNCI.* **1995,** *87*, 1767–1776.

Gowda, B. *Vanaspathi Kosha*; Kalpatharu Research Academy: Bangalore, 2004; p 67.

Gupta, B. L. *Forest flora of Chakrata*; Publisher Forest Research Institute Press: Dehradun, 1945.

Gürbüz, I.; Ustün, O.; Yesilada, E.; Sesik, E.; Kutsal, O. Antiulcerogenic Activity of Some Plants Used as Folk Remedy in Turkey. *J. Ethnopharmacol.* **2003,** *88*, 93–97.

Handique, L.; Parkash, V. Nodular Micro–Endosymbiontic and Endomycorrhizal Colonization in *Elaeagnus latifolia* L.: A Noble Host of Symbiotic Actinorhizal and Endomycorrhizal Association. *IJSR*. **2014,** *3*, 1713–1717.

Harekrishna, B.; Dipak, K. R.; Priyanka, S.; Sankar, P.; Ajay, M. Green Synthesis of Silver Nanoparticles Using Latex of *Jatropa curcas*. *Colloids Surfaces A: Physicochem. Eng. Aspects* **2009,** *339*, 134–141.

Hazarika, T. K.; Marak, S.; Mandal, D.; Upadhyaya, K.; Nautiyal, B. P.; Shukla, A. C. Underutilized and Unexploited Fruits of Indo–Burma Hot Spot, Meghalaya, North–East India: Ethno–Medicinal Evaluation, Socio–Economic Importance and Conservation Strategies. *Genet. Resour. Crop Evol.* **2016,** *63*, 289–304. https://doi.org/10.1007/s10722-015-0248-0.

Hokmabad, V. R.; Davaran, S.; Aghazadeh, M. Alizadeh, E.; Salehi, R.; Ramazani, A. Effect of Incorporating *Elaeagnus angustifolia* Extract in PCL–PEG–PCL Nanofibers for Bone Tissue Engineering. *Front. Chem. Sci. Eng*. **2019,** *13*, 108–119.

Hosseinzadeh, H.; Ramezani, M.; Namjo, N. Muscle Relaxant Activity of *Elaeagnus angustifolia* L. Fruit Seeds in Mice. *J. Ethnopharmacol.* **2003,** *84*, 275–278.

Institute of Medicine. *Dietary Reference Intakes for Thiamin, Riboflavin, Niacin, Vitamin B6, Folate, Vitamin B12, Pantothenic Acid, Biotin, and Choline*; The National Academies Press: Washington, DC, 1998; p 592. https://doi.org/10.17226/6015.

Ismail, F.; Farah, N. T.; Memon, A. N. Determination of Water Soluble Vitamin in Fruits and Vegetables Marketed in Sindh Pakistan. *Pak. J. Nutr.* **2013,** *12*, 197–199.

Jahn, W. Review: Chemical Aspects of the Use of Gold Clusters in Structural Biology. *J. Struct. Biol.* **1999,** *127*, 106–112.

Jannathul, F. M.; Lalitha, P.; Shubashini, K. S. Novel Synthesis of Silver Nanoparticles Using Leaf Ethanol Extract of *Pisonia grandis* (R.Br). *Der Pharma Chemica*, **2012,** *4*, 2320–2326.

Kadir, M.; Kuerban–jiang, B. Research on fat and protein compositions in seeds of wild *Elaeagnus angustifolia* from Xinjiang Uigur Autonomous Region. *Med. Plant*, **2011,** *2*, 51–53.

Kedtien, P.; Joomwong, A. Study on Total Phenolic Contents and Antioxidant Activities in Wild Olive (*Elaeagnus latifolia)* Fruit. *Acta Hortic.* **2018,** *1213*, 429–432. DOI: 10.17660/ActaHortic.2018.1213.63.

Kekuda, P. T. R.; Nethravathi, H. R.; Vinayaka, K. S.; Thippeswamy, N. B. Studies on Antibacterial and Anthelmintic Activity of *Elaeagnus kologa* SCHLDL. *J. Pharm. Res.* **2009,** *2*, 1885–1887.

Kessler, K. J. Destruction of *Gnomnoia leptostyla* Perithecia on Juglansniga Leaves by Microarthropodes Associated with *E. umbellata* Litter. *Mycologia* **1990,** *82*, 387–390.

Khan, S. U.; Khan, A.; Shah, A. A.; Liu, P. T.; Hui, L. F.; Liu, Z. Heavy Metals Content, Phytochemical Composition, Antimicrobial and Insecticidal Evaluation of *Elaeagnus angustifolia. Toxicol. Ind. Health* **2016,** *32*, 154–161. DOI: 10.1177/0748233713498459.

Khomdram, S.; Arambam, S.; Devi, G. S. Nutritional Profiling of Two Underutilized Wild Edible Fruits *Elaeagnus pyriformis* and *Spondias pinnata. Ann. Agric. Res.* **2014,** *35*, 129–135.

Kim, S. C.; Chang, K. U.; Park, D.; Kim, M. C.; Song, C. H.; An-Chung-Sun, S. D. Isolation of Symbiotic Frankia Eulk 1 Strain from Root Nodule of *E. umbellata. Korean J. Bot.* **1993,** *36*, 177–182.

Kirsh, V. A.; Mayne, S. T.; Peters, U.; Chatterjee, N.; Leitzmann, MF.; Dixon, L. B.; Urban, D. A.; Crawford, E. D.; Hayes, R. B. A Prospective Study of Lycopene and Tomato Product Intake and Risk of Prostate Cancer. *Cancer Epidemiol. Biomarkers Prev.* **2006,** *15*, 92–98.

Kohlmeier, L.; Kark, J. D.; Gomez–Garcia, E.; Martin, B. C.; Steck, S. E. Lycopene and Myocardial Infection Risk in the EURAMIC Study. *Am. J. Epidemiol.* **1997,** *146*, 618–626.

Krebs, M. O.; Bellon, A.; Mainguy, G.; Jay, T. M.; Frieling, H. One–Carbon Metabolism and Schizophrenia: Current Challenges and Future Directions. *Trends Mol. Med.* **2009,** *15*, 562–570.

Krupinsky, J. M.; Frank, A. B. Effects of Water Stress on *Tubercularia* Canker of Russian Olive. *Montana State Univ. Coop. Ext.* **1986,** *117*, 171–172.

Kumar, V.; Said, P. P. Physico-Chemical Characterization *Elaeagnus latifolia* from Sikkim Himalayas. *Int. J. Pure App. Biosci.* **2018,** *6*, 361–366. DOI: http://dx.doi.org/10.18782/2320–7051.6386.

Lachowicz, S.; Bieniek, A.; Gil, Z.; Bielska, N.; Markuszewski, B. Phytochemical Parameters and Antioxidant Activity of New Cherry Silverberry Biotypes (*Elaeagnus multiflora* Thunb.). *Eur. Food Res. Technol.* **2019,** *245*, 1997–2005.

Lee, M. S.; Lee, Y. K.; Park, O. J. Cherry Silver Berry (*Elaeagnus multiflora*) Extracts Exert Antiinflammatory Effects by Inhibiting COX–2 and Akt Signals in HT–29 Colon Cancer Cells. *Food Sci. Biotechnol.* **2010,** *19*, 1673–1677.

Li, L. H.; Baek, I. K.; Kim, J. H.; Kang, K. H.; Koh, Y. S.; Jung, Y. D.; Cho, C. K.; Choi, S. Y.; Shin, B. A. Methanol Extract of *Elaeagnus glabra*, a Korean Medicinal Plant, Inhibits HT1080 Tumor Cell Invasion. *Oncol. Rep.* **2009,** *21*, 559–563.

Liao, C. R.; Ho, Y. L.; Huang, G. J.; Yang, Y. S.; Chao, C. Y.; Chang, Y. S.; Kuo, Y. H. One Lignanoid Compound and Four Triterpenoid Compounds with Anti–Inflammatory Activity from the Leaves of *Elaeagnus oldhamii* Maxim. *Molecules* **2013,** *18*, 13218–13227.

Liao, C. R.; Kuo, Y. H.; Ho, Y. L.; Wang, C. Y.; Yang, C. S.; Lin, C. W.; Chang, Y. S. Studies on Cytotoxic Constituents from the Leaves of *Elaeagnus oldhamii* Maxim. in Non–Small Cell Lung Cancer A549 Cells. *Molecules* **2014,** *19*, 9515–9534.

Mahmood, L. The Metabolic Processes of Folic Acid and Vitamin B12 Deficiency. *J. Health Res. Rev.* **2014,** *1*, 5–9.

Marshillong, K. L. *Qualitative Phytochemical Study of Bioactive Compounds from Natural Source (Elaeagnus latifolia and Flemingia vestita) and Screening of Antimicrobial and Antifungal Activity*. PhD Thesis, Assam Bon Bosco University, 2017.

Matthews, V. *The New Plantsman*, Vol. 1. Royal Horticultural Society, 1994.

Meeploy, M.; Niwaspragrit, C. Biological Activities and GABA Content of Ma–Lod (*Elaeagnus latifolia* Linn.). International Conference on Engineering, Applied Sciences and Technology, IOP Publishing, IOP Conf. Series: *Mater. Sci. Eng.* **2019,** *639*, 1–9. DOI: 10.1088/1757–899X/639/1/012043.

Merculieff, Z.; Ramnath, S.; Sankoli, S. M.; Venkataramegowda, S.; Murthy, G. S.; Ceballos, R. M. Phytochemical, Antioxidant and Antibacterial Potential of *Elaeagnus kologa* (Schlecht.) Leaf. *Asian Pac. J. Trop. Med.* **2014,** *7S1*, S599–S602.

Minorsky, P. V. The Hot and the Classic. *Plant Physiol.* **2002,** *130*, 1077–1078.

Mirhydar, H. *Encyclopedia of Plants: Indications of Plants in the Prevention and Treatment of Diseases*, Vol. 2; Islamic Farhang: Tehran, 1998; pp 163–164.

Murphy, C. J. Sustainability as an Emerging Design Criterion in Nanoparticle Synthesis and Applications. *J. Mater. Chem.* **2008,** *18*, 2173–2176.

Murty, P. P.; Rao, G. M. N.; Rao, D. S. Nutritional and Medicinal Values of Some Wild Edible Fruits of Andhra Pradesh, India. *J. Sci.* **2015,** *5*, 385–393.

Nalwa, H. S. *Hand Book of Nanostructural Materials and Nanotechnology*; Academic Press New York, 2000, pp 1–5.

Natanzi, M. M.; Pasalar, P.; Kamalinejad, M.; Dehpour, A. R.; Tavangar, S. M.; Sharifi, R.; Ghanadian, N.; Rahimi–Balaei, M.; Gerayesh–Nejad, S. Effect of Aqueous Extract of *Elaeagnus angustifolia* Fruit on Experimental Cutaneous Wound Healing in Rats. *Acta Med. Iran.* **2012,** *50*, 589–596.

Natarajan, R. K.; Ekambaram, N.; Nayagam, A. A. J.; Gurunagarajan, S.; Muthukumar, Manimaran, A. *Elaeagnus indica* Mediated Green Synthesis of Silver Nanoparticles and its Potent Toxicity Against Human Pathogens. *Glob. J. Pharmacol.* **2013,** *7*, 222–231.

Nath, A.; Swer, T. L.; Deka, B. C.; Patel, R. K. Nutritional Status and Value Addition in Sohshang Fruit (*Elaeagnus latifolia*). *Beverage Food World* **2011,** *38*, 46–47.

Nazir, N.; Zahoor, M.; Nisar, M. A Review on Traditional Uses and Pharmacological Importance of Genus *Elaeagnus* Species. *Botan. Rev.* **2020,** *86*, 247–280.

Nikniaz, Z.; Ostadrahimi, A.; Mahdavi, R.; Ebrahimi, A. A.; Nikniaz, L. Effects of *Elaeagnus angustifolia* L. Supplementation on Serum Levels of Inflammatory Cytokines and Matrix Metalloproteinases in Females with Knee Osteoarthritis. *Complement. Ther. Med.* **2014,** *22*, 864–869.

Nima, P.; Astalakshmi, A.; Ganesan, V. A Comparative Account on the Green Synthesis of Silver Nanoparticles Using Leaves and Pods of *Peltophorum pterocarpum* (DC.) Backer ex Heyne. *J. Bioprocess Technol.* **2013,** *98*, 308–317.

Nima, P.; Ganesan, V. Differential potentialities of pod broth of *Indigofera linifolia* (L.F) Retz. in fabricating noble metal nanoparticles. *Int. J. Curr. Res.* **2015,** *6*, 9504–9510.

Nonaka, S.; Arai, C.; Takayama, M.; Matsukura, C.; Ezura, H. Efficient Increase of γ–Aminobutyric Acid (GABA) Content in Tomato Fruits by Targeted Mutagenesis. *Sci. Rep.* **2017,** *7*, 1–14. https://doi.org/10.1038/s41598-017-06400-y

Obón, C.; Rivera, D.; Alcaraz, F.; Attieh, L. Beverage and Culture. "Zhourat", a Multivariate Analysis of the Globalization of a Herbal Tea from the Middle East. *Appetite* **2014,** *79*, 1–10.

Okmen, G.; Turkcan, O. A Study on Antimicrobial, Antioxidant and Antimutagenic Activities of *Elaeagnus angustifolia* L. Leaves. *Afr. J. Tradit. Complement Altern. Med.* **2014,** *11*, 116–120.

Pandey, Y.; Upadhyay, S.; Bhatt, S. S. Phyto–Chemical Constituent of Some Wild Edible Fruits of Sikkim Himalaya. *J. Pharmacogn. Phytochem.* **2018,** *7*, 1045–1047.

Panja, S.; Chaudhuri, D.; Baban, G. N.; Le, M. H.; Mandal, N. In Vitro Assessment of Phytochemicals, Antioxidant and DNA Protective Potential of Wild Edible Fruit of *Elaeagnus latifolia* Linn. *Fruits*, **2014,** *69*, 303–314.

Paschke, M. W.; Dawason, J. O.; David, M. B. Soil Nitrogen Mineralization Under Black Walnut Interplanted with *Elaeagnus umbellata* or Black Alder. *J. Plant Soil*, **1989,** *118*, 33–42.

Patel, R. K.; Singh, A.; Deka, B. C. Soh–Shang (*Elaeagnus latifolia*): An Under–Utilized Fruit of North East Region Needs Domestication. *ENVIS Bull.: Himalayan Ecol.* **2008,** *16*, 1–2.

Patel, R. K.; Verma, V. K.; Deshmukh, N. A.; Jha, A. K.; Nath, A.; Patel, R. S. Underutilized Fruits of Meghalaya: An Option for Food and Nutritional Security. In: *Horticulture for Economic prosperity and Nutritional Security in 21st Century*; Hazarika, T. K., Nautiyal, B. P., Eds.; Westville Publishing House: New Delhi, 2013, pp 256–263.

Patel, S. Plant Genus *Elaeagnus*: Underutilized Lycopene and Linoleic Acid Reserve with Permaculture Potential. *Fruits* **2015,** *70*, 191–199.

Paulsamy, S.; Kumar, P. S.; Kumar, A. M. A.; Kumar, P. S. *Elaeagnus kologa* Schlecht.–An Under Utilized Edible and Endemic Fruit Plant in Nilgiris, the Western Ghats. *Indian J. Nat. Prod. Resour.* **2010,** *1*, 258–260.

Phanjom, P.; Sultana, A.; Sarma, H.; Ramchiary, J.; Goswami, K.; Baishya, P. Plant–Mediated Synthesis of Silver Nanoparticles Using *Elaeagnus latifolia* Leaf Extract. *Digest J. Nanomater. Biostruct.* **2012,** *7*, 1117–1123.

Pinto, M. P.; Santos, C. N.; Henriquesa, C.; Lima, G.; Quedas, F. Lycopene Content and Antioxidant Capacity of Portuguese Watermelon Fruits. *Electron. J. Environ. Agric. Food Chem.* **2011,** *10*, 2090–2097.

Pourcel, L.; Moulin, M.; Fitzpatrick, T. B. Examining Strategies to Facilitate Vitamin B_1 Biofortification of Plants by Genetic Engineering. *Front. Plant Sci.* **2013,** *4*, 1–8. https://doi.org/10.3389/fpls.2013.00160.

Pumnuan, J.; Sannongmueang, T.; Inyod, T.; Insung, A. Effectiveness of Bastard Oleaster (*Elaeagnus latifolia*) Extracts Against the Nymph of Mealybug (*Phenacoccus manihoti*). *Acta Hortic.* **2019b,** *1259*, 127–134. DOI: 10.17660/ActaHortic.2019.1259.21.

Pumnuan, J.; Sannongmueang, T.; Inyod, T.; Insung, A. Insecticidal Properties of Bastard Oleaster (*Elaeagnus latifolia*) Extracts Against Diamondback Moth (*Plutella xylostella*). *Acta Hortic.* **2019a,** *1259*, 135–142. DOI: 10.17660/ActaHortic.2019.1259.22.

Qureshi, R. A.; Ghufran, M. A.; Sultana, K. N.; Ashraf, M.; Khan, A. G. Ethnobotanical Studies of Medicinal Plants of Gilgit District and Surrounding Areas. *Ethnobot. Res. Appl.* **2006,** *5*, 115–122.

Ramezani, M.; Hosseinzadeh, H.; Daneshmand, N. Antinociceptive Effect of *Elaeagnus angustifolia* Fruit Seeds in Mice. *Fitoterapia*, **2001,** *72*, 255–262.

Rauwel, P.; Küünal, S.; Ferdov, S.; Rauwel, E. A Review on the Green Synthesis of Silver Nanoparticles and Their Morphologies Studied via TEM. *Adv. Mater. Sci. Eng.* **2015,** *2015*, 1–9. https://doi.org/10.1155/2015/682749.

Roomi, M. W.; Kalinovsky, T.; Monterrey, J.; Rath, M.; Niedzwiecki, A. In Vitro Modulation of MMP–2 and MMP–9 in Adult Human Sarcoma Cell Lines by Cytokines, Inducers and Inhibitors, *Int. J. Oncol.* **2013,** *43*, 1787–1798.

Rymbai, H.; Deshmukh, N. A.; Roy, A. R.; Roy, S. S.; Jha, A. K. Floral Morphology of *Elaeagnus latifolia* L. *Indian J. Hort.* **2017,** *74*, 340–345. DOI: 10.5958/0974–0112.2017.00068.8.

Rymbai, H.; Deshmukh, N. A.; Talang, H. D.; Assumi, S. R.; Devi, M. B.; Mawlein, J.; Jha, A. K. Evaluation of Fruit Characteristics of *Elaeagnus latifolia* L. in the North Eastern Hill Region, India. *Int. J. Minor Fruits Med. Aromatic Plants* **2020,** *6*, 38–42.

Rymbai, H.; Patel, R. K.; Deshmukh, N. A.; Jha, A. K.; Verma, V. K.; Roy, A. R. Physical and Bio–Chemical Profiling of sohshang (*Elaeagnus latifolia* Linn) Fruit in the Northeastern Hills Region, India. In: *Proceeding of National Seminar on 'Strategies for Conservation, Improvement and Utilization of Underutilized Fruits' Organized by Central Horticultural Experiment Station and Society for Promotion of Horticulture, Bangalore, India* on 4–5th December, 2014; p 368. DOI: 10.13140/RG.2.2.23349.42720.

Rymbai, H.; Roy, A. R.; Deshmukh, N. A.; Jha, A. K.; Shimray, W.; War, G. F.; Ngachan, S. V. Analysis Study on Potential Underutilized Edible Fruit Genetic Resources of the Foothills Track of Eastern Himalayas, India. *Genet. Resour. Crop Evol.* **2016,** *63*, 125–139.

Sabir, M. S.; Ahmad, D. S.; Hussain, I. M.; Tahir, K. M. Antibacterial Activity of *Elaeagnus umbellata* (Thunb.) a Medicinal Plant from Pakistan. *Saudi Med. J.* **2007,** *28*, 259–263.

Sabir, S.M; Riaz, K. Morphological, Biochemical and Elemental Analysis of *Elaeagnus umbellata*, a Multipurpose Wild Shrub from Pakistan. *J. Appl. Hortic.* **2005,** *7*, 113–116.

Saboonchian, F.; Jamei, R.; Hosseini, S. S. Phenolic and flavonoid Content of *Elaeagnus angustifolia* L. (Leaf and flower). *Avicenna J. Phytomed.* **2014,** *4*, 231–238.

Sachin, S.; Anupama, P.; Meenal, K. Biosynthesis of Silver Nanoparticles by Marine Bacterium, Idiomarine *sp.* PR58–8. *Bull. Mater. Sci.* **2012,** *35*, 1201–1205.

Sahan, Y.; Dundar, A. N.; Aydin, E.; Kilci, A.; Dulger, D.; Kaplan, F. B.; Gocmen, D.; Celik, G. Characteristics of Cookies Supplemented with Oleaster (*Elaeagnus angustifolia* L.) Flour. I. Physicochemical, Sensorial and Textural Properties, *J. Agric. Sci.* **2013,** *5*, 160–168. DOI: 10.5539/jas.v5n2p160.

Sahan, Y.; Gocmen, D.; Cansev, A.; Celik, G.; Aydin, E.; Dundar, A. N.; Dulger, D.; Kaplan, H. B.; Kilci, A.; Gucer, S. Chemical and Techno–Functional Properties of Flours from Peeled and Unpeeled Oleaster (*Elaeagnus angustifolia* L.). *J. Appl. Bot. Food Qual.* **2015,** *88*, 34–41. DOI: 10.5073/JABFQ.2015.088.007.

Saifuddin, N.; Wong, C. W.; Yasimura, A. A. N. Rapid Biosynthesis of Silver Nanoparticles Using Culture Supernatant of Bacterial with Microwave Irradiation. *E–J. Chem.* **2009,** *6*, 61–70.

Sakamura, F.; Suga, T. Changes in Chemical Components of Ripening Oleaster Fruits. *Phytochem.* **1987,** *26*, 2481–2484.

Sarwar, A. G. Medicinal Plant Genetic Resources of Bangladesh–Genera Represented by Single Species and Their Conservation Needs. *JMPS* **2015,** *3*, 65–74.

Sasikumar, J. M.; Patharaj, J.; Adithya, E. S.; Paul, H. C.; Shamna, R. Antioxidant Capacity and Phenolic Content of *Elaeagnus kologa* Schlecht. an Underexploited Fruit from India. *Free Radicals Antioxidants* **2012,** *2*, 28–35.

Sastry, M.; Ahmad, A.; Khan, M. I.; Kumar, R. Microbial Nanoparticle Production. In: *Nanobiotechnology*; Niemeyer, C. M., Mirkin, C. A., Eds.; Wiley–VCH: Weinheim, Germany, 2004; pp 126–135.

Savadi, S.; Vazifedoost, M.; Didar, Z.; Nematshahi, M. M.; Jahed, E. Phytochemical Analysis and Antimicrobial/Antioxidant Activity of *Cynodon dactylon* (L.) Pers. Rhizome Methanolic Extract. *J. Food Qual.* **2020,** 10. https://doi.org/10.1155/2020/5946541.

Schwalbe, R.; Steele–Moore, L.; Goodwin, A. C. *Antimicrobial Susceptibility Testing Protocols*; CRC Press: New York, 2007.

Seal, T. Nutritional Composition of Wild Edible Fruits in Meghalaya State of India and Their Ethno–Botanical Importance. *Res. J. Bot.* **2011,** *6*, 58–67. DOI: 10.3923/rjb.2011.58.67.

Seal, T.; Chaudhuri, K.; Pillai, B. Water Soluble Vitamin Estimation in Five Wild Edible Fruits Consumed by the Tribal People of North–Eastern Region in India by High Performance Liquid Chromatography. *IJCS* **2017,** *5*, 1576–1584.

Seal, T.; Pillai, B.; Chaudhuri, K. Nutritional Potential of Wild Edible Fruits, Traditionally Used by the Local People of Meghalaya State in India. *Indian J. Nat. Prod. Resour.* **2014,** *5*, 359–364.

Sharma, H. K.; Chhangte, L.; Dolui, A. K. Traditional Medicinal Plants in Mizoram, India. *Fitoterapia* **2001,** *72*, 146–161.

Shaw, D. L. The Design and Use of Living Snow Fences in North America. *Agric. Ecosyst. Environ.* **1988,** *22–23*, 351–362.

Si, C.; Zhang, Y.; Zhu, Z.; Liu, P. T.; Hui, L. F.; Liu, Z. Antibacterial Activity of *Elaeagnus angustifolia* Bark. *Planta Med.* **2010,** *76*, 88.

Soares, N. D.; Machado, C. L.; Trindade, B. B.; Lima, I. C.; Gimba, E. R.; Teodoro, A. J.; Takiya, Ch.; Borojevic, R. Lycopene Extracts from Different Tomato–Based Food Products Induce Apoptosis in Cultured Human Primary Prostate Cancer Cells and Regulate TP53, *Bax* and *Bcl–2* Transcript Expression. *Asian Pac. J. Cancer Prev.* **2017,** *18*, 339–345. DOI: 10.22034/APJCP.2017.18.2.339.

Sternberg, G. *Elaeagnus umbellata* in Illinois Conservation Practice. Report. 111. Department of Conservation, Virginia, 1982; pp 251–278.

Sundriyal, M.; Sundriyal, R. C. Underutilized Edible Plants of the Sikkim Himalaya: Need for Domestication. *Curr. Sci.* **2003,** 85.

Suwanaruang, T. Analyzing Lycopene Content in Fruits. International Conference on Inventions & Innovations for Sustainable Agriculture 2016, ICIISA, 2016. *Agric. Agric. Sci. Procedia* **2016,** *11*, 46–48.

Suwannasom, N.; Kao, I.; Pruß, A.; Georgieva, R.; Bäumler, H. Riboflavin: The health benefits of a forgotten natural vitamin. *Int. J. Mol. Sci.* **2020,** *21* (3), *950*. DOI: 10.3390/ijms21030950.

Tahiliani, A. G.; Beinlich, C. J. Pantothenic Acid in Health and Disease. *Vitamins Hormones* **1991,** *1*, 165–228.

Takahashi, H.; Sumi, M.; Koshino, F. Effect of Gamma–Aminobutyric Acid (GABA) on Normotensive or Hypertensive Rats and Men. *Jpn. J. Physiol.* **1961,** *11*, 89–95.

Takahashi, H.; Tiba, M.; Iino, M.; Takayasu, T. The Effect of γ–Aminobutyric Acid on Blood Pressure. *Jpn. J. Physiol.* **1995,** *5*, 334–341.

The Plant List, Version 1.1; Published on the Internet, 2013. http://www.theplantlist.org

Tuskan, G. A.; Laughlin, K. Windbreak Species Performance and Management Practices as Reported by Montana and North Dakota Landowners. *J. Soil Water Conserv.* **1991,** *46*, 225–228.

USDA.NCC Carotenoid Database for U. S. Foods. 1998. http://www.nal.usda.gov/fnic/foodcomp/Data/car98/car98.htm

USDA–United States Department of Agriculture. *US Food Supply Database*; Center for Nutrition and Policy Promotion: Beltsville, 2003. http://www.ars.usda.gov/ba/bhnrc/ndl.

Vinayaka, K. S.; Kekuda, T. R. P.; Nethravathi, H. R.; Thippeswamy, N. B.; Sudharshan, S. J. Free Radical Scavenging and Insecticidal Activity of *Elaeagnus kologa* Schldl. *Drug Invention Today* **2009,** *1*, 74–77.

Wan, L.; Tan, H. L.; Thomas–Ahner, J. M.; Pearl, D. K.; Erdman, J. W.Jr.; Moran, N. E.; Clinton, S. K. Dietary Tomato and Lycopene Impact Androgen Signaling–and Carcinogenesis–Related Gene Expression During Early TRAMP Prostate Carcinogenesis. *Cancer Prev. Res. (Phila)*. **2014,** *7*, 1228–1239. DOI: 10.1158/1940-6207.CAPR-14-0182.

Wang, S. Y.; Bowman, L.; Ding, M. Variations in Free Radical Scavenging Capacity and Antiproliferative Activity Among Different Genotypes of Autumn Olive (*Elaeagnus umbellata*). *Planta Med.* **2007,** *73*, 468–477.

Willner, I.; Basnar, B.; Willner, B. Nanoparticle–Enzyme Hybrid Systems for Nanobiotechnology. *FEBS J.* **2007,** *274*, 302–309.

Wu, M. C.; Hu, H. T.; Yang, L.; Yang, L. Proteomic Analysis of Up–Accumulated Proteins Associated with Fruit Quality During Autumn Olive (*Elaeagnus umbellata*) Fruit Ripening. *J. Agric. Food Chem.* **2011,** *59*, 577–583.

Yingthongchai, P.; Naphrom, D.; Smitamana, P. Assessment of Genetic Diversity in *Elaeagnus latifolia* L. by Inter–Simple Sequence Repeat (Issr) Markers. *Int. J. Agric. Technol.* **2014b,** *10*, 791–802.

Yingthongchai, P.; Naphrom, D.; Smitamana, P. Genetic Diversity of *Elaeagnus latifolia* L. as Revealed by Random Amplified Polymorphic DNA (RAPD) Markers. *J. Agric. Technol.* **2014a,** *10*, 177–187.

Zarger, T. G. Forestation of Surface Mines for Wildlife. In: *Trees for Reclamation Symposium Proceedings*; Gen. Tech. Rep. NE-61; U. S. Department of Agriculture, Forest Service, Northeastern Forest Experiment Station: Broomall, 1980; pp 71–74.

CHAPTER 3

Swertia chirayita: A Critically Endangered High-Value Medicinal Plant of the Himalayan Region

JITENDRA KUMAR SHUKLA[1], BHARTI KASHYAP[1], KISHOR BASOR[1], PREETI DHAKAL[2], EVANYLLA KHARLYNGDOH[1], PARDEEP KUMAR BHARDWAJ[3], and PULOK KUMAR MUKHERJEE[3]

[1]*Institute of Bioresources and Sustainable Development, Sub-centre: Meghalaya, Upper Shillong, Meghalaya, India*

[2]*Agricultural Technology Management Agency (ATMA), Mangan, North Sikkim, Sikkim, India*

[3]*Institute of Bioresources and Sustainable Development, Meghalaya, Takyelpat, Imphal, Manipur, India*

ABSTRACT

Swertia chirayita belonging to the family Gentianaceae is one among the top 32 prioritized medicinal plants by the National Medicinal Plants Board (NMPB), New Delhi, Government of India is globally renounced due to its medicinal importance. *S. chirayita* is a critically endangered (CR) plant growing at an altitude of 1200–3000 m above mean sea level (msl) in the temperate Himalayas. Depleting population of *S. chirayita* from its natural wild habitats has necessitated its conservation through ex situ cultivation. With the increasing demand and CR status of *S. chirayita*, the development of mass multiplication techniques through different regeneration pathways becomes imperative. Conservation through cultivation is the mantra for fetching the growing demand for CR medicinal plants of high demand

High-Value Plants: Novel Insights and Biotechnological Advances. Thangjam Anand Singh, Prakash Kumar Sarangi, and Ch. Brajakishore Singh (Eds.)
© 2025 Apple Academic Press, Inc. Co-published with CRC Press (Taylor & Francis)

along with conservation in natural habitats. However, there is a lack of understanding of how this species can be cultivated. The present script emphasizes on the various facets of *S. chirayita* form properties and uses for cultivation; requirements of soil type; propagation techniques; nursery and field preparation; transplanting and cultural practices; insect; pest; diseases prevalence and threats; commercial products and cost–benefit analysis of cultivation; etc. This information is basic for the globally widespread industrial requirements of *S. chirayita* and for its conservation through a cultivation program.

3.1 INTRODUCTION

The great Himalaya region encompasses high mountain ranges endowed with diverse flora and fauna and extends for 2500 km from east to west over five countries, namely, China, Bhutan, Nepal, India, and Pakistan. It contains the most widespread and uneven highest peaks on Earth including over a hundred mountains exceeding 7200 m (23,600 ft) in elevation including the highest, Mount Everest, with the largest areas covered by glaciers. This region plays an important role in the global atmospheric movement, biodiversity, water resources, and the hydro-ecological balance, apart from the esthetic landscape and provision of other ecosystem services. Himalayan ranges are rich repositories of biodiversity and water and provide ecosystem goods and services on which, both regional and global communities depend (Hamilton, 2002; Korner, 2004; Viviroli and Weingartner, 2004). Due to unique topography and landscape, diverse climatic conditions along the altitudinal ascent, and diversified environmental habitats ranging from tropical forest, and grassland to alpine meadows with vast and diverse natural resources, the Himalayas encompasses a diversity of medicinal and aromatic plants. However, unfortunately, some of the bioresources are unscientifically exploited for various purposes, particularly medicinal plants. Most of the medicinal plants are being unscientifically extracted for drug and pharmaceutical industries, traditional use, and research purposes from wild populations. These activities affect adversely the existence of a number of medicinal plants, particularly those of high commercial value due to which the original populations of many of these valuable resources have been vanishing. Chiraita [Botanical name: *Swertia chirayita* (Roxb. ex Fleming) Karsten; Family: Gentianaceae] (Fig. 3.1a) is among the several medicinal plant species growing in the Himalayan region. It is a medicinal plant of temperate Himalaya growing at an altitude of 1200–3000-m msl from the state of Kashmir in India to Bhutan and at 1200–1500 m above msl in the Khasi hills of Meghalaya state (Kirtikar and Basu, 1975; Clarke, 1885; Blatter, 1984; Tandon et al., 2010). The plant has an erect 2–3-ft long stem, rounds up to the middle, and four-angled upper branches (Fig. 3.1a–c). Flowers are small, stalked, green–yellow, tinged with purple color, rotate, and tetramerous. Petals have a pair of nectarines covered with fringed scales (Fig. 3.1d, e).

FIGURE 3.1 *S. chirayita*: (a) Mature plant, (b) flowering branch, (c) upper four-angled branch, (d) & (e) flower variations.

S. chirayita is medicinally used since ancient times because of the presence of bitter glycosides, triterpenoids, iridoids, xanthones, and flavonoids (Negi et al., 2011; Kumar and Staden, 2016). The swertiamarin, amarogentin (secoiridoid glucosides), and mangiferin, a xanthone C-glucoside are the main phytochemicals of the bitter-tasting plant. *S. chirayita* has two plus years life cycle, in the first and second years, the plant remains in a rosette form, and in the later part of second year to the third year, the plant rises up (erect stage) to bear flowers and fruits before the harvesting. Erect stage of plant remains for 6–7 months from June to December (Fig. 3.2a–c)

FIGURE 3.2 Life cycle of S. *chirayita* showing rosette stage in first and second year (a, b) and erect stage in third year (c).

Commercially the plants of *S. chirayita* are harvested during the erect stage. *S. chirayita* being a critically endangered (CR) medicinal herb is now on the verge of extinction. Although the extinction of species is a natural process, the current rate of extinction through human interventions is estimated approximately 100–1000 times faster than the natural speed of extinction (Chapin et al., 2000). *S. chirayita* population depletion is widely observed across its natural range. Since 2004, the export of wild-harvested *S. chirayita* is banned by the Government of India, prohibiting the export of *S. chirayita* plants, plant portions, and their extracts and derivatives obtained from the wild with exception of formulations, by the Director General of Foreign Trade (Khanal et al., 2014; Cunningham et al., 2018). In view of the increasing demand and CR status of *S. chirayita* the need to develop techniques for its mass multiplication through different regeneration pathways becomes imperative. Conservation through cultivation is a new mantra for fetching the growing demand of CR medicinal plants of high demand along with conservation in natural habitats. However, there is a lack of understanding of how this species can be cultivated (Rabgyal et al., 2021). The present script emphasizes on the various facets of *S. chirayita* form properties and uses for cultivation; requirements of soil type; propagation techniques; nursery and field preparation; transplanting and cultural practices; insect; pest; diseases prevalence and threats; commercial products; and cost–benefit analysis of cultivation; etc. This information is basic for the globally widespread industrial requirements of *S. chirayita* and for its conservation through a cultivation program.

3.2 PROPERTIES AND USES

It is one among the 32 highly prioritized medicinal plants by NMPB (Kala and Sajwan, 2007). Indian Pharmacopoeia List 1946 recognized *S. chirayita* as an official drug and as the most reputed plant of the Indian system of medicine (Ayurveda) (API, 2011; Handa, 1952). Significant biological activities of *S. chirayita* include hepatoprotective, antihepatotoxic, anti-inflammatory, antiarthritic, anticarcinogenic, antibacterial, antifungal, antileprosy, hypoglycemic, antimalarial, antifeedant, anthelmintic, analgesic antioxidant and central nervous system (CNS) depressant (Negi et al., 2011). The traditional use of *S. chirayita* is described as a home remedy for various ailments including intestinal worms in children, indigestion, cough, cold, asthma, liver diseases, headache, boils, scabies, and malarial fever (Pradhan and Badola, 2008; Joshi and Joshi, 2008; Pradhan and Badola, 2012a). *S. chirayita* is mentioned as annual (Kirtikar and Basu, 1975; Bentley and Trimen, 1981; Chauhan, 1999) or a biennial (Garg, 1987; Shreshtha and Joshi, 1992; Grierson and Long, 1999; Joshi and Dhawan, 2005; Nautiyal and Nautiyal, 2004; Rijal, 2009) or triennial (Pradhan and Badola, 2012a) or flowering once in the third year, that is, pluriannual (Edwards, 1993; Shah, 2008; Raina et al., 2013). It is used against fever, cold, cough, pneumonia, stomach pain, gastritis, diabetes, hypertension, hepatitis, liver diseases, bronchial asthma, dyspepsia, intestinal worm, general debility, arthritis, diarrhea, dysentery, spasm, pain, boils , and scabies, and is used as a blood purifier.

3.3 CULTIVATION

The importance of conservation of high priority, endangered medicinal plant *S. chirayita* through ex situ cultivation and development of agrotechniques had been optioned by various workers (Purohit et al., 2013; Tabassum et al., 2012; Badola and Pradhan, 2011; Phoboo and Jha, 2010; Pradhan and Badola, 2010a, b; Joshi and Joshi, 2008; Joshi, 2008; Joshi and Dhawan, 2005). Farmers hesitate to take up its cultivation due to its slow-growing nature (Basnet, 2001), long gestation/cultivation period of 3 years (Abrol et al., 2012: Badola and Pal, 2002), poor germination of seed, and difficulty handling seedlings because of very small size, etc. (Joshi and Dhawan, 2005; Raina et al., 1994; Chaudhuri et al., 2007). Whereas unsustainable and over harvesting from the wild for fetching high pharmaceutical demand is causing the depletion of the natural population of *S. chirayita*. Rai et al. (2000) mentioned that *S. chirayita* plant collected from Sikkim Himalaya is very high, amounting to 3,000,000 plants per year. International Union for Conservation of Nature (IUCN) categorized *S. chirayita* as "critically endangered" (Joshi and Dhawan, 2005). The declining population of *S. chirayita* from its natural habitat had made it essential to take up its ex situ cultivation, for conservation and to fetch the growing demand of Pharmaceutical Industries (Pradhan and Badola, 2010a, b; 2011; Joshi and Joshi, 2008; Badola and Pal, 2002). Ex situ cultivation is an appropriate method for the in situ conservation of *S. chirayita* (Badola and Pradhan, 2011). For the ex situ cultivation, it is important to understand the growth and adaptability performance and metabolite variation of *S. chirayita*. Many locations in the Northeastern states of Indian harbor a very congenial environment for the commercial cultivation of *S. chirayita*. However, ex situ cultivation of *S. chirayita* especially at lower altitudes has not been tried.

3.3.1 SOIL TYPE

S. chirayita can be grown in loamy to sandy-loam, friable, porous, well-drained soil rich in carbon and humus with pH 4.7–5.5. Cultivation of *S. chirayita* on the north-facing slopes getting morning and afternoon inclined sun rays is favorable (Bhattarai and Shreshta, 1996; Joshi, 2012). It generally prefers to grow on north-facing sloppy and moist shady places (Gaur, 1999). However, research introspections for location-specific climatic conditions, soil profile, soil composition, and microenvironment are required (Pradhan and Badola, 2012b; Shukla et al., 2017; Cunningham et al., 2018).

3.3.2 PROPAGATION

Natural regeneration of plants takes place by biologically mature seeds collected during November (Fig. 3.3 a, b). Under proper storage conditions (low moisture; cool and aerated conditions) seeds of *S. chirayita* remain viable for about 18 months. Farmers hesitate to take up the plants' cultivation due to one of the reasons of poor seed germination and difficulty handling seedlings because of very small size, etc. (Badola and Pal, 2002; Joshi and Dhawan, 2005; Raina et al., 1994; Chaudhuri et al., 2007). Shukla et al. (2017) suggested seeds can be harvested from mature plants cultivated at higher altitudes for utilization for cultivation at lower altitudes. Up to 70–80% germination and 60–70% seedling emergence can be achieved in nursery/field conditions (Fig. 3.3c).

3.3.3 NURSERY PREPARATION

Seed should be sown between February to April (December to February at lower altitudes 800–1200 m), into moist, fertile nursery beds of 1 m × 10 m dimension and 15–20 cm height (Mukherjee, 2013; Shukla et al., 2017). For nursery–bed preparation, sieved forest top soil, sand and vermicompost in 1:2:1 ratio is utilized. Approximately 200 g of seeds are required for raising planting stock for 1 ha (spacing size of approximately 45 cm × 45 cm) [In case of sowing as broadcast method 1 kg of seed is required for cultivation in 20 ropani/1 ha (1 ropani = 500 sq. m) land]. In nursery-bed seeds are sown @ 1g/sqm for which 1g seed is mixed homogeneously with fist full (approx. 100 g) of sterile ash and sown in 1 m^2 area of nursery bed. The seeds are covered with thin layer of soil not more than double the size of seed. Seed sown beds are mulched for better germination and mulching is removed as seeds start germinating after 15 to 30 days of sowing (Fig. 3.3d). Shading by shade net is required to protect the seed bed from heavy rain, direct sunlight, hailstorm, frost, etc. (Fig. 3.3e). Moisture is maintained in seeded bed by spraying water but drenching is avoided.

FIGURE 3.3 Propagation of *S. chirayita:* a and b- Seeds of *S. chirayita* (with 1000× seed view under a microscope), c- Germinating seedlings, d- Mulched seed bed, e- Plantlets in bed.

3.3.4 LAND PREPARATION

The land is prepared by weed removal, plowing 2–3 times in crisscross directions, followed by harrowing and planking for porous, fine tilting and moisture conservation of soil. The addition of vermicompost @ 3.75 tonnes/

ha and forest litter @ 2 tons/ha at the time of bed preparation is recommended (Das, 2008). For plantlet transplantation, a 15 cm raised bed having a top 10-cm surface dibbled with vermicompost and forest litter is utilized, with proper drainage to avoid water stagnation. In the case of direct sowing in the field by broadcast method, a mixture of seed and sterile ash is sown in the field, and after the establishment of seedlings desired spacing is maintained by thinning/transplanting. However, Gaire et al., (2019) reported that production doubles when seedlings are transplanted from the nurseries rather than directly broadcasting seeds in the field.

3.3.5 TRANSPLANTING

Seedlings of 6–8 cm with five to six leaves are transplanted in pre-prepared raised beds at 45 cm × 45 cm spacing in May–June after pre-monsoon shower or else in July–August at three to four leaves (Manual Chirata Cultivation, 2018). For transplanting, plantlets are uprooted with adhering soil, kept in moist gunny bags, shifted to the nearest planting site, and transplanted (Fig. 3.4a).

3.3.6 CULTURAL PRACTICES

Regular weeding and soil work are carried out without damaging the roots of plantlets in both nursery and field beds. Nursery beds and field plantation should be irrigated periodically as and when required weekly or fortnightly,

FIGURE 3.4 Growth and harvest of S. *chirayita: a*- Transplanted plants, b- Mature crop, c- Harvested crop.

till plants flower. Proper drainage systems by digging channels around fields and beds should be ensured to avoid stagnation of water.

3.3.7 HARVESTING AND YIELD

The traditional medicine systems use the whole plant (Bhargava et al., 2009; Tabassum et al., 2012), wherein the entire plant, including its roots, is uprooted before seed set, depriving the species of its natural regeneration potential (Aleem and Kabir, 2018). *S. chirayita* is harvested after 2 years growth period with the start of the third year, after seed maturation in November–December (Fig. 3.4b). The whole plant is uprooted cleaned, sun-dried, bundled, and packed in jute bags (Fig. 3.4c). However, the yield of *S. chirayita* can also be taken starting from the first year, when the plants are in the rosette stage while maintaining the plants for further growth under nursery or field conditions. Leaves harvested during the first and second years contain higher swertiamarin content compared to stems harvested in the third year. This maximizes the yield by two more harvestings during the gestation period of the crop. Shukla et al., (2017) mentioned the projected per hectare dry leaf yield of the 1-year-old ex situ cultivated plant at a lower altitude to about 280 kg.

Farmers usually sundry the harvested plant material. In the absence of enough sunlight plant material is stacked over the house attic by farmers, which creates favorable conditions for fungus and bacteria growth leading to low-quality products (FRDD, 2008). Post-harvested crops if inappropriately dried or handled, lead to most of the leaf's loss through decay or shedding. Farmers can suffer a loss of 30% of produce from harvesting to selling because of inappropriate drying, packaging, and storage (FRDD, 2008). Biomass and metabolite concentration also depends on the growth stage of the plant. No significant difference in plant biomass harvested during the flowering and senescence stage was observed, however, a detailed study of the active metabolite at the two growth stages is emphasized (Bhatt et al., 2006). Crop yields around 1000 kg/ha dry stem after 3 years (Fig. 3.4c). The seeds are air dried and stored in cloth-lined bags at aerated, cool places to be used as germplasm for next season crop. Seeds if properly stored remain viable for 1 year (Fig. 3.3a, b).

In the analysis of herbal samples, Ayurvedic Pharmacopeia of India (API) norms are very important. The plants raised through good agricultural practices and optimally and timely harvested qualify for the API norms. Barakoti et al. (2012) mentioned wild populations, in general, contain a higher bitter principle, which varies depending on agroecological growing conditions. Phoboo et al. (2010) reported the maximum content of these phytochemicals (amarogentin, mangiferin, and swertiamarin) in a mixture of inflorescence and leaf as compared to stem and root. No significant difference in the amounts of photochemical content in wild and cultivated plant parts of *S. chirayita* was observed (Phoboo et al., 2010). Shukla et al., (2017) reported that the leaves with high swertiamarin content can be sustainably harvested from the plants within 1 year of growth period.

3.3.8 COST: BENEFIT (B:C) RATIO

Based on the value-chain analysis and cost-benefit assessments, *S. chirayita* cultivation has been shown to be profitable (Cunningham et al., 2018) An emergent condition for ex situ cultivation of *S. chirayita* felt by many workers; the study conducted by Shukla et al. (2017) provide important information such as *S. chirayita* can be cultivated at a lower altitude of 800 ± 10 m above msl. Second, the yield of *S. chirayita* can be taken from the first year onward while maintaining the plants for further growth under nursery or field conditions. This maximizes the yield by two more harvestings during the gestation period of the crop (Shukla et al., 2017).

The yield of *S. chirayita* comes to around 1000 kg/ha dry stem after in third year (Fig. 3.4c) Dry herbage is sold at around ₹350 per kg amounting to ₹3,50,000/ha. The cultivation cost comes to around ₹100,000 for 2 years thus giving a B:C ratio of 1:2.5 (*Note:* B:C ratio may vary depending upon the yield, prevailing input cost and market rates, etc.).

3.3.9 INSECT, PESTS AND DISEASES

Metabolite content and other specifications as per API norms also depend on the storage conditions, post-harvest handling, geographical location of plant growth, climatic conditions during the harvest, etc. of the produce (Khanal et al., 2014). No specified insects/pests and diseases have been observed on *S. chirayita*. However, white grubs as cosmopolitan soil pests may affect the growth of plants under nursery and field conditions (Fig. 3.5a, b). The population of white grub can be controlled by the application of neem cake (40–50 kg/ha) in soil. Alternately, neem oil can also be used as a repellent inhibiting egg laying, growth, and feeding. Neem oil is mixed with water and sprayed onto affected areas. Adult beetles emerging from the soil from May to October can be killed by a user-friendly low-cost light trap for efficient mass trapping of beetles. Fungal leaf spots appear especially in the rosette stage; however, it does not adversely affect the yield of the crop. Diseased leaves or plants can be safely uprooted and destroyed by burning away from the field (Fig. 3.5c).

FIGURE 3.5 Insect, pest and disease observation in S. *chirayita* plantation site: a, b: White Grub affecting root of plant and c: Fungal leaf spots.

3.3.10 COMMERCIAL PRODUCTS

Herbal medicines such as Ayush-64, Diabecon, Mensturyl syrup, Mahasudarshan churn, Kiratatiktadi kvath, Sarva-jwar-hara-lauha, Chandra Prabati, Melicon, etc. are widespread traded commercial products.

3.3.11 THREATS

The ex situ cultivation at the lower altitude may face many bottlenecks. Major among them is initial germination of seed and slow seedling growth, competition with weed and subsequent loss of seedlings during weed removal. Non-uniform seedlings' emergence of *S. chirayita* is a big constraint in commercial cultivation (Pradhan and Badola, 2011). *S. chirayita* grows slower than simultaneously growing weed. Large-grown weed, when uprooted, brings out the small seedling of *S. chirayita* too. Therefore, proper weed removal at an appropriate stage to avoid competition is important. Another problem faced during cultivation at lower altitudes is an attack of pests on plants. The sails and sludge attacked the lower leaves during the night which later on fed on upper young and growing leaves too. Harvesting of fully grown mature lower leaves is useful as a control measure of snails and sludge. Water stagnation or moisture accumulation during monsoon is detrimental to the *Swertia* plant. For the removal of water and accumulating moisture, more study in terms of design, direction, and slope of bed is required through different agrotechniques, so that plants could reach the flowering and fruiting stage when cultivated at lower altitudes. Bhattarai and Shreshta (1996) mentioned north facing gentle slopes getting morning and afternoon-inclined sun rays are favorable for *S. chirayita* cultivation. Further, the main threat faced by *S. chirayita* plants are their premature harvesting deprives their natural regeneration. Second, low seed germination percentage and viability, long crop gestation periods, and delicate field handling pose threats to low interest in cultivation. Third, lack of large-scale commercial plantations, destructive harvesting, and loss of habitat are other threats.

3.4 CONCLUSIONS

S. chirayita is endangered, industrially highly demanded, and among the top 32 prioritized plants by the NMPB because of its medicinal importance. Indian Pharmacopoeia List 1946 recognized *S. chirayita* as an official drug and as the most reputed plant of the Indian system of medicine (Ayurveda). Significant biological activities of *S. chirayita* include hepatoprotective, antihepatotoxic, anti-inflammatory, antiarthritic, anticarcinogenic, antibactertial, antifungal, antileprosy, hypoglycemic, antimalarial, antifeedant, anthelmintic, analgesic antioxidant, and CNS depressant. The specificity of the niche environment at an altitude of 1200–3000 m msl in the temperate Himalayas makes its availability scarce leading to harvesting from the wild. Depleting the population of *S. chirayita* from its natural wild habitats has necessitated its conservation through ex situ cultivation for fetching the growing demand of CR medicinal plants of high importance. Ex situ cultivation is an appropriate method for the in situ conservation of *S. chirayita*. However, there is a lack of understanding and know-how of *S. chirayita* cultivation. Factors such as slow growing nature, long gestation/cultivation period of 3 years, poor germination of seed, and difficulty handling seedlings because of very small size, etc. make farmers hesitant to take up its cultivation. *S. chirayita* can be grown in loamy to sandy loam, friable, porous, well-drained soil rich in carbon and humus with pH 4.7–5.5. Seed should be sown between February and April (December to February at lower altitudes 800–1200 m), into moist, fertile nursery beds of 1 m × 10 m and 15–20 cm height. Seedlings of 6–8 cm with five to six leaves are transplanted in raised beds at 45 cm × 45 cm spacing in May-June after a pre-monsoon shower or else in July–August. Regular weeding and soil work is carried out without root damage with periodical irrigation as and when required (weekly or fortnightly) till flowering ensuring proper drainage to avoid water stagnation. *S. chirayita* is harvested after 2 years of growth period with the start of the third year, after seed maturation in November–December. The yield of *S. chirayita* can also be taken starting from the first year when the plants are in the rosette stage while maintaining the plants for further growth under nursery or field conditions. The yield of *S. chirayita* comes to around 1000 kg/ha dry stem in the third year giving a B:C ratio of 1:2.5. The main threats faced by the natural population depletion of *S. chirayita* is due to deprivation of natural regeneration because of premature harvesting. Low seed germination percentage and viability, long crop gestation periods, and delicate field handling lower the farmer's interest in cultivation. Besides, the lack of large-scale commercial plantations, destructive harvesting, and loss of habitat are other threats.

ACKNOWLEDGMENTS

The authors are grateful to the Director of the Institute of Bioresources and Sustainable Development (IBSD), Imphal, Manipur for his support and encouragement in publishing the work. The above work was funded by the Department of Biotechnology, Government of India under the IBSD Core Fund.

KEYWORDS

- ***Swertia chirayita***
- **endangered plant**
- **ex situ cultivation**
- **conservation**

REFERENCES

Abrol, E.; Vyas, D.; Koul, S. Metabolic Shift from Secondary Metabolite Production to Induction of Anti-Oxidative Enzymes During NaCl Stress in *Swertia chirata* Buch.-Ham. *Acta Physiologiae Plantarum* **2012,** *34*, 541–546.

Aleem, A.; Kabir, H. Review on *Swertia chirata* as Traditional Uses to Its Pyhtochemistry and Phrmacological Activity. *J. Drug Deliv. Therap.* **2018,** *8*(5-s), 73–78.

API. *Kiratatikta*. In: *The Ayurvedic Pharmacopoeia of India*, Vol. 1 (1). Delhi, India, 2011; pp 98–100 (ISBN 11-901151-3-8).

Badola, H. K.; Pal, M. Endangered Medicinal Plant Species in Himachal Pradesh. *Curr. Sci.* **2002,** *83*, 797–798.

Badola, H. K.; Pradhan, B. K. Economic Viability of Cultivation of *Swertia chirayita*, a High Value Endangered Medicinal Herb in Himalaya. Zeitschrift fur Arznei- & Gewurzpflanzen. *J. Med. Spice Plants*. **2011,** *16*, 118–124.

Barakoti, T. P.; Tiwari, N. N.; Yonzon, M. Quality of Chiretta (*Swertia chirayita*) in Cultivated and Wild Samples Collected from Different Districts of Nepal. *Nepal J. Sci. Tech.* **2012,** *13*, 57–62.

Basnet, D. B. Evolving Nursery Practices and Methods of Cultivation of High Value Medicinal Plant *Swertia chirayita* Ham. *Environ. Ecol.* **2001,** *19*, 935–938.

Bentley, R.; Trimen, H. *Medicinal Plants*, Vol. 4; International Book Distributors: Dehradun, India, 1981; p 306

Bhargava, S.; Rao, P. S.; Bhargava, P.; Shukla, S. Antipyretic Potential of *Swertia chirata* Buch Ham. root Extract. *Sci. Pharm.* **2009,** *77* (3), 617–624.

Bhatt, A.; Rawal, R. S.; Dhar, U. Ecological Features of a Critically Rare Medicinal Plant, *Swertia chirayita*, in Himalaya. *Plant Species Biol.* **2006,** *21*, 49–52.

Bhattarai, K. R.; Shreshta, K. K. Ecology Study of on Chiraito in Northern Gorkha. *J. Nat. History Museum*. **1996,** *15*, 13–16.

Blatter, E. *Beautiful Flowers of Kashmir (Vol. II)*; International Book Distributors: Dehradun, India, 1984.

Chapin, Iii, F. S.; Zavaleta, E. S.; Eviner, V. T.; Naylor, R. L.; Vitousek, P. M.; Reynolds, H. L.; Hooper, D. U.; Lavorel, S.; Sala, O. E.; Hobbie, S. E.; Mack, M. C. Consequences of Changing Biodiversity. *Nature* **2000,** *405* (6783), 234–242.

Chaudhuri, R. K.; Pal, A.; Jha, T. B. Production of Genetically Uniform Plants from Nodal Explants of *Swertia chirata* Buch.-Ham. ex Wall.- and Endangered Medicinal Herb. *In Vitro Cell, Dev. Biol. Plant*, **2007,** *43*, 467–472.

Chauhan, N. S. *Medicinal and Aromatic Plants of Himachal Pradesh*; Indus Publishing Company: New Delhi, 1999.

Clarke, C. B. *Verbenaceae*. In: *The Flora of British India*, Vol. 4; Hooker, J. D., Ed.; London, 1885; pp 560–604.

Cunningham, A. B.; Brinckmann, J. A.; Schippmann, U.; Pyakurel, D. Production from Both Wild Harvest and Cultivation: The Cross-Border *Swertia chirayita* (Gentianaceae) Trade. *J. Ethnopharmacol.* **2018,** *225*, 42–52.

Das, A. *Agro-Techniques of Selected Medicinal Plants*, Vol. 1; National Medicinal Plants Board, Government of India, TERI Press, 2008; pp 189–192.

Edwards, D. M. *The Marketing of Non-Timber Forest Product from the Himalaya: The Trade Between East Nepal and India*; Rural Development Forestry Network Paper 15b, Overseas Development Institute: London, 1993; p 24.

FRDD. Guidelines for Resource Assessment and Management of Chirata Swertia chirayita. Forest Resources Development Division, Department of Forest, Ministry of Agriculture, Royal Government of Bhutan, *2008*, 22. http://www.snvworld.org/sites/www.snvworld.org/files/images/guidelines_for_resource_assessment_and_management_of_chirata_swertia_chirayita_-_2008.pdf

Gaire, D.; Yadav, B. K.; Kattel, R. R. Value Chain Analysis and Community-Based Strategies of Chirayita *(Swertia chirayita)* in Eastern Nepal. *Int. J. Environ Sci. Nat. Res.* **2019,** *20* (1), 556027. DOI: 10.19080/IJESNR.2019.20.556027

Garg, S. Gentianaceae of the North West Himalaya (a Revision). In: Bioscience International, Monograph17; Today and Tomorrow's Publication Co.: New Delhi, 1987; pp 183–184.

Gaur, R. D. *Flora of the District Garhwal*, North West Himalaya (with Ethnobotanical Notes). Transmedia: Srinagar, Garhwal, 1999; p 408.

Grierson, A. J. C.; Long, D. G. *Flora of Bhutan* (Ed. Long DG), Vol.2(2) Published by Royal Botanical Garden Edinburgh, UK and the Royal Government of Bhutan, 1999; pp 626.

Hamilton, L. S. *Why Mountain Matters. World Conservation: IUCN Bull.* **2002,** *1*, 2002.

Handa, K. L. A Modified Method for the Assay of *Swertia chirata. Indian J. Pharm Sci.* **1952,** *14*, 87–88.

Joshi, P.; Dhawan, V. *Swertia chirayita*—An Overview. *Curr. Sci.* **2005,** *89*, 635–640.

Joshi, K.; Joshi, A. *Swertia* L. (Gentianaceeae) in Nepal Himalaya: Checklist, Phytogeography, Ethnobotany and Conservation Status. *Ethnobot. Leaflets* **2008,** *12*, 361–372.

Joshi, K. *Swertia* L. (Gentianaceeae) in Nepal: Ethnobotany and Agenda for Sustainable Management. *Ethnobot. Leaflets*. **2008,** *12*, 1–6.

Joshi, K. *A Field Manual on Nursery Management and Cultivation of Chiraita (Swertia chirayita)*; Royal Government of Bhutan, Ministry of Agriculture and Forests, Social Forestry and Extension Division, Department of Forests and Park Services: Thimphu, 2012.

Joshi, P.; Dhawan, V. *Swertia chirayita*—An Overview. *Curr. Sci.* **2005,** *89*(4), 635–640.

Kala, C. P.; Sajwan, B. S. Revitalizing Indian Systems of Herbal Medicines by the National Medicinal Plant Board Through Institutional Networking and Capacity Building. *Curr. Sci.* **2007,** *93*, 797–806.

Khanal, S.; Shakya, N.; Nepal, N.; Pant, D. *Swertia chirayite*: The Himalayan Herb. *Int. J. Appl. Sci. Biotechnol.* **2014,** *2*, 389–392.

Kirtikar, K. R.; Basu, B. D. *Indian Medicinal Plants*, Vol. III, 2nd ed.; Bishen Singh Mahender Pal Singh: Dehradun, India, 1975; pp 1663–1668.

Korner, C. Mountain Biodiversity, Its Causes and Function. *AMBIO: J. Human Environ.* **2004,** *33* (13), 11–17.

Kumar, V.; Staden, J. V. A Review of *Swertia chirayita* (Gentianaceae) as a Traditional Medicinal Plant. *Front. Pharmacol.* **2016,** *6*, 1–14. http://dx.doi.org/10.3389/fphar.2015.00308.

Manual Chirata Cultivation; Medicinal and Aromatic Plants Programme, Agriculture Research and Development Centre, Department of Agriculture: Thimphu, 2018, ; p 44. http://www.doa.gov.bt/wp-content/uploads/2020/03/Chirata-Cultivation-Manual-1.pdf

Mukherjee, D. Improved Agricultural Practices for High Value Medicinal Plants Under Mid to High Altitude Situation. *J. Crop Weed* **2013,** *9* (1), 201–206.

Nautiyal, M. C.; Nautiyal, B. P. *Agrotechniques for High Altitude Medicinal and Aromatic Plants*; Bishen Singh Mahendra Pal Singh: Dehradun, India, 2004.

Negi, J. S.; Singh, P.; Rawat, B. Chemical Constituents and Biological Importance of *Swertia*: A Review. *Curr. Res. Chem.* **2011,** *3* (1), 1–15.

Phoboo, S.; Jha, P. K. Trade and Sustainable Conservation of *Swertia chirayita* (Roxb. ex Fleming) h. Karst in Nepal. *Nepal J. Sci. Technoi.* **2010,** *11*, 125–132.

Phoboo, S.; Pinto, M. D.; Bhowmik, P. C.; Jha, P. K.; Shetty, K. Quantification of Major Phytochemicals of *Swertia chirayita*, a Medicinal Plant from Nepal. *Ecoprint*, **2010,** *17*, 59–68.

Pradhan, B. K.; Badola, H. K. Ethnomedicinal Plant Use by Lepcha Tribe of Dzongu Valley, Bordering Khangchendzonga Biosphere Reserve, in North Sikkim, India. *J. Ethnobiol. Ethnomed.* **2008,** *4*, 22. DOI: 10.1186/1746-4269-4-22. http://www.ethnobiomed.com/content/4/1/22

Pradhan, B. K.; Badola, H. K. *Swertia chirayita*, a High Value Endangered Medicinal Herb: Potential in North-East India. *Ecotone* **2010a,** *2* 24–27.

Pradhan, B. K.; Badola, H. K. Chemical Stimulation of Seed Germination in Ex Situ Produced Seeds in *Sweria chirayita*, a Critically Endangered Medicinal Herb. *Res. J. Seed Sci.* **2010b,** *3*, 139–149.

Pradhan, B. K.; Badola, H. K. Assessment of Seedling Emergence and Vigour for Quality Planting Material in Thirteen Populations of *Swertia chirayita*—A High Value Endangered Medicinal Herb, Using Substrate Combinations. Zeitschrift für Arznei- and Gewürzpflanzen. *J. Med. Spice Plants*, Germany **2011,** *16* (1), 34–41.

Pradhan, B. K.; Badola, H. K. *Effect of Storage Conditions and Storage Periods on Seed Germination in Eleven Populations of Swertia chirayita: A Critically Endangered Medicinal Herb in Himalaya*, 2012a; pp 1–9. http://dx.doi.org/10.1100/2012/128105. DOI: 10.1100/2012/128105

Pradhan, B. K.; Badola, H. K., Effects of Microhabitat, Light and Temperature on Seed Germination of a Critically Endangered Himalayan Medicinal Herb, *Swertia chirayita*: Conservation Implications. *Plant Biosyst. Int. J. Deal. Asp. Plant Biol.* **2012b,** *146* (2), 345–351.

Purohit, V. K.; Bahuguna, Y. M.; Tiwari, D.; Tiwari, A.; Andola, H. C.; Negi, J. S.; Chauhan, R. S. *Swertia chirayita* (Roxb. ex Fleming) Karsten on the Verge of Extinction in the Himalayan Region. *Curr. Sci.* **2013,** *104*, 161–162.

Rabgyal, J.; Bajgai, Y.; Pelden, K.; Tobgay, T.; Rinchen, S. Assessment of Nursery Methods and Manures for Cultivation of Chirayita (*Swertia chirayita* Buch-Ham.) in Lauri Gewog, Bhutan, *Bhutanese J. Agric.* **2021,** *4* (1) 137–151.

Rai, L. K.; Prasad, P.; Sharma, E. Conservation Threat to Some Important Medicinal Plants of the Sikkim Himalaya. *Biol. Conserv.* **2000,** *93*, 27–33.

Raina, R.; Johri, A. K.; Srivastava, I. J. Seed Germination Studies in *Swertia chirayita* L. *Seed Res.* **1994,** *22*, 62–63.

Raina, R.; Patil, P.; Sharma, Y. P.; Rana, R. C. Reproductive Biology of *Swertia chirayita*—A Temperate Critically Endangered Medicinal Plant. *Cryologia: Int. J. Cytol. Cytosyst. Cytogen.* **2013,** *66*, 12–22.

Rijal, D. P. Taxonomic Study of Some Medicinally Important Species of *Swertia* L. (Gentianaceae) in Nepal. *Botanica Orientalis—J. Plant Sci.* **2009,** *6*, 18–24.

Shah, I. A. Studies on Seed Germination and Seed Set in *Swertia chirayita* (Roxb. ex. Flem.) Karst. MSc (Forestry) Thesis, Dr.Y. S. Parmar University of Horticulture & Forestry, Nauni, Solan (HP), India, 2008.

Shreshtha, J. N.; Joshi, S. D. Tissue Culture Techniques for Medicinally Important Herbs—*Orchis incarnate* and *Swertia chirayita*. *Banko Janakari* (*J. Forest, Info. Nepal*), **1992,** *3*, 24–26.

Shukla, J. K.: Dhakal, P.; Uniyal, R. C.; Paul, N.; Sahoo, D. Ex Situ Cultivation at Lower Altitude and Evaluation of *Swertia chirayita*, a Critically Endangered Medicinal Plant of Sikkim Himalayan Region, India. *South African J. Bot.* **2017,** *109*, 138–145.

Tabassum, S.; Mahmood, S.; Hanif, J.; Hina, M.; Uzair, B. An Overview of Medicinal Importance of *Swertia chirayita. Int. J. Appl. Sci. Technol.* **2012,** *2* (1), 298–304.

Tandon, P.; Kumaria, S.; Kayang, H. Conservation of Medicinal and Aromatic Plants of Northeast India. In: *Medicinal Plants in Changing Environment*; Ahmad, A., Siddiqi, T. O., Iqbal, M., Eds.; Capital Publishing Company: New Delhi, India, 2010. pp 203–212.

Viviroli, D.; Weingartner, R. Hydrological Significance of Mountains: From Regional to Global Scale. *Hydrol. Earth Syst. Sci.* **2004,** *8* (6), 1016–1029.

CHAPTER 4

Genus *Garcinia* in Northeast India: Folklore Applications, Scope, and Opportunities

J. C. BORA [1] and D. D. JOSHI[2]

[1]*Associate Professor, Institute of Advanced Studies in Science and Technology, Guwahati, Assam, India*

[2]*Amity Herbal Consortium, Amity University Uttar Pradesh, Noida, UP, India*

ABSTRACT

Garcinia, a flavoring fruit is rich in flavonoids, organic acids, and various other nutritional ingredients. Hydroxy citric acid (HCA), a potential supplement for weight management and anti-obesity agent has been identified in the fruits of genus *Garcinia* L. Various *in vivo* studies have contributed to the understanding of the anti-obesity effects of Garcinia/HCA as it regulates the serotonin level and uptake of glucose along the enhance fat oxidation. State Assam, India holds massive fruit diversity, and genus *Garcinia* is one of these. The documentation of unexplored potentials of a genus and the collection of fragmented information inspires the conservation and propagation practices. Garcinia-based local knowledge (e.g., as an edible fruit, medicine, wood, ornamental plant, and extracting oil, etc.) has been influenced by the needs of different ethnic communities. There has been a common urge in every research article to update the indigenous knowledge with modern molecular science; highlight the same is another objective of the present text. The collective information on the field distribution, estimated usage intensities, a prediction for potential distribution, and art of the traditional management will help the entrepreneurs in the future for innovation.

High-Value Plants: Novel Insights and Biotechnological Advances. Thangjam Anand Singh, Prakash Kumar Sarangi, and Ch. Brajakishore Singh (Eds.)
© 2025 Apple Academic Press, Inc. Co-published with CRC Press (Taylor & Francis)

4.1 INTRODUCTION

Garcinia has been in use for centuries in Asian countries for culinary purposes as a condiment and flavoring agent in place of tamarind or lemon and to make meals more tasteful. Dried rinds of *G. cambogia* fruits with salt and other edible organic acids are used to lower the pH and adjust the bacteriostatic effect for fish curing. The assay of hydroxy citric acid (HCA) varies in different species of Garcinia. Researches indicate that Garcinia seems to reduce the body glucose absorption rate into the intestines after eating, and helps to decrease blood sugar levels, an indication that these phytochemicals can improve glucose metabolism, a goal associated with diabetic patients attempting to get it under control. Herbals and their value-added forms are used as food supplements by people, globally. Traditional evidence sheds light on the consumption of Garcinia fruits as an effective strategy for disease treatment and health maintenance. Currently, the nutraceutical industry is flourishing, and interest in establishing scientific credibility has attained importance for many industries and scientists.

Garcinia is mainly used as the fruit (with the rind, pulp as well as seeds) along with flowers, floral nectar, pollen, leaves, roots, and the associated micro-microbes. The ecological anthropology of traditional Garcinia products shows that the useful properties of edible parts were discovered by the concerned communities by "trial and error" and "cause and effect" strategies perhaps during the food gathering era. Products like food colorants, juice, brined solution, butter, and oil were prepared without understanding their bio-chemical composition. These traditional products and the associated knowledge for their properties may lead to the chemist searching for novel biomolecules. Several of these have applications as bioactive, ingredients, and part of the food industry, and may have many opportunities for the producers to supply the best quality fruits or any other plant parts in demand. Value addition of primary products is the best opportunity for Garcinia producers. The extractable yellow pigment of Garcinia fruits is a natural food color, which has an assay of 2.4% of the fruit biomass, and fruits juice can be fermented to produce high quality Garcinia wine. The weight-loss effects of Garcinia or its active constituents, the HCA, are a matter of study over the last few decades. Its efficacy also has been proven for ameliorating obesity-related side-effects such as inflammation, insulin resistance, and stress.

4.1.1 DESCRIPTION, DISTRIBUTION, AND TAXONOMY

The genus *Garcinia* L. (family; Clusiaceae) comprises about 250 species, globally (Nimanthika and Kaththriarachchi, 2010; Rogers and Sweeney, 2007). Garcinia has a wide range of tropical habitats, often in areas of great biological and ethnic diversity, including tropical Asia, Southern Africa, Northeast Australia, and tropical America. The genus is characterized by a brownish–yellow gum resin in the trunk, leaves, and fruits. The gum resin is used as purgative or cathartic (Li, 1979). *Garcinia* species are well known for being traditionally used as fresh fruit, medicine, timber, and ornamentation (Iwu and Igboko, 1982; Yapwattanaphun et al., 2002) as well as important animal food sources (Wadsworth et al., 2005). Some species are used as important plant resources by local communities as an integral part of the culture and family based economy. Garcinia fruits are berries with fleshy and juicy mesocarps, which in most species can be eaten raw. The most significant species in the genus is the purple mangosteen (*G. mangostana* L.), which is presumed originated in Southeast Asia or Indonesia and has been largely remained indigenous to Cambodia, Malay, Myanmar, Peninsula, Thailand, and Vietnam (Akao et al., 2008). Fruits of Garcinia spp. are used as a spice (the exocarp of *G. multiflora*) to flavor and color the famous "Bun-rieu soup" in Vietnam, and of *G. gummigutta*, in Kerala, India, as a spice called kodumpulli (CSIR, 1956).

Though Garcinia has a rich diversity, only a few species have been well-known for unique and delicious taste, so there is a need to identify the potential and to promote the fruit species by all possible means. Traditional knowledge documentation is one of the possible methods, the source for entrepreneurs for innovative ideas to identify potentials of indigenous crops and develop a process of economic values. However, if documentation was not done, the knowledge is really under threat to extinction. Documentation is the way for the survival of knowledge, and conservation as well. Thus, the present text aims to collect the Garcinia-based knowledge, uses, and practices from Assam, India.

Sharma and Sanjappa (1993) have reported that the genus *Garcinia* is known for edible fruits, and medicinal properties in the tropical regions but a few species also thrive in relatively low rainfall areas. There are about 200 species of the genus native to South Asia from southern Thailand and Peninsular Malaysia to Indonesia, distributed in the Southeast Asian region (Mabberley, 2005). Roberts (1984) and Maheswhari (1964) have separately reported about 35 species in India, of which 15 species are from Northeast India, and 9 species from the undivided Assam (Maheshwari, 1982; Roberts,

1984). Further, eight species have been reported from Sonitpur districts of Assam (Kar et al., 2008). In India, *G. combogia* is found naturally in the western peninsular coastal regions and the adjoining Western Ghats [in the states of Goa, Karnataka, Maharashtra, Kerala, and parts of Eastern India (West Bengal, Assam, and North-eastern Hill regions)], but grown in other parts of peninsular India (Padhye et al., 2009). Fruits of some Garcinia species are the richest sources of red pigments in the plant kingdom. The fruit and syrup of *G. indica* are very popular in "Konkan" region and have antioxidant and antibacterial properties (Negi et al., 2008).

Garcinia spp. are also cultivated either for vegetables, traditional medicines, or other domestic uses such as timber, firewood, and landscaping, etc. The cultivated species are *G. atroviridis*, *G. cowa*, *G. hombroniana, G. lanceaefolia*, *G. mangostana, G. morella*, and *G. prainiana*. *Garcinia* species are known to produce edible fruits and *G. magostana* is often considered as most famous fruit, "the Queen of Fruits in Thailand." Young leaves of the few species of *Garcinia* are eaten, cooked by some tribes in Northeast India (Jain and Dam, 1979). The fruits of *G. cambogia*, *G. indica,* and *G. atroviridis* are the main natural source of the HCA. The latex of *G. cowa* is used in Thai folk medicines as an anti-fever agent (Pattalung et al., 1994). In the traditional system, a few *Garcinia* species are used after childbirth medication, for menstrual problems and dysentery, and treat fever (Burkill, 1935), HIV (Rukachaisirikul et al., 2003), and cancer (Nabandith et al., 2004).

Garcinia has been used as Ayurvedic medicine (Lim, 2012). *G. cambogia* decoction is purgative and used in the treatment of intestinal worms, bilious digestive conditions, dysentery, rheumatism, and tumors. For veterinary applications, it is used to treat mouth disease in cattle (Jena et al., 2002). Its fruit rind is used in rickets and enlargement of the spleen and healing of bone fractures (Khare, 2007). In Southeast Asian countries, it is used in folkloric medicine to treat cough, dandruff, earache, stomach pains associated with pregnancy, and throat irritation (Permana et al., 2005). The dried fruit of *G. atroviridis* is used to enhance blood circulation, treat coughs, used as a laxative, and as an expectorant. With vinegar, its fruits are used as a lotion and rubbed over the abdomen of women after confinement (Lim, 2012).

4.2 PHYTO-PHARMACOLOGY

Traditionally, Garcinia fruit is commonly known as the herb for weight loss, suppress appetite, reduce food intake, and increasing energy expenditure. The imbalance between energy intake and expenditure generates obesity, the leading cause of premature death in the world. Other potential risks of obesity include cardiovascular disease, diabetes, cancer, and hormonal imbalances in women, leading to infertility. Therefore, supplementation with safe, and source-out from natural products like Garcinia, in addition to a healthy diet are preferred. Diabetes mellitus, the most common disorder is increasing rapidly in urban societies. The use of plants in traditional medical practice has a long-drawn history and remains the mainstay of primary healthcare even for diabetes. *G. quaesita*, commonly known as "Goraka" is an endemic fruit of Sri Lanka, and a natural source of HCA, which has the potential of weight loss. The fruit is popular to curbs appetite and treats obesity. Traditionally, it is collected from wild sources like food, medicines, and various commodities. Additionally, the fruit of Garcinia is known as a potent cardio-protective in the traditional medical practices of Sri Lanka. These effects are attributed to a variety of phyto-constituents present in Garcinia fruits (Munaweera, 2016).

The composition of Goraka (*G. quaesita*) has not been properly studied yet. But some studies show that the rind of Goraka fruit assays 16–30% (on a dry basis) of HCA, an ingredient reported causing weight loss in humans without stimulating the central nervous system. HCA was found to act in three different ways toward obesity and fat burning, namely (i) It reduces the conversion of carbohydrates into fat by inhibiting the activity of the enzyme ATP citratelyase, (ii) HCA reduces appetite, (iii) HCA also enables to burn glucose before it is stored as fat. HCA is a competitive inhibitor of ATP-citratelyase, an extra-mitochondrial enzyme involved in the initial steps of *de novo lipogenesis*. HCA reduces the transformation of citrate into acetyl coenzyme A, a necessary step for the formation of fatty acids in the liver. This increase low-density lipoprotein (LDL) receptor activity in liver cells that can pull LDL cholesterol out of circulation. This is a key factor in weight control and reduction of obesity using Goraka's HCA as an antagonist to citric acid (Munaweera, 2016).

HCA increases the production of hepatic glycogen due to activation of gluco-receptors, resulting in the feeling of fullness and reduced appetite. Experimental results reveal that HCA generates the metabolic response to appetite-suppressing signals, and communicates to the brain via 5-hydroxy-tryptophan, a precursor of serotonin. There are not many more precise and targeted research and clinical trials, so the applicability of *G. quaesita* is still problematic. *G. cambogia* is another species from Sri Lanka which has been more commonly used for weight loss, generally available as supplements, but concerns and adverse effects are reported with this species as well. Thus, the use of *Garcinia* species has been still not fully concluded to be effective toward weight loss and obesity (Munaweera, 2016).

Studies on *G. Quaesita* are limited, phytochemical analyses of other *Garcinia* species reveal that most of the fruits contain flavonoids, tannins, saponins, and cardiac glycosides. The reduction of the glucose concentration could be that the bi-flavanoids, which promotes the entry of glucose into the cells, stimulation of glycolytic enzymes and glycogenic enzymes or inhibiting the glucose 6-phosphatease in liver and subsequently reducing the release of glucose in the blood. Diabetes is also often associated with increased dyslipidemia. Lack of insulin in diabetic patients is known to be associated with increased synthesis of cholesterol and insulin resistance in diabetic patients could increase the hepatic uptake of fatty acids released by lipolysis of adipose tissues, the intra-hepatic synthesis of triglycerides, and the overproduction and secretion of VLDL (very-low-density lipoprotein) particles that in turn leads to increased plasma levels of TG (triglyceride). For this condition, phytochemicals in Garcinia fruits were found to be effective in reducing serum total cholesterol, LDL cholesterol, triglycerides, VLDL cholesterol. This reduction is found a dose-related property. The underlying mechanism of lipid lowering effect of these could be by inhibition of lipid absorption due to the presence of saponins and tannins or inhibition of cholesterol esterase, activation of fatty acid synthase, acetyl CoA carboxylase, and production of TG precursors such as acetyl CoA and glycerol phosphate. Flavonoids can also be playing a role in this process. Flavonoids are used in the reduction of lipids by inhibiting hepatic HMG-CoA reductase. This evidence suggests that Garcinia fruits are effective toward diabetes mellitus in a dose-dependent manner. Experimental studies reveal that there is a drop in blood sugar levels in diabetic patients after using garcinia or its supplements (Munaweera, 2016).

HCA (Fig. 4.1) is a principal acid of fruits of the genus *Garcinia. G. Cambogia* DSER fruits have adenosine 59-triphosphate (ATP) citrate lyase (EC 4.1.3.8) inhibition activity. The enzyme catalyzes the extramitochondrial cleavage, result in a decrease in body weight (Watson et al., 2000).

Recently, more clinical trials have been conducted to elucidate the functional effects of Garcinia/HCA, supplementation to promote human health. Studies have revealed that the metabolic functions for HCA, Garcinia extract rich in HCA, and/or garcinol (Fig. 4.2) reduces blood lipids, suppress the appetite, and reduces food intake (Saito et al., 2005). Further, there is scientific evidence that intake of HCA and HCA-SX alone does not produce any adverse effects and a dietary dosage of up to 2800 mg/day in humans

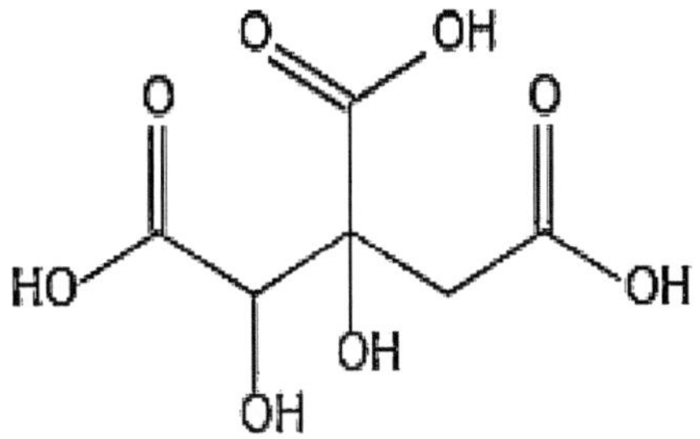

FIGURE 4.1 (-) -Hydroxy citric acid .

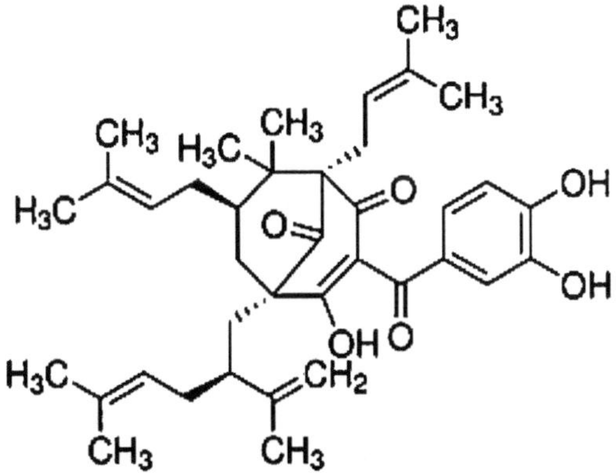

FIGURE 4.2 Garcinol.

(Deshmukh et al., 2008; Downs et al., 2005). Though, Garcinia/HCA supplements are efficient tools for various health problems especially obesity (Fig. 4.3) (Pandey, 2001). It is difficult to evaluate the effectiveness of single agents when the combination products are tested. In addition, awareness of the safety and efficacy of the Garcinia supplements available in the market is to be carefully analyzed for the risks and benefits of the dietary supplements vs. whole Garcinia fruits.

Currently, the world is in a health transition as death is giving way to new epidemics of non-communicable disorders such as cancer, cardiovascular diseases, and diabetes. Infections are the major cause of diseases in urban life. The trend of increasing obesity and obesity-related comorbidity and mortality is one of the concerns in the last few decades (WHO, 2004; Pandey, 2001; IASO, 2012). World Health Organization (WHO) data for the overweight reveals the body problems such as hyperlipidemia, hypertension, coronary heart disease, diabetes, cancer, disease of the gall bladder, osteoarthritis,

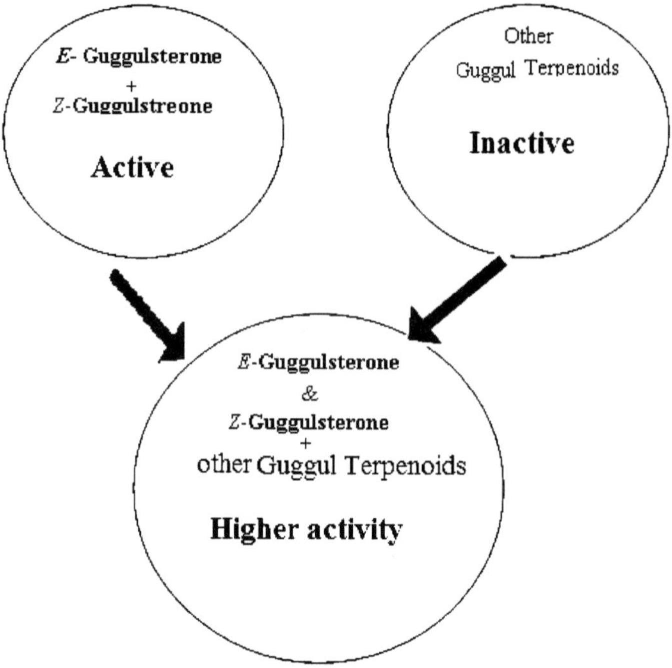

FIGURE 4.3 Diagrammatic sketch for synergistic effect of different ingredients in a formulation using the standardized extract of *Commiphora mukul.*

Source: Adapted from Pandey (2001)

shortage of breath, abnormal dilation of the veins, backache, and even psychological problems (WHO, 2000). In the market, there are a few drugs to ameliorate or prevent obesity, but costs, efficacy, and side effects are also matter of great concern, for example, Orlistat, Phentermine, Rimonabant, and Sibutramine, which are licensed for weight management purposes, but have some adverse effects (Bray and Greenway, 2007). For example, Phentermine has been reported to cause dry mouth, insomnia, headache, dizziness, fatigue, and palpitation (US FDA, 2010 ; US FDA, 2012). In the year 2010, Food and Drug Administration (FDA) had announced the market withdrawal of Meridia (Sibutramine) due to its risk of serious cardiovascular events (US FDA, 2010; US FDA, 2012). Several ethnobotanical studies have reported the bio-prospecting surveys on the positive application of herbs in the treatments for obesity (Heber, 2003). A myriad of health effects has been attributed to genus *Garcinia* such as anti-obesity effects (Preuss et al., 2004), antiulcerogenic (Deore et al., 2011), anti-oxidative (Yonei et al., 2008), antidiabetic (Wielinga et al., 2005), antimicrobial (Mackeen et al., 2000), antifungal (Mackeen et al., 2002), anti-inflammatory (Mazzio and Soliman, 2009), and anticancer effects (Mackeen et al., 2000).

Garcinia fruits are used in the preparation of traditional fermented food, "Ngari," in Manipur, India. Due to the increasing domestic demand for "Ngari" some traditional producers have started using urea to speed up the fermentation process. Recently, introduced use of plastic containers and polythene bags are to be studied before using these as an accessory utensil to be used for the fermentation process, as may cause allergic/toxic effects on human health. Over the years, there were no systematic and deliberate efforts made by researchers for the introduction of production technology through hygienic, utilization of pure microbial starter cultures, and optimum process control. Still, the local manufacturers are depending on the age-old traditional art. Recent works comparing the effect of a different starter culture of indigenous micro-flora have shown a significant difference in the taste, aroma, and chemical composition of the traditional fermented food using Garcinia fruits (Holzapfel, 2002). There has been a growing interest world over to explore search and collect microbial germplasm in search of the gene pool, which can help biotechnologist to develop value-added products for human welfare. Development of value-added products by selecting microorganisms, genetic improvement, process improvement, raw material improvement, etc., will lead to the industrialization of these food products. An increasing number of recent scientific articles report the positive effects of fermented foods on various aspects of human health and diseases through different mechanisms (Sarangthem and Singh, 2003). Development of novel fermented foods, with clinically documented health-promoting properties, can be marketed at a lower price.

4.3 TRADITIONAL MANAGEMENT OF *GARCINIA*

4.3.1 GARCINIA IN CHINA

The use of *Garcinia* spp. can be traced to ancient China, the "Compendium of Materia Medica" (ca.1592–1596). In China, there are 19 species of *Garcinia*, of which 13 are endemic. A total of 16 species occur in Yunnan and Guangxi (14 in Yunnan and 8 in Guangxi), *G. Oligantha* Merr. occur in Guangdong and Hainan, and *G. linii* C. E. Chang and *G. Subelliptica* Merr. occur in Taiwan (Li et al., 2007). Liu et al. (2016) have conducted an investigation to record the traditional knowledge of Garcinia used by different ethnic groups in Yunnan and Guangxi, and assessing the traditional management modes, and using Garcinia collections to predict the potential range and suitable areas for future utilization through cultivation. The free listing technique was used in the Zhugen village (Guangxi), to collect the traditional know-how on *Garcinia* genus from the region. The natives largely rely on natural resources; hence, rich in traditional knowledge and had a long history of cultivating Garcinias in or near their home gardens. A total of nine species were traditionally used in the village from the 16 targeted in the study. Traditional uses, local names, and economic and ethnobotanical applications were recorded from interviewees. The usages had not been well recorded previously, though, *Garcinia* spp. has been used by different ethnic groups (such as Dai, Zhuang, Hani, Jinuo, Yi, Dulong, and other local people) and developed different utilization modes and management of the wild Garcinia resources (Liu et al., 2016).

As economic plant species are commonly cultivated in rural areas and are essentially important for the rural economy (Joshi et al., 2015). Local people are generally supposed to support and participate in conservation initiatives if addresses on direct benefits from such efforts (Zou et al., 2010). As per Liu et al. (2016), Zhuang, Jinuo, and Dai people have had conventions of cultivating *Garcinia* for a long time in China. Different ethnic groups have similar methods for retrieving trees or saplings from nearby mountains after taste of fruits from wild trees and choose the one they prefer, and then transplant the saplings to their home garden or backyard. There were no reports for people using seeds to cultivate *Garcinia saplings. Saplings acquire stability in 2*–3 years and are enough mature to bear fruits. Most households cultivate two to five trees in traditional home gardens to provide for themselves, while in Fangchenggang and Hechi, Guangxi, some households cultivate about 1–10 hector of *Garcinia* trees in or near their home garden to serve for local people in and near the villages both in fresh fruits and pickles made from the fruit mesocarp. The fruit price ranges from 6 to 10 RMB/kg (0.9–1.5 US dollars/kg), and pickled mesocarp ranges from 12 to 16 RMB/kg (1.8–2.4 US dollars/kg) (Liu et al. 2016).

Jinuo people in Xishuangbanna, Yunnan, cultivate *G. xanthochymus* in traditional agroforestry systems on sunny slopes as the shade tree for tea (*Camellia sinensis* Linn.) plants. The area around *G. xanthochymus* is covered by hay on the ground to prohibit the growth of weeds and to cushion the falling of the fruit during harvest. Patterns of use among species also reflect species traits. In Fangchenggang, southern most Guangxi, local Zhuang people cultivate all three of the local species; *G. bracteata*, *G. multiflora*, and *G. oblongifolia*. It has been reported that there is no taste difference in fruits from all the three species but explained that the sequential phenologies of the three species provide fresh fruit for a longer time: *G. bracteata* usually has fruits in Aug.–Oct., *G. oblongifolia* in Sept.–Dec., and *G. multiflora* in Oct.–Dec. (Liu et al., 2016).

4.3.1.1 STUDIES TO PREDICT THE DISTRIBUTION OF GARCINIA

There are different kinds of models to study a species for distribution and predict potential distribution, including "Classification and Regression Trees" (CART). Liu et al. (2016) have collected specimen samples of nine *Garcinia* spp. which were retrieved, together with the distributions, and mapped during the ethnobotanical surveys. All data were coordinated on the map of Yunnan and Guangxi to locate the suitable distribution altitude. It was 200–400 m in most parts of Guangxi, while in Yunnan, most species were distributed from 400 to 1500 m in valleys in the southern part of the province. In North Yunnan where the altitude is higher, there were a few specimen records were belonging to three species, *G. multiflora*, *G. pedunculata*, and *G. esculenta,* all found in Gongshan, were distributed in low elevations along with the upstream of the Lancangjiang River (Liu et al., 2016).

4.3.1.2 ETHNOBOTANICAL UTILIZATION AND PHYTO-HARMACOLOGY

Medicinal chemistry of natural products reveals the active compounds of species and has reasonable explanations for usage patterns. Multi-disciplinary-signals approach for more clues to the ethno-botanists for more insight into the understanding of traditional knowledge and indigenous usage. Liu et al. (2016) have checked the traditional uses of Garcinia against the phytochemistry literature. Several phyochemical studies reveal that *Garcinia* spp. has been in practice to cure inflammation and relieve pains. *G. yunnanensis* and *G. xanthochymus* were observed for scanty phytochemical investigation data, and their medicinal use seems not fully recognized (Liu et al., 2016).

Liu et al. (2016) have conducted field investigations to understand traditional knowledge and traditional management for *Garcinia.* Niche modeling was used to explore more suitable areas for cultivating *Garcinia* in the future. Traditionally, *Garcinia* is most frequently used as edible fruits, followed by medicinal, wood, and ornamental use. Liu et al. (2016) have stressed that the local people recognized their own ethnotaxa and observed several ethnotaxa for *G. multiflora*. All these ethnotaxa were distinguished using morphological characteristics (Liu et al., 2016).

Liu et al. (2016) have further stressed that men generally knew more traditional use modes than women. One reason for this may be that men go to the forest more frequently than women to collect wild food and medicine and tend to range farther from home. Interviewees over the age of 50 years tended to have much more traditional knowledge. Those in the group of 30–50 years of age occasionally appeared unsure, mistook one species for another, and were more likely to use the same name for two different ethnotaxa. Most of the people under 30 years old have left to find jobs in urban areas. Few remain in the villages (Liu et al., 2016). Liu et al. (2016) have finally stressed for the systematic documentation of indigenous knowledge, a resource is of great significance (Liu et al., 2016).

Studies by Liu et al. (2016) have concluded that *Garcinia* species are well-known resources in China and useful various ways by local people in Yunnan and Guangxi, China. It is used as edible fruit, medicine, wood, ornamental, and for extracting oil. The knowledge is diversified and dependent on sex, age, and ethnic group. Liu et al. (2016) have narrated that *Garcinia as* edible fruits and other use modes were highly appreciated by the indigenous people in Yunnan and Guangxi, in China. They had collected wild populations from the mountains, cultivated in traditional home gardens, and sold in markets. They have great numbers of natural individuals and a large potential distribution area for cultivation. Some phytochemical studies have been done to help to explain the traditional uses of *Garcinia.* Liu et al. (2016) have further stressed for future studies to focus on phytochemical screening of less studied, but traditionally used species, as well as taxonomic investigations of *Garcinia* ethnotaxa (Liu et al. 2016).

4.3.2 GENUS GARCINIA IN AFRICA

Many African *Garcinia* species have been reported to be used in traditional medicine for many centuries. Some of them have been studied and reported to have many ethnomedical uses. Among the effects reported in the treatment of head and abdomen pains, fever, ulcers, impotence, throat and bronchial infections, venereal diseases, diarrhea, hepatitis, asthma rheumatism, arthritis, cancer, liver cirrhosis, and cough. Literature survey reveals that *G. kola* grows in many Western African countries, to be the most studied *Garcinia* species in all aspects, including ethno-medical use, pharmacology, and its phytochemistry. Sometimes, it is referred to as a "wonder plant" because every plant part is medicinally important (Dalziel, 1937). Ethno medically, many Garcinia plants are used as decoction, infusion, or as juice. Most extracts are prepared with cold or hot water and are applied for the treatment of toothache, inflammations, wound-healing, jaundice, ulcers, dysentery, aphrodisiac, fever, sleeping sickness, venereal diseases, liver cirrhosis, arthritis, and respiratory tract diseases. Other Garcinia plants are used as chewing sticks, fertilization stimulants, and aid childbirth while some fresh or dried fruits are used as food. For instance, *G. livingstoneii* is commonly known as an "African Mangosteen" as it produces very tasty and delicious fruits. The powdered root of the plant is used as an aphrodisiac (Arnold and Gulumian, 1984). Many of the Garcinia plants are known for different traditional uses while a few have not been reported for any ethnomedical use (Magadula and Mbwambo, 2014). A survey to locate the Tanzanian *Garcinia* species indicated that some of the species are threatened for extinction due to habitat destruction as *G. acutifolia* formerly reported to grow in Pugu Forest, coastal region, which has been reported to be wiped-out from the area (Bamps et al., 1978).

4.3.3 GARCINIA IN BANGLADESH

Garcinia is an important predominant fruit of the North-eastern region of India, having several medicinal properties and being used against various diseases in folk medicine. Negi et al. (2008) have reported that the extracts of Garcinia fruits have pronounced inhibitory effects against the Gram-positive bacteria the extracts from G. *Pedunculata* (Jacob et al., 2015). Mazumdar et al. (2016) have studied *G. Pedunculata* from the Sylhet region of Bangladesh (local name; Toikor) as a natural preservative a better option to extend the shelf-life of bread and to ensure health safety. The dried powder of *G. pedunculata* fruits was studied for bread preservation. Yeast and mold count, total viable count (TVC), and moisture content were measured every other day (Day 0–10) during the study. Results for nutritive analysis of bread revealed that it was in the range of 7.41–8.95% protein, 4.16–5.86% fat, 50.49–55.77% carbohydrate, 0.91–1.31% minerals, and 1.08–1.24% fiber. The moisture range of the bread was found at 29.86–33.74%. Among different treatments

T1 having 1% Toikor powder and T2 treatments having clove powder, 0.5% showed the highest shelf life with minimum yeast and mold count at the end of storage. Clove and Toikor powder can be used as a natural preservative to make the bread long-lasting. Finally, it was concluded that Toikor powder combinations (0.2 and 0.3%) affect the quality characteristics of bread and provide better protection against microbial spoilage (Jacob et al., 2015).

The research study was concluded with the need for clinical evaluation of different extracts and compounds isolated from the species using *in vivo* models, and further studies on Garcinia should focus on a comprehensive phytochemical analysis of the species and to more accurately define its biological activities to co-relate with phytochemical components. Additionally, new biological evaluations are still needed to scientifically validate some of its ethno pharmacological applications so that the species can be used as a future resource for disease treatment and management, including the toxicity of the species on humans.

4.3.4 PROXIMATE INFORMATION ON GARCINIA

4.3.4.1 PHYTOCHEMISTRY OF GARCINIA

Studies have revealed that Garcinia rind contains fat (1.4%), moisture (80.0 g/100 g), protein (1%), tannin (1.7%), pectin (0.9%), and total sugars (4.1%). Leaves of the plant are reported to contain 1.24 g fiber, 75% moisture, 17.2 g of carbohydrates, 2.3 g of protein, 0.5 g of fat, 15.14 mg of iron, 250 mg of calcium, 10 mg of ascorbic acid, and 18.10 mg of oxalic acid (Sheela et al., 2004). Garcinia seeds are very rich in oleic, stearic, and stearic triglycerides (Dushyantha et al., 2010). The plant also contains HCA-lactone and citric acid, but in minor quantities (Jayaprakasha and Sakariah, 2002).

4.3.4.2 INDUSTRIAL USE

Fruit rinds of Garcinia are commercially used to prepare concentrated syrups. The syrup on appropriate dilution is used as a cool health drink, especially as an off-season drink. In Goa, India, local community uses the rinds to prepare wine. Dried rinds are powdered and marketed to prepare acidulants for traditional curries (Nayak et al., 2010).

4.3.4.3 TRADITIONAL MEDICINAL USES

Fruits and leaves of Garcinia are sour, astringent, thermogenic, constipating, and digestive. The preparations from Garcinia rinds are used to treat rheumatic pains and bowel complaints. The fruit is anti-helminthic and cardio-tonic. The juice (sherbet) made from the rind is used for piles, hemorrhoids, colic problems, ulcers, inflammations, treat sores, dermatitis, diarrhea, dysentery, ear infection, to facilitate digestion and to prevent over perspiration or hyper perspiration. Garcinia fruits are a type of natural antacid. The rind of Garcinia fruit with yogurt and salt is used against the burning sensation and relieve gastric ulcerations (Perlin, 2009; Sahasrabudhe and Deodhar, 2010).

4.3.4.4 FREE RADICAL SCAVENGING PROPERTIES

Free radicals, consisting of reactive oxygen species (ROS) and reactive nitrogen species (RNS), in excess cause damage to DNA, lipids, proteins, and other biomolecules. The antioxidants are necessary to prevent the formation/ nullify the harmful effects of the ROS and RNS. Garcinia products existing in the market as concentrated syrup, cold aqueous, and hot (boiled) aqueous extract are effective in scavenging the free radicals (Preuss et al., 2004; Del Rio D et al., S 2005).

4.3.4.5 ANTI-AGING ACTIVITIES

Sahasrabudhe and Deodhar (2010) have observed that Garcinia rind extract by methanol, ethyl acetate, and water fraction of the methanolic extract have anti-hyaluronidase and anti-elastase activities. Aging is a natural process and the development of wrinkles and sagging of the skin is an indication of it. The activity of the elastase enzyme decreases the elasticity of the skin, and gradually causes sagging, with age. The concentration of hyaluronic acid decreases and makes the skin dry and wrinkled. Skin exposure to the sun and detrimental chemicals also hasten the process of aging and wrinkling and application of antioxidant-rich cosmaceuticals are known to retard the process (Baliga and Katiyar, 2006). Sahasrabudhe and Deodhar (2010) have noted that the ethyl acetate fraction showed significant hyaluronidase inhibition even at 25 μg/mL (a low concentration), while the aqueous fraction was effective against both elastase and hyaluronidase at concentration 90 μg/mL. Garcinia pigments have UV light absorbing properties, reveals the utility in skincare management (Bhat et al., 2005).

4.3.4.6 GASTRO-PROTECTIVE EFFECTS

Studies have shown that the oral administration of garcinol (40–200 mg/kg) reduced the indomethacin-induced gastric ulceration in rats. The optimal effects were noted at 200 mg/kg and the protective results were better than that of cetraxate-HCl, a positive control. Garcinol has been reported effective in reducing water immersion-induced gastric ulceration (23°C), and effects were similar to the cetraxate-HCl, a positive control (Yamaguchi et al., 2000). A peptic ulcer is a multi-factorial disease and affects a significant number of the global population.

4.3.4.7 ANTI-DIABETIC EFFECTS

Garcinia rind aqueous extract (100 mg/kg and 200 mg/kg) on oral treatment for 4 weeks to streptozotocin-induced type 2 diabetic rats have been reported effective in decreasing blood glucose in both cases of fasting and postprandial, respectively [68] (King and Loeken 2004).

Diabetes, a disease as old as mankind, is characterized by chronic hyperglycemia, a report has revealed that endocrine disorder has affected nearly 5% of the global population. Chronic hyperglycemia is a cause of secondary complications that are more dangerous than hyperglycemia, so it requires constant medical attention and care of the patients (King and Loeken, 2004).

4.3.4.8 ANTINEOPLASTIC AND CHEMO-PREVENTIVE EFFECTS

Studies on garcinol, iso-garcinol, and xanthochymol, extracted from Garcinia have properties to affect the growth and proliferation of four human leukemia cell lines. The growth inhibitory effects of iso-garcinol and xanthochymol were more potent than that of garcinol (Braide, 1993).

4.3.4.9 LIPID-LOWERING EFFECT

Malabar Tamarind (*G. cambogia*) is seen abundantly in the evergreen forests of Konkan in South India. Many traditional recipes in Kerala are used to treat hepatitis, laryngitis, reducing the LDL, and mouth disease. The bi-avonones are the most dominant components in most Garcinia species (Waterman and Hussain, 1983).

4.3.4.10 ANTI-OBESITY ACTIVITY

In the Ayurvedic system of medicine, Garcinia is used to treat illnesses related to obesity. Multiple studies have revealed that hydroxy citric acid (also known as Garcinia acid), possesses anti-obesity effects. Studies have shown that consumption of hydroxyl citric acid reduces appetite, inhibits fat synthesis, lipogenesis, decreases food intake, and reduces body weight (Edirisinghe et al., 2015). WHO reports have been cited for deciphered the increasing number of obese people, globally, and sub-sequently increase in the incidence of cancer, cardiovascular diseases, diabetes, and digestive diseases. Obesity is associated with other health risks as diabetes mellitus, heart disease, high B.P., etc. WHO has reported that globally with more than 1 billion adults is over-weight. The main cause of obesity is the consumption of more calories from food than needed. Garcinia fruits are a rich source of HCA, the active ingredient of the fruit that aids in weight-loss by inhibiting fat production and suppressing appetite. *G. cambogia* extract is used as a popular ingredient in many weight loss supplements (Nguyen and El-Serag, 2010). Garcinia fruits, a rich source of HCA, are an active agent that aids in weight loss by inhibiting fat production and suppressing appetite. Citrine is another ingredient of Garcinia fruits; it inhibits an enzyme that helps the body synthesize fat for storage in adipose tissue. HCA inhibits lipogenesis, lowers the production of cholesterol and fatty acids, increases the production of glycogen in the liver, promotes energy, suppresses appetite, and activates thermogenesis (Shara et al., 2003).

Many studies have been pointing to strong possibilities to used *G. Cambogia* extract as a rich source of ingredients for weight loss supplements, but the evidence is not conclusive. Despite without conclusive evidence, it has been cited in many weights loss supplements for losing weight (Preuss et al., 2004). *G. cambogia* is a small tropical sour fruit, but the rind is used as a spice in Indian cooking. The active ingredient HCA, unlike caffeine, is not a stimulant. Neither is it an appetite suppressant, so does not directly work on the nerve centers of the brain. HCA helps to maintain the required body energy and improves the signaling system to the brain when it has eaten enough. In overweight people, the response to the brain is often delayed, causing delay to stop eating (Ohia et al., 2002). HCA-rich supplements are most suitable for people who overeat. At the stage of stress or anxiety, it gives the same calming effect that they get from food. Generally, a person does not need to eat any special foods when taking HCA, and there are no forbidden foods. The selection of healthy foods with Garcinia/HCA, a concerned person loses weight and fat slowly but steadily without really trying. An obese patient does not need to be hungry for weight loss. *G. cambogia* fruit/extract rich in HCA in the diet of the patient is usually combined with chromium. Chromium is a mineral that is often lacking in our diets, and as we get older, chromium deficit can contribute to weight gain and possibly diabetes. HCA and chromium seem to work in a positive synergistic mode to regulate blood sugar levels. However, already diabetes person should see a medical doctor before taking any products containing chromium. The use of HCA-rich supplements in pregnancy or feeding mother reveals best to be cautious just like with all other weight loss supplements (Jones and Andrews, 2007), though HCA has some significant advantages over many other weight loss supplements ingredients as it has been extensively tested in research laboratories on both human and animal. There were no side-effects have been reported, and it is considered a natural weight loss supplement (Edirisinghe et al., 2015).

4.3.5 HCA IN WEIGHT LOSS PROCESS AND REACTION MECHANISM

In normal conditions, orally taken carbohydrates (glucose, fructose, and galactose) are converted by the body into ATP (energy) and the excess carbohydrates that could not be immediately used are converted into glycogen, a storage form of carbohydrates. It is deposited in muscles and liver. When the natural glycogen stores are reasonably full, additional carbohydrates are then converted into excess of extra-mitochondrial Acetyl-CoA required for fatty acid synthesis using ATP Citrate lyase enzyme. (-)- hydroxy citric acid [(-)-HCA] is the principal acid of fruit rinds of Garcinia fruits. (-)-HCA is a potent inhibitor of ATP citrate lyase, which reduces the availability of acetyl-CoA units required for fatty acid synthesis and lipogenesis of a lipogenic diet (i.e., diet rich in carbohydrates). The added glycogen load in the liver stimulates a long-lasting neuro-signal from the liver to the brain to satiety (satisfaction) and helping to suppress appetite (Ishihara et al., 2000). Inhibition of citrate lyase aids aerobic endurance and owing to a substantial increase in glucose uptake by working muscle, and an increase in hepatic glucose output. Garcinia/HCA have not been known for any adverse effects in healthy adults, but children, pregnant, and lactating women, those diagnosed with diabetes mellitus, and people with Alzheimer's or other forms of dementia disease are advised to take it under strict medical supervision or not to touch (Asghar et al., 2007).

4.3.6 THERAPEUTIC DOSAGES

Garcinia/HCA supplements are available as OTC (over the counter) products in various forms (including tablets, capsules, powders, extracts, and even snack bars) in markets, globally. These supplements are generally standardized for a fixed percentage of HCA. The usual dosage for Garcinia is 300–500 mg tablets 3 times daily taken half an hour before meals with water (Stevens et al., 2005). Asish et al. (2008) have made a comparative study among *Garcinia* species for the (-) HCA content with HPLC and acid–base titration and concluded that the titration results are slightly higher as it gives the presence of total acids in the sample. Percentage difference in the (-)-HCA of the two methods varies between 0.09 and 0.879 in leaves and 0.15 and 3.35 in case of fruits. Among the four species of *Garcinia* (*G. gummi-gutta, G. indica, G. Tinctoria,* and *G. cowa*) studied G. *tentoria* was found to have very low HCA in leaves (0.013%) and fruits (0.07%). Further, it was revealed that fruit rinds have more amount of HCA (3.63–4.62%, as such basis) in comparison with leaves (2.07–2.44%). The magnitude of HCA in leaves and fruit rinds followed in the order of *G. gummi-gutta*>*G. indica*>*G. cowa*>G. *tinctoria* irrespective of the methods studied (Asish et al., 2008).

Studies on various animal models and humans have been conducted for the safety of Garcinia/HCA, but no serious effects have been reported in any of the reports, and all results were comparable to the placebo. Dose-dependent studies have also concluded that HCA is safe under the experimental conditions employed (Asghar et al. 2007; Stevens et al., 2005).

4.3.7 GARCINIA AS FUNCTIONAL FOOD

Historically, Garcinia has been used to treat gastric ulcers. Currently, garcinol is reported to lower acidity and protect the gastric mucosa. It is a benzophenone derivative, has assay 2–3% in the fruit rind. It has been reported as a potent inhibitor of histone acetyltransferase and is considered as potential anti-cancer agent. It has been characterized for anti-oxidative activity, chelating activity, free radical scavenging activity, and antiglycation activity. It is assumed that garcinol is a potent antioxidant and a glycation inhibitor under specified conditions. Hydroxyl (OH^-) radicals are regarded as the most dangerous ROS, and, therefore, garcinol is expected to be useful for preventing diseases caused by these radicals, such as stress-induced gastric ulcers and non-steroidal anti-inflammatory drug-induced gastric. The mechanism of its anti-ulcer activity is not clear, garcinol may scavenge ROS on the surface of gastric mucosa, supposed to protecting cells from injury (Ananthakrishnan and Rameshkumar, 2017).

Fat extracted from Garcinia fruits solidify organic liquids in 12 h at room temperature but at 4°C the process completes in just 1 h. This property of the fat has wide applications useful in commercial and industrial purposes (Reddy and Raghuveer, 2018). Garcinia fruits juice has about 4% sugars, it is used to produce wine by fermentation. Garcinia wine also known as kokum wine is prepared in Goa, India, using the traditional method (Obolskiy et al., 2009). However, the procedure has been found to wine for poor quality wine and has a scope for innovation.

The dried exudates from Garcinia bark are used as a coloring agent and in several traditional medicines. The phytochemical investigations reveal that bark exudates as a rich source of bioactive secondary metabolites such as caged xanthones (Reddy and Raghuveer, 2018). The literature is scanty about various amino acids, vitamins, proteins, enzymes, glycol-conjugates, lectins, polyphenolics, etc., from the genus Garcinia.

Several products of *G. Cambogia* or its derivatives had been patented and commercialized. A total of 66 patents have been applied to Garcinia or HCA with the US Patent and Trademark Office (USPTO) since 1976. These patents are on different aspects including HCA-rich supplements from Garcinia rind, HCA and its value-added products and their applications. Most of the patents are related to *G. cambogia/atroviridis* and HCA derived from Garcinia on obesity and weight loss [84] (Chuah et al. 2012). Many of the food supplements having Garcinia/HCA marketed for weight management are the combination of active ingredients rather than containing a single agent (Downs et al., 2005). Generally, it is difficult to evaluate the effectiveness of a single ingredient with multi-ingredient combination products. The awareness for safety and efficacy of the weight management supplements on the marketed products is necessary to analyze the risks and benefits of the dietary supplements. Thus, scientific investigations on the potential health-promoting benefits of herbals as a diet supplement is a prerequisite (Deshmukh et al., 2008). Currently, the traditional healers last hope and choice at the illness when it is untreatable by modern treatment and became a mysterious illness. As for traditional cultural knowledge, there is a need for more public awareness to document it. This may help communities to update the gaps, if any and connect it with modern molecular science (Deshmukh et al., 2008).

4.3.8 *FLORAL BIOLOGY OF THE GENUS GARCINIA*

A detailed knowledge of morphology and floral biology of the plants is helpful for a purposeful breeding program. Te-chato (2007) has conducted a study on floral and fruit morphology of some species in *Garcinia* (*G. cowa, G. speciosa, G. atrouredis, G. mangostana, G. dulcis,* and *G. schomburgkiana*) and observed some differences and similarities. Flowers of all the species are composed of the male and female flowers. Female flowers are solitary, and it occurs in single or occasionally in a cluster. All the species in the genus comprise of four petals and four sepals, which were different in size and color. Many stamens both filamentous and sessile anthers are observed in both male and female flowers (Te-Chato, 2007).

Neog (1990) carried out an experiment on *G. cowa* (Kuji Thekera) and reported that the flowering period of the plant from April to May and anthesis occurs from 2 am to 4 am and dehiscence completed within 3–4 h of anthesis. The pollen fertility has been reported 75%. Baruah and Borthakur (2012) conducted a study on morphology and ethnobotany of six species of *Garcinia* like *G. pedunculata, G. paniculata, G. morella, G. cowa, G. lanceaefolia,* and *G. Xanthachymus* and reported that all the six species are evergreen in nature. Flowers are polygamous in nature. The flowering in *G. Pedunculata* is throughout the year. Flowering time of *G. cowa* is from March to April and that of *G. Lanceaefolia* and *G. Xanthochymus* are March–May.

Richards (1990) studied on the phenology, pollination biology, and fertilization of *G. Hombrniana* Pierre, which is pollinated by trigona bees. Nectar is restricted to the large discaid stigma, which also captures and hydrates pollen. The wet stigma and binucleate pollen suggest that *Garcinia* are from hermaphrodite plants with a gametophytic self-incompatibility system. On stigmas, nectar is secreted early on three or four successive days. On male pistillodes, nectar is secreted when anthers dehisce, on the second morning after anthesis. Pollen is most viable when freshly collected, but some pollen is viable for 4 days after collection (Richards, 1990).

Sasithorn *et al.* (2007) have reported a significant morphological and anatomical variation among female and hermaphrodite flower of *G. atrovirides*. Hermaphrodite flower has a long filament, and they produce fertile pollen grains, whereas females produce pollen less anthers. The hermaphrodite flowers have more flowers per inflorescence than female flowers, but they gradually drop off before fruit setting (Sasithorn et al., 2007).

Parthasarathy and Nandakishore (2014) have studied nine species (*G. indica, G. gummi-gulta, G. cowa, G. Subelliptica,* and *G. Mangostana* from Western Ghats, *G. pedunculata, G. Lanceaefolia* and *G. Kydia* from

Northeast Himalayas and *G. Xanthochymus* from both the ecosystem) for morphological characterization and observed that *Garcinia* species are evergreen and having an opposite decussate foliar arrangement. They have simple petiolate leaves. The petiols are so small that they look like sessile. Venation of the leaves is pinnate in all the cases, but venation is different. In the case of *G. Cowa* mid-vein raised abaxially, however, they observed that flowering initiate in summer in NE region (April–June) while in Western Ghats it is spring (Dec.–March). Flowers are mostly sessile except *G. Kydia* which has pedicels of 4–7 mm length. Flowers are actinomorphic except *G. subelliptica*. Flower buds are round and have variations in flower color anthers are sessile celled and longitudinally dehiscent. Fruits are fleshy berry with smooth fleshy pericarp with high diversity in shape, size, and color (Parthasarathy and Nandakishore, 2014).

Devi et al. (2012) have studied on genetic diversity of *G. indica* (Kokum) in Goa and reported that male and female flowers are in separate plants. Male trees have typical flowers with long pedicels, stamens are fertile, but pistils were absent in this type. Female flowers are globose in shape with well-developed ovary with the stigmatic surface has eight distinct lobes. The study shows that there is a great variation in flowering and fruiting. Flowering spread for 8–9 weeks and fruiting season lasted for 2 months. Flower initiations take place during first week of December and last week of November and 50% flowering occurs 1 month after the first initiation of flower (Devi et al., 2012).

The ethno-botanical survey indicated that local people or people of Assam had knowledge about *Garcinia* species (*G. pedunculata, G. cowa, G. Lanceaefolia,* and *G. xanthochymus*) and they were known for their edible fruits and medicinal properties. It is commonly known as "Thekera" in Assamese and Assamese people have rich traditional uses of these fruits as medicine and in some social customs. A similar result was also observed in Assam by Baruah and Borthakur (2012) who had conducted ethno-botanical survey on different *Garcinia* species of Assam.Barukial and Sarmah (2011) also found that the fruits of all four species of *Garcinia* (i.e., *G. pedunculata, G. lanceaefolia, G. cowa,* and *G. xanthochymus*) have digestive and anti-dysenteric properties. They are also effective for diarrhea and jaundice (Barukial and Sarmah, 2011).

4.3.8.1 KARYOTYPE OF GENUS GARCINIA

Karyotype study revealed that chromosome number of *G. pedunculata, G. Cowa,* and *G. Lanceaefolia* was found to be $2n = 48$ whereas *G. Xanthochymus* had $2n = 72$ numbers of chromosome. Therefore, from the study it can be concluded that these *Garcinia* species could be used to enrich diet formulations and nutraceuticals, and also there is a scope for genetic assessment and breeding due to the presence of variability in different characters (Gogoi, 2015).

4.3.9 ASSAM; A HUB FOR GENUS GARCINIA

Dutta et al. (2017) have recorded a total of 10 species of *Garcinia* from different forest areas, patch vegetation in rural areas and homesteads of the valley of Assam, India. Among the 10 species recorded (Table 4.1), *G. dulcis, G. morella, G. sopsopia, G. spicata,* and *G. Xanthochymus* were found in wild, three species *G. pedunculata, G. cowa,* and *G. kydia* were found distributed in wild as well as under domesticated conditions. The distributions of these species have been found in the patches of rural areas. The species *G. Xanthochymus* was found throughout the entire valley with its distribution in the natural forest as well as patch vegetation. The species of *Garcinia* recorded in the study sites were medium size trees except for *G. lanceifolia* which is a small tree only found in the cultivated condition in the homestead gardens (Dutta et al., 2017).

TABLE 4.1 Distribution of Genus *Garcinia* in Assam, India (Dutta et al., 2017).

S No.	Species	Local name	Habitat
1	*Garcinia atroviridis* Griff. ex T. Anderson	–	It is wild and randomly distributed.
2	*G. cowa* Roxb. ex Choisy	Kuji Thekera	It is a wild species, localized in certain pockets, and in home gardens as a domesticated plant.
3.	*G. dulcis* (Roxb.) Kurz.	–	It is a wild species, not uniformly distributed.
4	*G. kydia* Roxb.	Kuji Thekera	It is a wild, localized in certain pockets, and in home gardens as a domesticated plant.
5.	*G. lanceifolia* Roxb.	Rupahi Thekera	It occurs at the home gardens as a domesticated plant.
6.	*G. Morella* (Gaertn) Desr.	–	It occurs in forest, but not uniform.
7.	*Garcinia pedunculata* Roxb. Ex Buch.-Ham.	Bor Thekera	Wild, localized in certain pockets and home gardens as domesticated plant.
8.	*G. sopsopia* (Buch.-Ham.) Mabb.	Mamoi Thekera	Wild, in certain pockets.
9.	*G. spicata* Hook.f.	–	It is a wild species, and occurs in certain pockets.
10.	*G. xanthochymus* Hook.f. ex. T. Anderson	Teportenga	It is a wild as well as domesticated as occurs in gardens also.

The association of four species of *Garcinia* (*G. pedunculata, G. kydia, G. cowa,* and *G. lanceifolia*) out-of-the ten recorded with other species of the plants of the study sites were found valuable for the people due to fruits. These species are economically important due to the various traditional usages of the whole fruits as well as the pericarp, and it is believed that the medicinal values of these dried sliced pericarps increase on storage.

TABLE 4.2 Ethnobotanical Usages of *Garcinia* Species in Assam, India (Dutta et al., 2017).

S No.	Name of species	Traditional uses
1.	*G. cowa*	Pericarps of the fruits are used as acidifying agents for curry. The fleshy outer covers of seeds of ripe fruits are eaten as raw. Dry-sliced pericarps of fruits are preserved for its medicinal value. It is used as medicine against dysentery and constipation. The juice prepared from the dry sliced pericarp is eaten as sherbet because of its sour test. The extract of dry-sliced fruit is used as medicine for high blood pressure and stomach disorder for human as well as domestic animals. The stem is used as pole for construction of traditional house.
2.	*G. kydia*	Pericarps of fruit are used as acidifying agents for curry. The fleshy outer cover of seeds of ripe fruits is eaten as raw. Dry-sliced pericarp of the fruits is preserved for its medicinal value, is used as medicine against dysentery and constipation. The juice prepared from the dry-sliced pericarp is eaten as sherbet because of sour test. The extract of dry-sliced fruit is used as medicine for high blood pressure and stomach disorder for human as well as domestic animals. The stem is used as pole for construction of traditional house, making Dhaki, a traditional rice husking implements, fuel wood, etc.
3.	*G. lanceifolia*	Leaves are taken as vegetables. Pericarps of fruits are eaten in raw and as pickles. The dry-sliced prepared from pericarps are used as acidifying agent for traditional curry, medicine against stomach trouble. The juice prepared from the dry-sliced pericarp is drink as sherbet during warmer days to get relief from hot.
4.	*G. pedunculata*	Fruits are used as acidifying agents for curry and pickles. The flashy outer covering of the seed coat is used for making cold drinks. Boiled unripe fruit is used as the medicine in dysentery. It is also eaten as with traditional food. The unripe fruits are used in Bohag Bihu festival. The wood is used for the construction of houses, making Dhaki, the traditional rice husking implements, fuel wood, etc.

Future scope: There is a fresh trend for use of food supplements and herbal medicines, as demand has shifted to use the natural sources for health benefits. Pharmacologists, pharmacognosists, microbiologists, botanists, and chemists are combing the earth for phytochemicals and lead from herbs and plants that could be developed for the treatment of various diseases. 16%

medicines from natural products were approved by the FDA during 2018 (De la Torre and Albericio, 2019), indicating the scope and search for new moieties from natural origin, globally. Natural products offer a potentially infinite source of chemical diversity unparalleled to any synthetic chemical collection or combinatorial chemistry approach. The natural product compounds can have astounding chemical structures that can lead to unexpected, alternative medicinal chemistry programs based on important biological targets. The genus *Garcinia* is among the list of plants to be studied in-depth as it has many un-deciphered molecules.

KEYWORDS

- **thekera**
- **diversity**
- **ethnopharmacology**
- **phyto-pharmacology**
- **perspective for innovation**

REFERENCES

Akao, Y.; NakagawaY, Yoshinori N. Anti-Cancer Effects of Xanthones from Pericarps of Mangosteen. *Int. J. Mol. Sci.* **2008,** *9*, 355–370.

Ananthakrishnan, R.; Rameshkumar, K. B. Phytochemicals and Bioactivities of *Garcinia indica* (Thouars) Choisy; A Review; In: *Diversity of Garcinia Species in the Western Ghats: Phytochemical Perspective*; Ramesh Kumar, K. B., Ed.; Chapter 9; JNTBGRI, 2017; pp 142–150.

Arnold, H. J.; Gulumian, M. Pharmacopoeia of Traditional Medicine in Venda. *J Ethnopharmacol.* **1984,** *12*, 35–74.

Asghar, M.; Monjok, E.; Kouamou, G.; Ohia, S. E.; Bagchi, D.; et al.Super CitriMax (HCA-SX) Attenuates Increases in Oxidative Stress, Inflammation, Insulin Resistance, and Body Weight in Developing Obese Zucker Rats. *Mol. Cell Biochem.* **2007,** *304*, 93–99.

Asish, G. R.; Parthasarathy, U.; Zachariah, T. J.; Gobinath, P.; Mathew, P. A.; George, J. K.; Saji, K. V.A Comparative Estimation of (-) Hydroxycitric Acid in Different Species of *Garcinia. Hort J.* **2008,** *21* (1), 26–29.

Baliga, M. S.; Katiyar, S. K. Chemoprevention of Photo-Carcinogenesis by Selected Dietary Botanicals. *Photochem. Photobiol. Sci.* **2006,** *5*, 243–253.

Bamps, P.; Robson, N.; Verdcourt, B. Flora of Tropical East Africa: Guttiferae, Crown Agents for Over sea Governments and Administrations, 1978; p 14.

Baruah, S.; Borthakur, S. K. Studies on Morphological and Ethno-Botany of Six Species of *Garcinia* L. (Clusiaceae) Found in the Brahmaputra Valley, Assam, India. *School Res. Libr.* **2012,** *2* (3), 389–396.

Barukial, J.; Sarmah, J. N. Ethno-Botanical Plants Used by the People of Golaghat district, Assam, India. *Intern. J. Med. Arom. Plants.* **2011,** *1* (3), 203–211.

Bhat, J. D.; Kamat, N.; Shirodkar, A. Compendium and Proceedings of 2nd National Seminar on GARCINIA (*Garcinia indica* Choisy); Goa University March, 4–5, 2005.

Braide, V. B. Anti-Inflammatory Effect of Kolaviron, a Biflavonoid Extract of *Garcinia kola. Fitoterapia* **1993,** *64* (5), 433–436.

Bray, G. A.; Greenway, F. L. Pharmacological Treatment of the Overweight Patient. *Pharmacol. Rev.* **2007,** *59* (2), 151–184.

Burkill, I. H. *A Dictionary of the Economic Product of the Malay Peninsula 1*. Reprint Edition (1966) Lumpur, Ministry of Agriculture and Co-Operatives, 1935.

Chuah, L. O.; Yeap, S. K.; Ho, W. Y.; Beh, B. K.; Alitheen, N. B. In Vitro and In Vivo Toxicity of *Garcinia* or Hydroxycitric Acid: A Review. *Evid.-Based Complement. Altern. Med.* **2012,** Article ID e197920.

CSIR. *The Wealth of India—Raw Materials*, Vol. IV; New Delhi, India, 1956.

Dalziel, J. M. *The Useful Plants of West Tropical Africa*; Crown Agents for the Colonies: London.

De la Torre, B. G.; Albericio, F. Review; The Pharmaceutical Industry in An Analysis of FDA Drug Approvals from the Perspective of Molecules. *Molecules* **2018,** *24*, 809–820. DOI: 10.3390/molecules24040809

Del Rio, D.; Stewart, A. J.; Pellegrini, N. A Review of Recent Studies on Malondialdehyde as Toxic Molecule and Biological Marker of Oxidative Stress. *Nutr. Metabol. Cardiovasc. Dis.* **2005,** *15*, 316–328.

Deore, A. B.; Sapakal, V. D.; Dashputre, N. L.; Naikwade, S. Antiulcer Activity of *Garcinia indica* Linn Fruit Rinds. *J. Appl. Pharm. Sci.* **2011,** *1*, 151–154.

Deshmukh, N. S.; Bagchi, M.; Yasmin, T.; Bagchi, D. Safety of a Novel Calcium/Potassium Salt of Hydroxyl Citric Acid (HCASX): I. Two-Generation Reproduction Toxicity Study. *Toxicol. Mech. Methods***2008,** *18* (5), 433–442.

Devi, S. P.; Balamohan, T. N.; Thangam, M.; Kumar, J. A.; Ramachandrudu, K.; Korikanthimath, V. S. A Study on Diversity and Distribution of Kokum (*Garcinia indica* (Choisy) (Thouars) Using DIVA-GIS in Goa with Respect to Fruit Characters. *Indian J Hort.* **2012,** *69* (2), 156–162.

Downs, B. W.; Bagchi, M.; Subbaraju, G. V.; Shara, M. A.; Preuss, H. G.; Bagchi, D. Bio-Efficacy of a Novel Calcium Potassium Salt of (-)-hydroxy citric acid. *Mutat. Res.* **2005,** *579* (1–2), 149–162.

Dushyantha, D. K.; Girish, D. N.; Suvarna, V. C.; Dushyantha, D. K. Native Lactic Acid Bacterial Isolates of Garcinia for Preparation of Fermented Beverage. *E J. Acad. Res. Rev.* **2010,** *2*, 21–24.

Dutta, D.; Hazarika, P.; Hazarika, P. Distribution and Diversity of *Garcinia* L. in Upper Brahmaputra Valley, Assam. *Int. J. Curr. Res.* **2017,** *9* (10), 59644–59655.

Edirisinghe E, A.; H. E.; Nawarathna, S. B.; Marapana R, A.; U. J.; Jaysinghe, J. Extraction, Crystallization, and Preservation, Pelletizing and Quantification of Hydroxyl Citric Acid from *Garcinia combogia. IJIRT.* **2015,** *2* (4), 2–10.

Gogoi, B. Morpho-Biochemical Characterization of *Garcinia* Species of Assam. Dept. of Horticulture, Faculty of Agriculture, Assam Agri Univ, Jorhat-7**85013,** Assam. Registration No. 190 of 2007 (Roll No. 10-AD-20). Ph.D. Thesis, 2015.

Heber, D. Herbal Preparations for Obesity: Are They Useful? *Primary Care* **2003,** *30* (2), 441–463.

Holzapfel, W. H. Appropriate Starter Culture Technologies for Small-Scale Fermentation in Developing Countries. *Int. J. Food Microbiol.* **2002,** *75*, 197–212. http://www.jntbgri.res.in/downloads/garcinia/chapter_9.pdf (accessed on 26 Mar 2019).

IASO (International Association for the Study of Obesity). % Global Prevalence of Adult Obesity (BMI ≥ 30 kg/m^2): Country Rankings, 2012. http://www.iaso.org/site media/uploads/Global prevalence of adult obesity Ranking by country 2012.pdf (accessed on 8 Mar 2019).

Ishihara, K.; Oyaizu, S.; Onuk, K.; Lim, K.; Fushiki T. Chronic (-)-hydroxycitrate Antiglycation activity of Garcinol from *Garcinia* and Promotes Lipid Oxidation During Exercise in Mice. *J. Nutr.* **2000,** *130*, 2990–2995.

Iwu, M.; Igboko, O. Flavonoids of *Garcinia kola* Seeds. *J. Nat. Prod.* **1982,** *45*, 650–651.

Jacob, K. M. P.; Ali, M. A.; Vishnu, H.; Shylaja, G.; Mythili, S.; Sathiavelu, A. Evaluation of Antibacterial and Antioxidant Activity of *Garcinia gummigutta. Int. J. Drug Dev. Res.* **2015,** *7* (3), 57–59.

Jain, S. K.; Dam, N. Some Ethnobotanical Notes from North-Eastern India. *Econ. Bot.* **1979,** *33*, 52–56.

Jayaprakasha, G. K.; Sakariah, K. K. Determination of Organic Acids in Leaves and Rinds of *Garcinia indica* (Desr.) by LC. *J. Pharm. Biomed. Anal.* **2002,** *28*, 379–384.

Jena, B. S.; Jayaprakasha, G. K.; Singh, R. P.; Sakariah, K. K. Chemistry and Biochemistry of (−)-Hydroxycitric Acid from Garcinia. *J. Agric. Food Chem.* **2002,** *50* (1), 10–22.

Jones, F. J.; Andrews A. H. Acute Liver Injury Associated with the Herbal Supplement Hydroxycut in a Soldier Deployed to Iraq. *Am. J. Gastroenterol.* **2007,** *102*, 2357–2358.

Joshi, N.; Siwakoti, M.; Kehlenbeck, K. Wild Vegetable Species in Makawanpur District, Central Nepal: Developing a Priority Setting Approach for Domestication to Improve Food Security. *Econ. Bot.* **2015,** *69*, 1–10.

Kar, A.; Borkakoti, S.; Borthakur, S. K. Extended Distribution of the Genus *Garcinia*L. in Sonitpur District, Assam, India. *Pleione* **2008,** *2* (2), 165–170.

Khare, C. P. *Indian Medicinal Plants: An Illustrated Dictionary*; Springer: Berlin, Germany, 2007.

King, G. L.; Loeken, M. R. Hyperglycemia-Induced Oxidative Stress in Diabetic Complications. *Histochem. Cell Biol.* **2004,** *122*, 333–338.

Li, J.; Robson, N. K. B.; Stevens, P. F. Clusiaceae Through Araliaceae. In: *Flora of China*; Wu, Z. Y., Raven, P. H., Hong, D. Y., Eds.; Science Press & Missouri Botanical Garden Press: Beijing and St. Louis, 2007.

Li, X. W. Guttiferae (Eds.; Flora Republicae Popularis Sinicae. Wu, C. Y., Li, X. W.); Beijing Academic Press: Beijing, 1979; pp 1–112.

Lim, T. K. *Edible Medicinal and Non-Medicinal Plants*, Vol. 2 Fruits; Springer: Heidelberg, Germany, 2012.

Liu, B.; Zhang, X.; Bussmann, R. W.; Hart, R. H.; Li, P.; Bai, Y.; Long, C. *Garcinia* in Southern China: Ethnobotany, Management, and Niche Modeling. *Econ. Bot.* **2016,** *70* (4), 416–430.

Mabberley, D. J. The Plant Book—A Portable Dictionary of the Vascular Plants, 2nd ed.; Cambridge University Press, 2005; p 293.

Mackeen, M. M.; Ali, A. M.; Lajis, N. H. et al. Antimicrobial, Antioxidant, Anti-Tumour-Promoting and Cytotoxic Activities of Different Plant Part Extracts of *Garcinia atroviridis* Griff. ex T. Anders. *J. Ethnopharmacol.* **2000,** *72* (3), 395–402.

Mackeen, M. M.; Ali, A. M.; Lajis, N. H.; Kawazu, K.; Kikuzaki, H.; Nakatani, N. Antifungal Garcinia Acid Esters from the Fruits of *Garcinia atroviridis*. *Zeitschrift fur Naturforschung C*. **2002,** *57* (3–4), 291–295.

Magadula, J. J.; Mbwambo, Z. H. *Garcinia Plant Species of African Origin: Ethnobotanical, Pharmacological and Phytochemical Studies*; New York, 2014. ISBN: 978-1-941926-10-9.

Maheswari, J. K. Taxonomic Studies on Indian Guttiferae III. The Genus *Garcinia* Linn. *Bull Bot Surv India*. **1964,** *6* (2–4), 107–135.

Mazumdar, R. M.; Faik Al, Nipa, M. N.; Shazada, N. E.; Das, S.; Bhuiyan, H. R. Effect of *Garcinia pendunculata* (Toikor) Fruit Powder and Natural Preservatives on the Shelf Life of Breads. *Int. J. Curr. Sci. Technol.* **2016,** *4* (2), 124–127.

Mazzio, E. A.; Soliman K. F. A. In Vitro Screening for the Tumoricidal Properties of International Medicinal Herbs. *Phytother. Res.* **2009,** *23* (3), 385–398.

Munaweera, R. *Garcinia in Diabetic and Fat Control*; University of Kelaniya, 2016. https://www.researchgate.net/publication/323614265 (accessed on 8 Mar 2019).

Nabandith, V.; Suzui, M.; Morioka, T.; Kaneshiro, T.; Kinjo, T.; Matsumoto, K.; Akao, Y.; Iinuma, M.; Yoshimi, N. Inhibitory Effects of Crude Alpha-Mangostin, a Xanthone Derivative, on Two Different Categories of Colon Pre-Neoplastic Lesions Induced by 1, 2-Dimethylhydrazine in the Rat. *Asian Pac. J. Cancer Prev.* **2004,** *5* (4), 433–438.

Nayak, C. A.; Rastogi, N. K.; Raghavarao, K. S. M. S. Bioactive Constituents Present in *Garcinia indica* Choisy and Its Potential Food Applications: A Review. *Int. J. Food Prop.* **2010,** *13*, 441–453.

Negi, P. S.; Jayaprakasha, G. K.; Jena, B. S. Antibacterial Activity of the Extracts from the Fruit Rinds of *Garcinica cowa* and *Garcinia pedunculata* Against Food Borne Pathogens and Spoilage Bacteria. *Food Sci. Technol.* **2008,** *41* (10), 1857–1861.

Neog, M. Studies on the Floral Biology, Growth and Development of Few Minor Species of Assam. M.Sc. (Agri) Thesis, Assam Agricultural University, Jorhat, India, 1990.

Nguyen, D. M.; El-Serag, H. B. The Epidemiology of Obesity. *Gastroenterol. Clin. North Am.* **2010,** *39*, 1–7.

Nimanthika, W. J.; Kaththriarachchi HS. Systematics of Genus *Garcinia* L. (Clusiaceae) in Sri Lanka. New Insights from Vegetative Morphology. *J. Natl. Sci. Foundation* **2010,** *38*, 29–44.

Obolskiy, D.; Pischel, I.; Siriwatanametanon, N.; Heinrich, M. 2009.*Garcinia mangostana*, a Phytochemical and Pharmacological Review. *Phytother Res.* **2009,** *23*, 1047–1065.

Ohia, S. E.; Opere, C. A.; LeDayAM, Bagchi, M.; Bagchi, D. et al. Safety and Mechanism of Appetite Suppression by a Novel Hydroxyl Citric Acid Extract (HCASX). *Mol Cell Biochem.* **2002,** *238*, 89–103.

Padhye, S.; Ahmad, A.; Oswal, N.; Sarkar FH.Emerging Role of Garcinol, the Antioxidant Chalcone from *Garcinia indica* Choisy and Its Synthetic Analogs. *J. Hematol. Oncol.* **2009,** *2*, 38. DOI: 10.1186/1756-8722-2-38.

Pandey, R. C. SOUVENIR. International Seminar on Medicinal Plants-Quality Standardization. June 9–10, 2001; pp 117–133.

Parthasarathy, U.; Nandakishore, O. P. Morphological Characterization of Some Important Indian *Garcinia* Species, 2014. http:/dx.doi.org/ 10.1155/2014/823705.

Pattalung, P. N.; Thongtheeraparp, W.; Wiriyachitra, P.; Taylor WC.Xanthones of *Garcinia cowa*. *Planta Medica.* **1994,** *60* (4), 365–365.

Perlin, D. S. Antifungal Drug Resistance: Do Molecular Methods Provide a Way Forward? *Curr. Opin. Infect. Dis.* **2009,** *22*, 568–573.

PermanaD, Abas, F.; Maulidiani, M. et al. Atrovirisidone, B.; A New Prenylated Depsidone with Cytotoxic Property from the Roots of *Garcinia atroviridis*. *Zeitschrift fur Naturforschung C.* **2005,** *60* (7–8), 523–526.

Preuss, H. G.; Bagchi, D.; Bagchi, M.; Rao, C.; V. S.; Dey, D. K.; Satyanarayana, S. Effects of a Natural Extract of (−)-hydroxy citric acid (HCA-SX) and a Combination of HCA-SX Plus Niacin Bound Chromium and *Gymnema sylvestre* Extract on Weight Loss. *Diab. Obes. Metabol.* **2004,** *6* (3), 171–180.

Preuss, H. G.; Rao, C. V.; Garis, R.; Bramble, J. D.; Ohia, S. E. et al. An Overview of the Safety and Efficacy of a Novel, Natural (-)-Hydroxyl Citric Acid Extract (HCASX) for Weight Management. *J. Med.* **2004,** *35*, 33–48.

Preussa, H. G.; Bagchi, D.; Bagchi, M.; Rao CVS, Satyanarayana, S, et al. Efficacy of a Novel, Natural Extract of (-)-Hydroxy Citric Acid (HCA-SX) and a Combination of HCA-SX, Niacin Bound Chromium and *Gymnema sylvestre* Extract in Weight Management in Human Volunteers: A Pilot Study. *Nutr. Res.* **2004,** *24*, 45–58.

Reddy, G. C.; Raghuveer, S. Non-Traditional Products from Kokum: Inland and Global Opportunities. *Int. J. Fauna Biol. Stud.* **2018,** *5* (6), 102–104.

Richards, A. J. Studies in Garcinia, Dioecious Tropical Forest Trees: The Origin of the Mangosteen (*G. mangostana* L.). *Botan. J. Linn. Soc.* **1990,** *103*, 301–308.

Roberts, E. *Vegetable Materia Medica of India and Ceylon*; Bishen Singh Mahendra Pal Singh: Dehra Dun, India, 1984.

Rogers, S. Z.; Sweeney PW. Two Distinctive New Species of Malagasy *Garcinia* (Clusiaceae). *System. Bot.* **2007,** *32*, 772–779.

Rukachaisirikul, V.; Pailee, P.; Hiranrat, A.; Tuchinda, P.; Yoosook, C.; Kasisit, J.; Taylor, W. C.; Reutrakul, V. Anti-HIV-1 Protostane Triterpenes and Digeranyl Benzophenone from Trunk, Bark and Stems of *Garcinia speciosa. Planta Medica.* **2003,** *69* (12), 1141–1146.

Sahasrabudhe, A.; Deodhar, M. Anti-Hyaluronidase, Anti-Elastase Activity of *Garcinia indica. Int. J. Bot.* **2010,** *6*, 1–10.

Saito, M.; Ueno, M.; Ogino, S.; Kubo, K.; Nagata, J.; Takeuchi, M. High Dose of *Garcinia cambogia* Is Effective in Suppressing Fat Accumulation in Developing Male Zucker Obese Rats, But Highly Toxic to the Testis. *Food Chem. Toxicol.* **2005,** *43*, 411–419.

Sarangthem, K.; Singh, T. N. Microbial Bioconversion of Metabolites from Fermented Succulent Bamboo Shoots into Phyto-Sterols. *Curr. Sci.* **2003,** *84* (12), 1544–1547.

Sasithron, P.; Bamroongrugsa, N.; Kanchanapoom, K.; Naualsri, C. An Evaluation of the Sexual System of *Garcinia artivorus* L. (Clusiaceae), Based on Reproductive Features. *Songklanakarin J. Sci. Technol.* **2007,** *29* (6), 1457–1468.

Shara, M.; Ohia, S. E.; Yasmin, T.; Zardetto-Smith, A.; Kincaid, A.; et al. Dose- and Time-Dependent Effects of a Novel (-)-Hydroxy Citric Acid Extract on Body Weight, Hepatic and Testicular Lipid Peroxidation, DNA Fragmentation and Histo-Pathological Data Over a Period of 90 Days. *Mol Cell Biochem.* **2003,** *254*, 339–346.

Sharma, B. D.; Sanjappa, M. *Flora of India* (Eds.; Balakrishnan, N. P.); Botanical Survey of India: Calcutta, 1993; pp 98–131.

Sheela, K.; Nath, K. G.; Vijayalakshmi, D.; Yankanchi, G. M.; Patil, R. B. Proximate Composition of Underutilized Green Leafy Vegetables in Southern Karnataka. *J. Hum. Ecol.* **2004,** *15*, 227–229.

Stevens, T.; Qadri, A.; Zein, N. N. Two Patients with Acute Liver Injury Associated with use of the Herbal Weight-Loss Supplement Hydroxycut. *Ann Intern Med.* **2005,** *142*, 477–478.

Te-chato, S. Floral and Fruit Morphology of Some Species in *Garcinia* spp. Songklanakarin. *J. Sci. Technol.* **2007,** *29* (2), 245–252.

US FDA (United States Food and Drug Administration). Meridia (Sibutramine): Market Withdrawal Due to Risk of Serious Cardiovascular Events, 2010. http://www.fda.gov/safety/medwatch/safetyinformation/safetyalertsforhumanmedicalproducts/ ucm228830.htm (accessed on 10th Mar 2019).

US FDA (United States Food and Drug Administration), Center for Drug Evaluation and Research. Endocrinologic and Metabolic Drugs Advisory Committee, 2012. http://www.fda.gov/downloads/AdvisoryCommittees/CommitteesMeetingMaterials/Drugs/EndocrinologicandMetabolicDrugsAdvisoryCommittee/UCM2241 80.pdf

Wadsworth, J. W.; Story, S. P.; Zhou, B. N.; Ogden, R. V.; Godbee, R. G.; Palu, A. K.; Jensen, C. J. *Garcinia mangostana* L. Enhanced Animal Food Product. U.S. Patent Application: USA, 2005.

Waterman, P. G.; Hussain RA.Systematic Significance of Xanthones, Benzophenones and Bi-avonoids in Garcinia. *Biochem. Syst. Ecol.* **1983,** *11*, 21–28.

Watson, P. M.; Commins, S. P.; Beiler, R. J.; Hatcher, H. C.; Gettys TW. Differential Regulation of Leptin Expression and Function in A/J vs. C57BL/6J Mice During Diet-Induced Obesity. *Am. J. Physiol. Endocrinol. Metab.* **2000,** *279*, E356–E365.

WHO. Obesity: Preventing and Managing the Global Epidemic, Report of a WHO Consultation. WHO Technical Report Series 894. World Health Organization, Geneva, Switzerland, 2000.

WHO. Expert Consultation, Appropriate Body-Mass Index for Asian Populations and Its Implications for Policy and Intervention Strategies. *Lancet.* **2004,** *363*, 157–163.

Wielinga, P. Y.; Wachters-Hagedoorn, R. E.; Bouter, B. et al. Hydroxy Citric Acid Delays Intestinal Glucose Absorption in Rats. *Am. J. Physiol.* **2005,** *288* (6), G1144–G1149.

Yamaguchi, F.; Ariga, T.; Yoshimura, Y.; Nakazawa, H. Anti-Oxidative and Antiglycation Activity of Garcinol from *Garcinia indica* Fruit Rind. *J. Agric. Food Chem.* **2000,** *48*, 180–188.

Yapwattanaphun, C.; Subhadrabandhu, S.; Sugiura, A.; Yonemori, K.; Utsunomiya, N. Utilization of Some *Garcinia* Species in Thailand. *Int. Symp. Trop. Subtrop. Fruits* **2000,** *575*, 563–570.

Yonei, Y.; Takahashi, Y.; Hibino, S.; Watanabe, M.; Yoshioka, T. Effects on the Human Body of a Dietary Supplement Containing L-Carnitine and *Garcinia cambogia* Extract: A Study Using Double-Blind Tests. *J. Clin. Biochem. Nutr.* **2008,** *42* (2), 89–103.

Zou, X. B.; Huang, F. Q.; Hao, L. M.; Zhao, J. W.; Mao, H. P.; Zhang, J. C.; Ren, S. Y. The Socio-Economic Importance of Wild Vegetable Resources and Their Conservation: A Case Study from China. *Kew Bull* **2010,** *65*, 577–582.

CHAPTER 5

Parkia timoriana (DC.) Merr.: An Important Economic Plant

MAIBAM BEEBINA CHANU and CH. BRAJAKISHOR SINGH

Plant Bioresource Division, Institute of Bioresources and Sustainable Development, Takyelpat, Imphal West, Manipur, India

ABSTRACT

Parkia timoriana (DC.) Merr is a flowering plant of the Fabaceae family which has many traditional medicinal uses. This plant is native to Assam and Manipur of India and different Asian countries. The aim of the present study is to compile the scattered information on morphology, distribution, traditional uses, edible uses, diseases of the plant, economic importance, pharmacological and phytochemical information of *Parkia timoriana*. The study included information from the electronic databases such as Pubmed, Web of Science, and Scopus up to December, 2020. Information on the traditional uses of *P. timoriana* toxicity study, pharmacological activities, phytochemicals have been provided in this research paper. *P. timoriana* is a valuable medical plant with many therapeutic importances. The study provided useful information on the various aspects of *Parkia timoriana*.

5.1 INTRODUCTION

Parkia timoriana (DC.) Merr. is a large tree up to 20–25 m high with spreading branches. This tree belongs to the Fabaceae family with many synonyms viz. *Parkia javanica* Merr and *Parkia roxburghii* G. Don. So far 34 species of *parkia* has been reported to be distributed throughout the world (Luckow and Hopkins, 1995; Neill, 2009). *Parkiatimoriana* (DC.) Merr is one of the most popular species of *Parkia* in the Indo-Pacific region.

High-Value Plants: Novel Insights and Biotechnological Advances. Thangjam Anand Singh, Prakash Kumar Sarangi, and Ch. Brajakishore Singh (Eds.)
© 2025 Apple Academic Press, Inc. Co-published with CRC Press (Taylor & Francis)

It is quite similar to *Parkia biglandulosa* and *Parkia speciosa* but differs in morphology. The present study would provide useful information on the distribution, vernacular names, morphology, traditional uses in different parts of the world, toxicological profile, diseases on *Parkia* tree, phytochemistry profile, reported pharmacological activities.

FIGURE 5.1 *Parkia timoriana* edible pod collected from local market.

5.2 DISTRIBUTION

Parkia timoriana (DC.) Merr. is native to Indo-Pacific region including Thailand, Malaysia, Myanmar, Indonesia, and northeastern states of India especially in Manipur and Assam.

5.3 VERNACULAR NAMES

Parkia timoriana (DC.) Merr. is called tree beans in English. Common names in different Indian regions and languages and other regions of the world include yongchak in Manipur, Manipur-urohi (Assamese), Manipuri seem (Bengali), supota/kharial (Hindi), Zongto (Mizo), Aoelgap (Garo), Bire-phang (Kachari), Themuk-arang (Mikir), Unkamn-pinching (Naga),

Riang in Thailand, petai hutan in Borneo, kedawung and peundung in Java, alai in Sumatra, petai kerayong and batai hutan in Malaysia, and kupang and amarang within the Philippines.

5.4 MORPHOLOGY

Parkia timoriana (DC.) Merr. is a large tree with spreading branches having cylindrical trunk and grayish brown bark with white spots. Its height at maturity varies approximately from 30–40 m or even more and spreads around 6–17 m. The leaf is bipinnate with 14–31 pairs of pinnae and 52–71 pairs of leaflets in each pinna. The inflorescence is head of racemose type with clusters of yellowish-white tiny flowers, hanging at top of peduncle up to 45 cm long during the month of June–July and fruiting starts after 10–15 days of flowering. Bracteoles are present at the base of every flower; it covers individual flowers in inflorescence during juvenile stages (Hopkins, 1994). In the early stage, the fruits are soft, delicate, and bright green in color and consists of bunches of pods up to 50 cm in length. The fruits mature during March–April, upon maturity pods turn black and comprise yellow dry powdery pulp in which the black seeds are inculcated (Longvah and Deosthale, 1998). When the seeds mature they have a hard, black seed coat. The seeds can be dried and stored.

5.5 EDIBLE USES

Parkia timoriana (DC.) Merr. is a valuable edible medicinal plant. In north-eastern India especially Manipur, *P. timoriana* is a delicious valuable food cuisine. Beginning from the flower up to the mature hard black pod every part of the fruit is edible. The pod is taken with dry fish to make “eromba,” “singju” a special dish of Manipur. People in Manipur buy fresh pod as costly as Rs. 100 per pod in early season. The dried seeds can be soaked in water and taken as various cuisines. It is also a common food in Thailand and Indonesia.

5.6 TRADITIONAL USES IN INDIA AND DIFFERENT PARTS OF THE WORLD

Traditional healers of Manipur, India used tender pods and bark to treat intestinal disorders, piles, dysentery, and diarrhoea (Khumbongmayum et al., 2005); decoction of bark is use to treat diabetes (Devi, 2011). Dimasa and Kachari tribes of Assam, India used paste of bark as plaster to treat eczema

(Rathi et al., 2012). Different ethnic communities of Mizoram, India use externally the paste of fruit to heal wounds and scabies; fruit and juvenile shoots are eaten to treat diarrhoea, dysentery and to get relief from food poisoning (Bhardwaj and Gakhar, 2005) and decoction of roots are used as antipyretic plants (Laha, 2015). Mizos, Garos, Kacharis, Nagas, and Mikir tribes in northeast, India used pods crushed in water as facewash and shampoo. Lotion made from bark and leaves is used in treatment of skin diseases and ulcer (Jena et al., 2018; Bhuyan, 1996). Decoctions of barks are used by locals of Gambia as a bath to get relief from fever and toothache and pulped bark is taken orally with lemon to heal wound and ulcer (Irvine, 1961). In Philippines, the bark is used in tanning and dyeing nets. The pods are used to cure stomach disorders and improve liver functions (Jena et al., 2018). In Indonesia, *P. timoriana* is used as a medicine to overcome stomach disorders such as bloating, cholera, intestinal inflammation and to expel intestinal worms. Dried and ground powder is mixed with honey or other mixes such as ginger to treat colic and colds (Hidayati et al., 2020). In Malaysia, the pods are used in headache, kidney disorder, diabetes, urinary tract infection, and hypertension (Samuel et al., 2010; Ong et al., 2011).

5.7 ECONOMICAL IMPORTANCE

Parkia pod is a favorite food item for every household in Manipur. The plant started its production at the age of 6 years but full bearing stage is after 10 years only. The life span of the plant is around 80–90 years or even more. In favorable season a full grown plant bears 10,000–15,000 pods. So, a single plant can yield approx. Rs. 100,000/-–1,50,000/- per annum. The cost of the dried seeds per kg ranges from Rs. 400 to Rs. 500. The tree can easily be propagated through seeds and 1 or 2 years old seedlings can be transplanted in the field. However, due to massive dying of *parkia* in recent years due to certain pathogens, Manipur is now importing *parkia* pods from the neighboring countries like Myanmar, Thailand, and Indonesia. For the plantation of *parkia*, maintenance is not required because being a legume it will also enrich soil through nitrogen fixation. Thus, large-scale plantation of *parkia* may save the drainage of economy to the neighboring countries.

5.8 DISEASES OF PARKIA

1. Vertillicium wilt disease is a soil-borne disease causing 80% mortality of seedlings caused by a soil fungus

called *Verticillium dahliae*. The disease can be managed by soil drenching with 0.2% Bavistin 50% (Kumar et al., 2012). The pathogen enters the plant through natural openings and wounds in the roots and grows up into the stem, where it affects the xylem system and blocks the conduction of water to the leaves.

2. Caterpillars of *Cadra cautella* infest the stored tree bean seeds while in the field and in properly sealed paper bags. The larvae of *C. cautella* feed uncontrollably on the green kernels. The larvae are mobile and make webs as they tunnel through the food, and excrete yellowish brown excreta near the point of entrance from where they start feeding from the surface of the kernels and continue inside. In case of heavy infestation, the whole inside of kernel was consumed, while the outer skin of the seed had tiny holes through which the larva entered rendering it unfit for consumption. Pupation occurred inside the crevices of the infested seeds and the paper bags. The adult moth appears after 2 weeks of pupation, but they do not feed. Adults spread infestation through eggs laid on seed surfaces. This larvae infestation leads to immature dropping of inflorescence and pod heads during the months of October–February (Thangjam, 2003).
3. *Botrydiplodia theobromae* is a fungal pathogen that leads to sudden death or decline of tree bean *Parkia* timoriana. Decline of tree bean can be identified by an initial wilting of half portion of the tree, excessive shedding of leaves, excessive gummosis, easy to peel off bark, gradual drying up leading to death, and secondary infestation by insects on dead trees. Moreover, tiny holes made by an unknown bark borer were frequently seen on the diseased tree wood which was suspected to act as a vector (Singh et al., 2018).

5.9 PHYTOCHEMICAL PROFILE

Phenolic compounds, flavonol glycoside, lactins, fatty acids, quassinoids, triterpenoid, phytosterols, and flavonoids are reported from *Parkia timoriana*. Presence of phenolic compounds, Epigallocatechin gallate and flavonol glycoside, hyperin has been reported from the pods of *P. roxburghii* (Sheikh et al., 2016). Kaur et al. (2005) isolated novel mannose/glucose specific lectins from seed extracts. Azizi et al. (2008) reported the presence of fatty acids undecanoic, myristic, palmitic, oleic, linoleic, elaidic, stearic, stearoic, lauric, arachidonic, and linoleic acids in the seeds of *P. timoriana*. Dinda et al. (2009) isolated Javanicoside A, Javanicoside A pentaacetate, Javanicoside B, and Javanicoside B hexaacetate along with ursolic acid and β-sitosterol from the methanol extract of leaf and stem bark of *P. timoriana*. Parkinol is a new compound isolated from leaves of *P. Timoriana* by Dinda et al. (2010). The specific phenolic composition of *Parkia roxburghii* extracts performed through HPLC and LC-MS/MS showed presence of gallic acid, caffeic acid, chlorogenic acid, ellagic acid, ferulic acid, and flavonoids like kaempferol, quercetin, diadzein, and genestein were present in low amount in some parts of the plant (Dubey et al., 2020). The presence of flavone compounds baicalein, quercetin, and chrysin from the *Parkia javanica* fruit extract was also reported by Das et al. (2017).

5.10 TOXICOLOGY STUDIES

Sheikh et al. (2016) performed the acute toxicities in albino mice as per OECD Guideline no. 423, Annexure 2c adopted by CPCSEA, Government of India. Ethylacetate fraction of *Parkia* roxbhurghii was found to be safe with no observed behavioral changes and no mortality up to a dose of 2000 mg/kg. Aisha et al. (2012) studied the cytotoxic effect of methanolic extract on normal human cell line and no acute cytotoxic effect was observed in normal cell line.

5.11 PHARMACOLOGICAL ACTIVITY

Different parts of *Parkia timoriana* have been reported to possess antimicrobial, antioxidant, antidiabetic, anticancer, antigonorrhoeal, immunomodulatory, and wound-healing activity. *Parkia* pods are reported to be effective in kidney disorder, urinary tract infection, hypertension, and headaches (Buring and Berg, 2013; Ong et al., 2011). Devi et al. (2007) reported the antibacterial and nematicidal activity of different parts of the tree. They reported that the aqueous extracts of seed, pod, and leaf were effective against various strains of bacteria like *Staphylococcus aureus, Streptococcus fecalis, Bacillus cereus,* etc., and also to nematode *Meloidogyne graminicol*. Das et al. (2017) reported the antimicrobial activity of the fruit on *Psuedomonas aeruginosa*. Sakunpak and Panichayupakaranant (2012) reported the antibacterial activity against gastrointestinal pathogenic bacteria *Shigella sonnei*. Saha et al. (2016) reported the bacteriostatic and bactericidal activity of the stem

TABLE 5.1 Phytochemicals Reported from *Parkia timoriana*.

Sl. No.	Plant species	Name of compounds	Chemical structure	Authors
1.	*Parkia roxburghii*	Epigallocatechin gallate	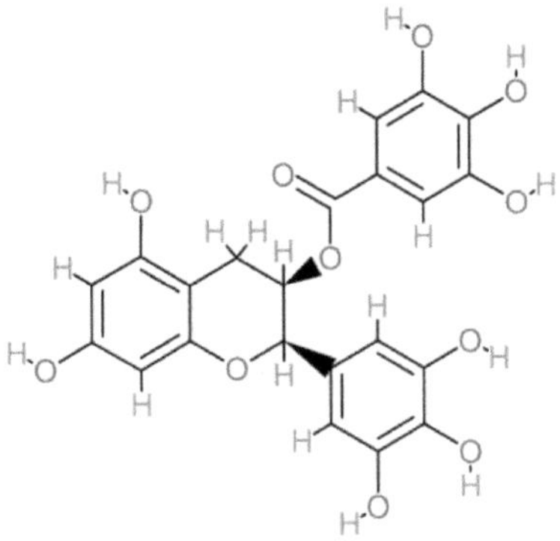	Sheikh et al., 2016
		Hyperin	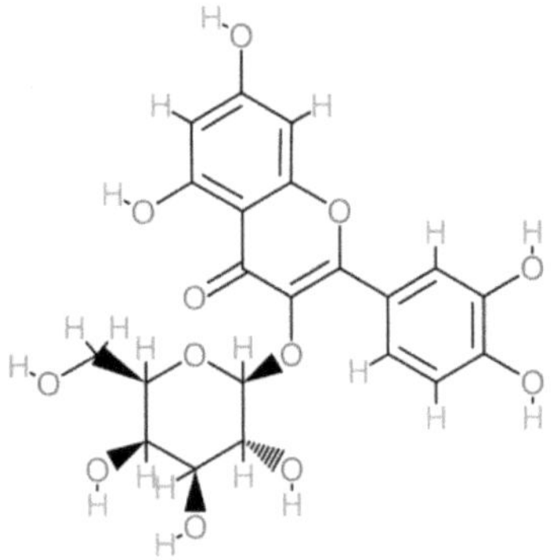	
2.	*Parkia timoriana*	Javanicoside A	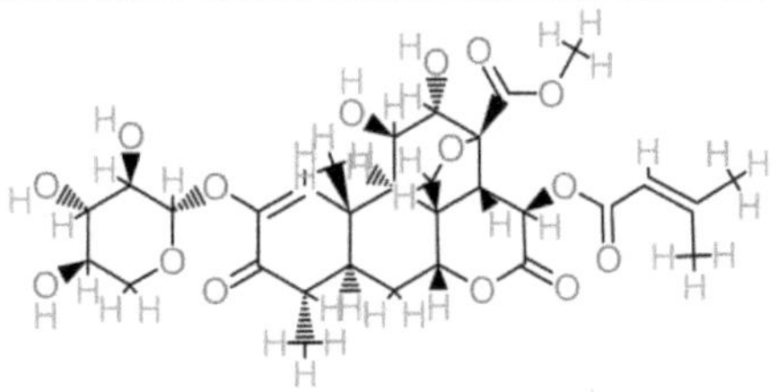	Dinda et al., 2009
		Hyperin	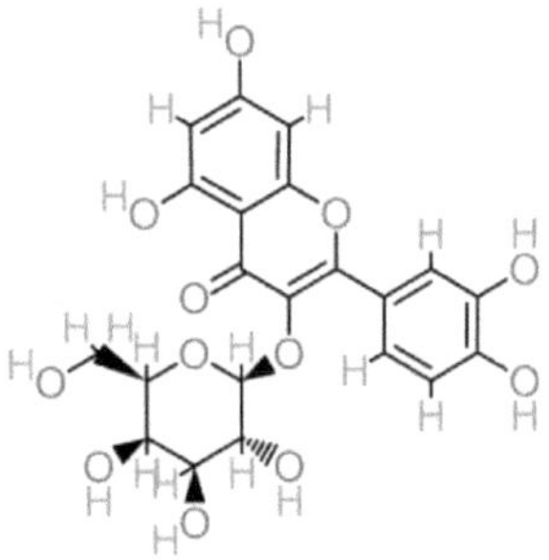	

TABLE 5.1 *(Continued)*

Sl. No.	Plant species	Name of compounds	Chemical structure	Authors
		Javanicoside B	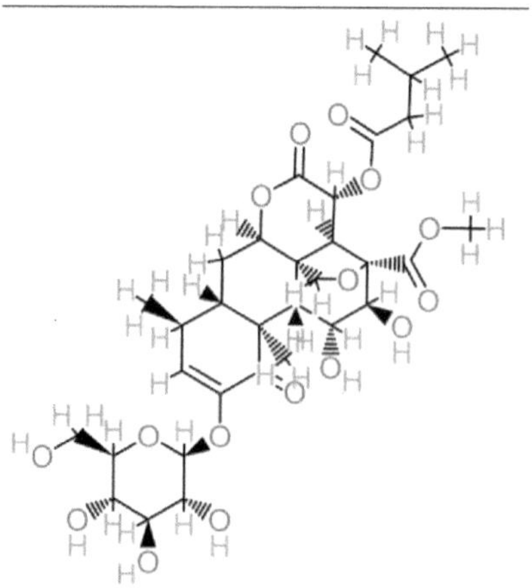	
		Ursolic acid	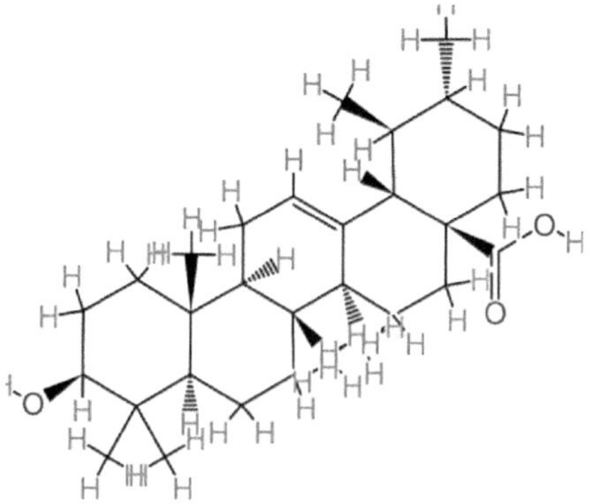	
		β-sitosterol	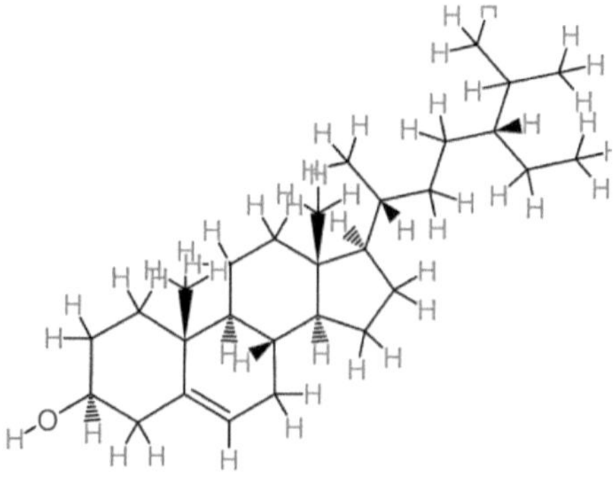	

TABLE 5.1 *(Continued)*

Sl. No.	Plant species	Name of compounds	Chemical structure	Authors
3.	*Parkia timoriana*	Parkinol (Trihexyphenidyl)	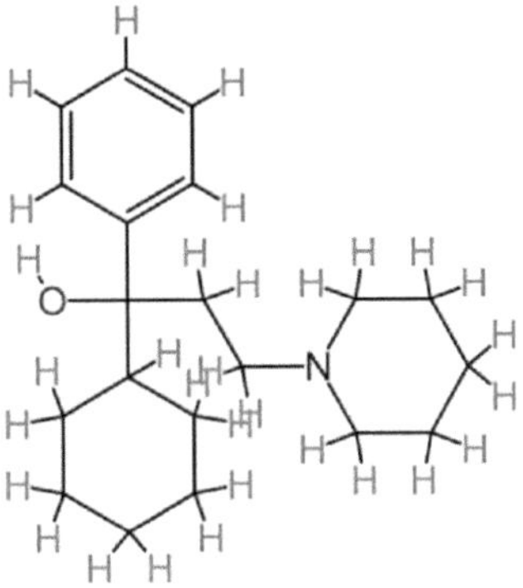	Dinda et al., 2010
4.	*Parkia timoriana*	Linoleic acids	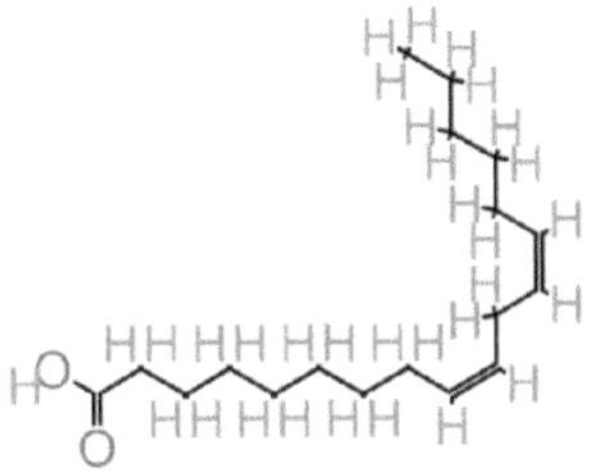	Azizi et al., 2008
		Undecanoic		
		Myristic		
		Palmitic acid	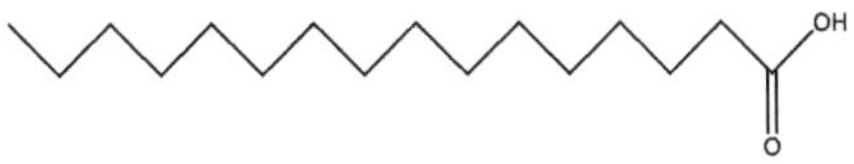	
		Oleic acid	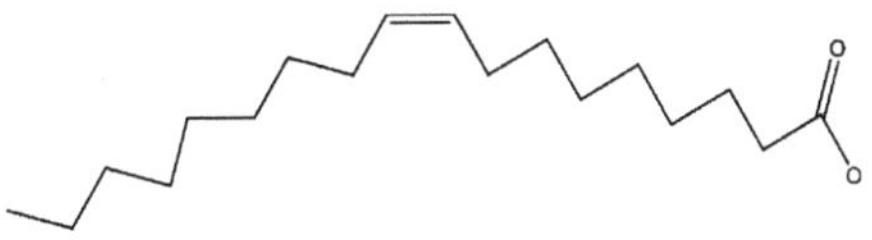	

TABLE 5.1 *(Continued)*

Sl. No.	Plant species	Name of compounds	Chemical structure	Authors
		Elaidic acid	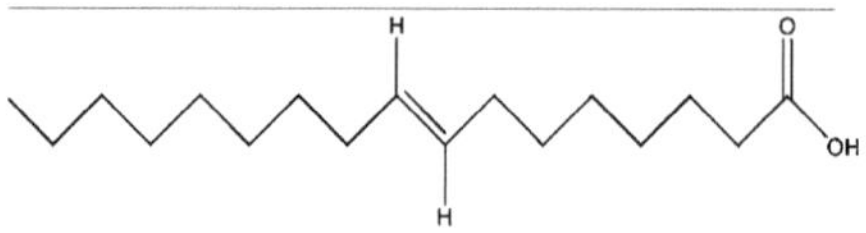	
		Stearic acid	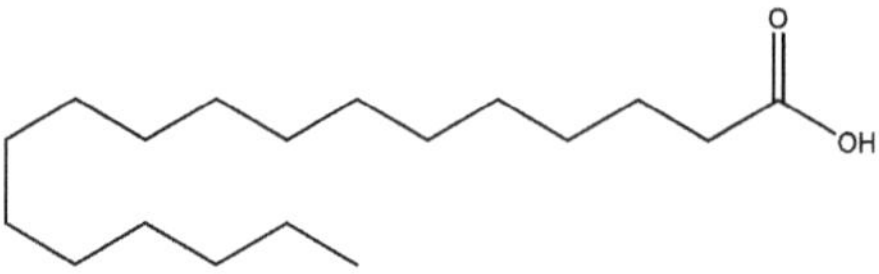	
		Lauric acid	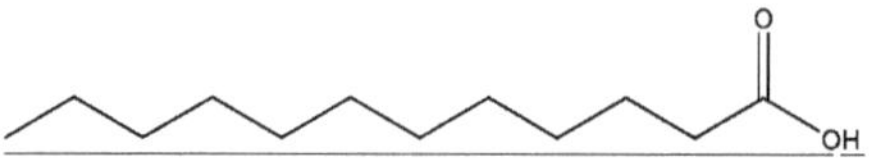	
		Arachidonic acid	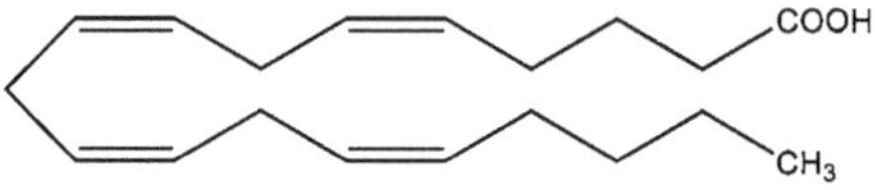	
5.	*Parkia roxburghii*	Gallic acid	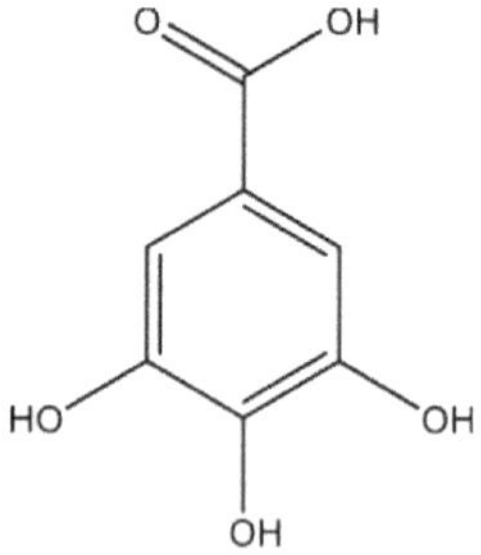	Dubey et al., 2020
		Caffeic acid	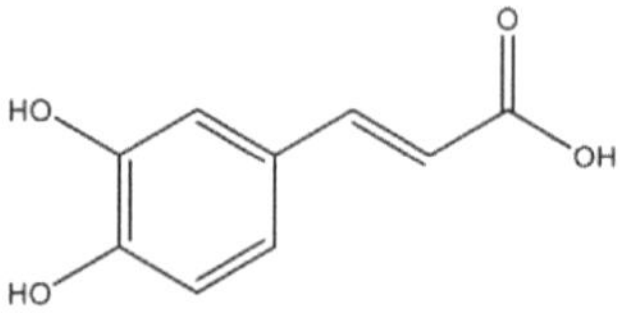	
		Chlorogenic acid	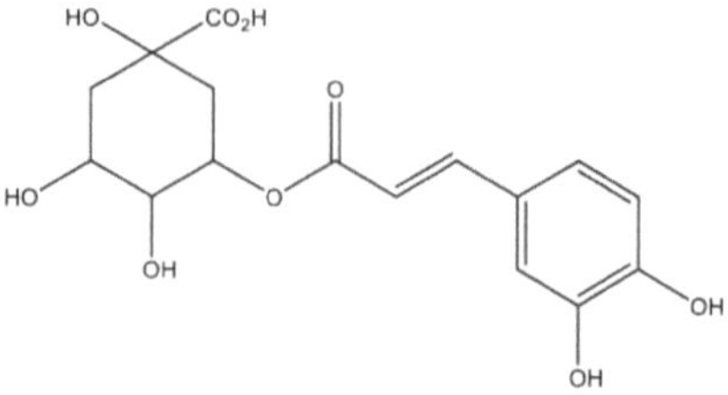	

TABLE 5.1 *(Continued)*

Sl. No.	Plant species	Name of compounds	Chemical structure	Authors
		Ellagic acid	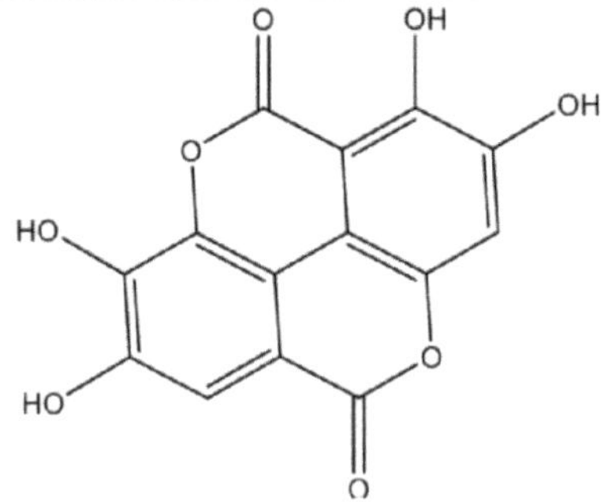	
		Ferulic acid	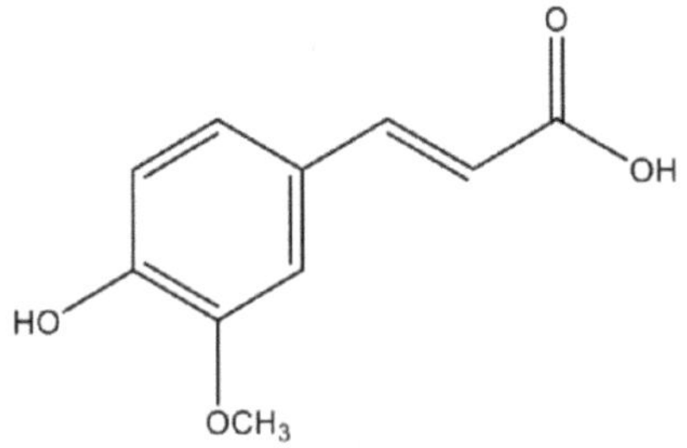	
		Kaempferol	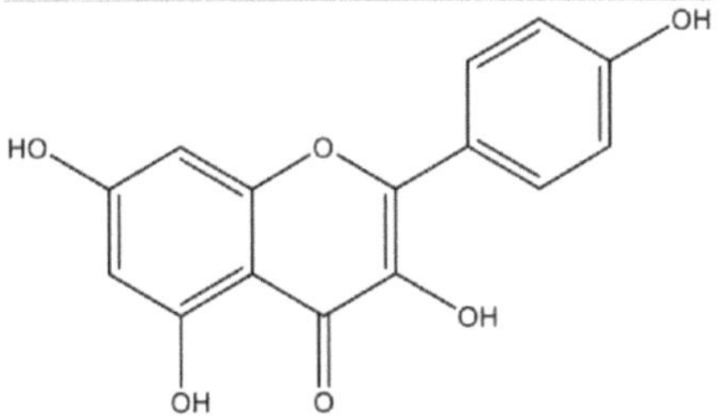	
		Quercetin	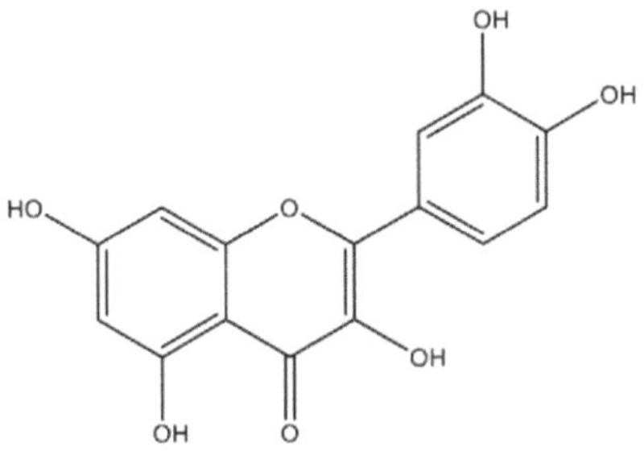	

TABLE 5.1 *(Continued)*

Sl. No.	Plant species	Name of compounds	Chemical structure	Authors
		Diadzein	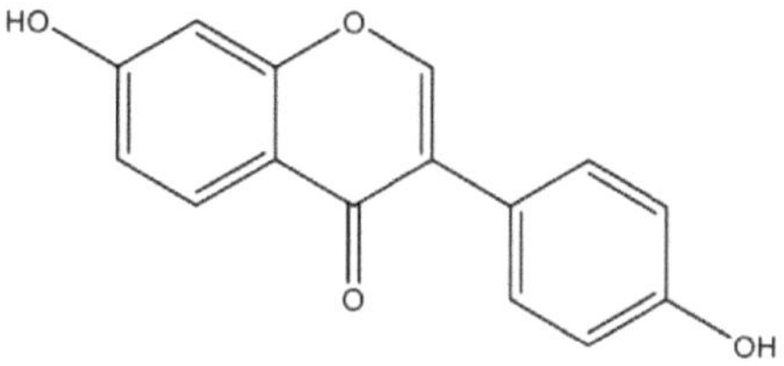	
		Genistein	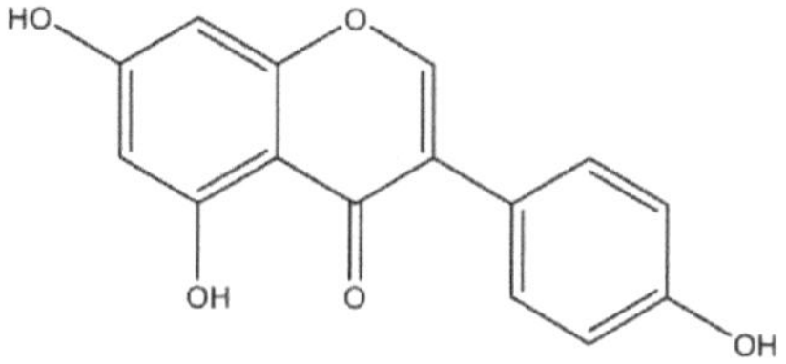	
6.	*Parkia javanica*	Quercetin	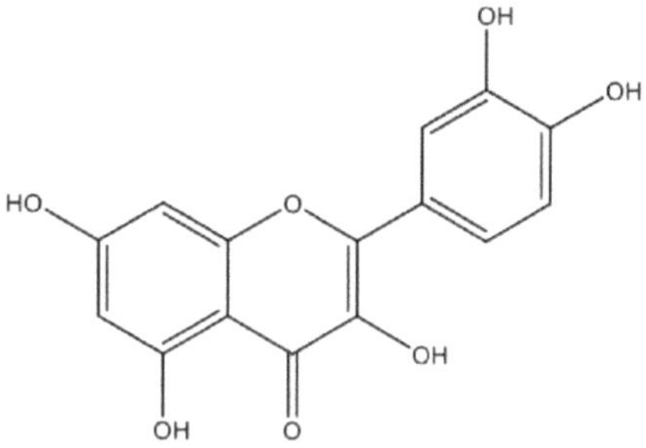	Das et al., 2017
		Chrysin	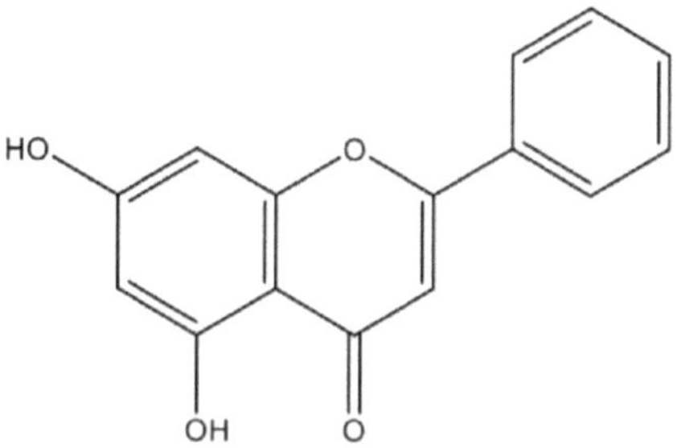	

bark on *Vibrio cholarae* strains. Saha et al. (2018) reported the bactericidal activity of the stem bark on *Enterobacter aerugenes*, *Pseudomonas aeruginosa*, and *Klebsiella pneumonia*.

Dubey et al. (2020) reported the antioxidant properties of the pods, leaves, seeds, and flower of *parkia*. Buathongjan et al. (2020) reported that the petins from pod husk and pod powder have antioxidant activities. Gan and Latiff (2011) reported that the freeze dried pod has antioxidant activity. Sheikh et al. (2016) have reported the antidiabetic and hepatoprotective activity of the edible pod of *parkia*. They reported that treatment with 5 mg/kg b.w. of the pod reduces the elevated blood glucose levels significantly.

Chanu et al. (2018) reported the anticancer activity of the seeds in HepG2 human liver cancer cells. Kaur et al. (2005) reported that the lectin isolated from *parkia* inhibited proliferation of HB98, B-cell hybridoma cell line. Khangembam et al. (2018) reported that the methanolic extract of *parkia* had antiproliferation activity in HeLa and MCF-7 cells. Patra et al. (2016) reported that the fruit extract has significant cytotoxicity activity in sarcoma-180 A549, AGS, and MDA-MB435S cancer cells.

Mullick et al. (2019) reported the antigonorrhoeal activity of the bark in multidrug-resistant clinical isolate as well as standard strain of *Neisseria gonorrhoeae*. Saha et al. (2018) reported the immunomodulatory activity of the stem bark of *parkia* on macrophage cells. They reported that treatment with the extract of *P. javanica* increased the expression of proinflammatory cytokines, IL-6, IL-8, IL-12, and TNF-α in mice macrophage RAW 264.7 cells and the maximum proliferation and migration of macrophage cells were observed at the dose of 20 μg/mL.

Sil et al. (2019) reported that the methanolic extract of *Parkia javanica* stem bark has potential wound-healing activity mediated via FAK/PI3K/Akt/NF-κB pathway and the extract stimulated proliferation and migration of both fibroblasts and keratinocytes significantly in celllines. They reported that wounded mice treated with the extract tropically fasten the excisional wound healing compared to placebo control group.

5.12 CONCLUSION

Parkia timoriana is a valuable medical plant with many therapeutic, economical, and consumable importances. The present study would encourage researchers to perform various studies on *P. timoriana* and validate the traditionally claimed uses. *P. timoriana* can be a good candidate for pharmaceutical industries in drug development and commercial production of drug and a good plant of economic importance.

TABLE 5.2 Pharmacological and Biological Activities Reported from *Parkia timoriana.*

Sl. No.	Species	Activity	Plants part used	References
1.	*Parkia roxburghii*	Antidiabetic and hepatoprotective activity	Ethlyacetate fraction of edible pod	Sheikh et al., 2016
2.	*Parkia javanica*	Anticancer activity	Methanolic extract of seeds	Chanu et al., 2018
3.	*Parkia javanica*	Apoptosis inducing activity	Methanolic extract of seeds	Khangembam et al., 2018
4.	*Parkia roxburghii*	Anti-proliferative activity	Lectin isolated from the plant	Kaur et al., 2005
5.	*Parkia javanica*	Anticancer activity	Methanol extract of fruit	Patra et al., 2016
6.	*Parkia roxburghii*	Antibacterial and nematicidal effect	Different part of plant	Devi et al., 2007
7.	*Parkia timoriana*	Antibacterial activity	Ethylacetate and methanol extract of seed	Sakunpak and Panichayupakaranant, 2012
8.	*Parkia javanica*	Antigonorrhoeal activity	Methanolic extract of *bark*	Mullick et al., 2019
9.	*Parkia timoriana*	Antioxidant activity	Pectins from pod husk and pod powder	Buathongjan et al., 2020
10.	*Parkia timoriana*	Antioxidant activity	Freeze dried and oven dried pod powder	Gan and Latiff, 2011a
11.	*Parkia javanica*	Antibacterial activity	Stem bark extract	Saha et al., 2016
12.	*Parkia javanica*	Antibacterial activity	Crude methanol extract of *Parkia javanica* stem	Saha et al., 2018
13.	*Parkia roxburghii*	Antioxidant activity	Pods, leaves, seeds and then flowers	Dubey et al., 2020
14.	*Parkia javanica*	Wound-healing activity	Methanolic extract of stem bark	Sil et al., 2019
15.	*Parkia javanica*	Immunomodulatory activity	Methanolic extract of stem bark	Saha et al., 2018

KEYWORDS

- ***Parkia timoriana***
- **medicinal plant**
- **pharmacological**
- **phytochemicals**

REFERENCES

Aisha, A. F. A.; Abu-Salah K. M.; Alrokayan, S. A.; Ismail, Z.; Majid, A. M. S. A. Evaluation of Antiangiogenic and Antioxidant Properties of *Parkia* speciosa Hassk. Extracts. *Pak. J. Pharm. Sci.* **2012,** *25*, 7–14.

Azizi, C. Y. M.; Salman, Z.; Norulain N. A. N.; Omar, A. K. M. Extraction and Identification of Compounds from *Parkia speciosa* Seeds by Supercritical Carbon Dioxide. *J. Chem. Natl. Res. Eng.* **2008,** *2*, 153–163.

Bhardwaj. S.; Gakhar, S. K. Ethnomedicinal Plants used by the Tribals of Mizoram to Cure Cuts and Wounds. *Indian J. Tradit. Knowl.* **2005,** *4*, 75–80.

Bhuyan, T. C. *Parkia* roxburghii, the tree bean. *Rain For News*, 1996; vol 1, pp 1–2.

Buathongjan, C.; Israkarn, K.; Sangwana, W.; Outrequina, T.; Gamonpilas, C.; Methacanon, P. Studies on Chemical Composition, Rheological and Antioxidant Properties of Pectin Isolated from Riang (*Parkia timoriana* (DC.) Merr.) pod. *Int. J. Biol. Macromol.* **2020,** *164*, 4575–4582.

Buring, A. B.; Berg, A. J. J. V. D. Freeze Dried Extract of *Parkia*, Preferably of *Parkia speciosa* Beans for Treatment of Diseases such as Diabetes Mellitus Type 2. US Patent, 2013, PCT/EP2012/000662.

Chanu, K. V.; Devi, L. G.; Srivastava, S. K.; Thakuria, D.; Kataria, M.; Telang, A. G. Phytochemical Analysis and Evaluation of Anticancer Activity of *Parkia javanica* Seeds. *Pharma Innov. J.* **2018,** *8*, 305–311.

Das, A.; Das, M. C.; Sandhu, P.; Das, N.; Tribedi, P.; De, U. C.; Akhterb, Y.; Bhattacharjee, S. Antibiofilm Activity of *Parkia javanica* Against*Pseudomonas aeruginosa*: A Study with Fruit Extract. *RSC Adv.* **2017,** *7*, 5497–5513.

Devi, A. P. Plants Used by Meitei Community of Manipur for the Treatment of Diabetes. *Assam Univ. J. Sci. Tech. Bio. Env. Sci.* **2011,** *7*, 63–66.

Devi, T. P.; Shakuntala, I.; Devi, G.; Nonglait, K. K. L.; Singha, L. B.; Pattanayak, A.; Rahaman, H. *Asian J. Microbio. Biotech. Env. Sci.* **2007,** *9*, 621–626.

Dinda. B.; Mohanta, B. C.; Debnath, S.; Ghosh, B.; Arima, S.; Sato, N.; Harigaya, Y. Iridoid Glucosides from Leaf and Stem Bark of *Parkia javanica. J. Asian Nat. Prod. Res.* **2009,** *11*, 229–235.

Dinda, B.; Mohanta, B. C.; Ghosh, P.; Sato, N.; Harigaya, Y. Chemical Constituents of *Parkia* javanica, *Alocasia indica* and *Premna latifolia. J. Indian Chem. Soc.* **2010,** *87*, 829–831.

Dubey, R. K.; Upadhyay G.; Singh, V.; Pandey, S. Antioxidant Potential and Free Radical Scavenging Activity of *Parkia roxburghii*, G. Don, A Lesser known Legumoinious Tree from North East India. *S. Afr. J. Bot.* **2020,** *131*, 454–461.

Gan, C. Y.; Latiff, A. A. Antioxidant *Parkia* Speciosa Pod Powder as Potential Functional Flour in Food Application: Physicochemical Properties' Characterization. *Food Hydrocoll.* **2011,** *25*, 1174–1180.

Hidayati, A. N.; Zuhud, E. A. M.; Andarwulan, N. Population Structure, Vegetation Composition and Economic Potentials of *Parkia timoriana* in Meru Betiri National Park, East Java, Indonesia. *Biodiversitas* **2020,** *21*, 203–210.

Hopkins, H. C. F. The Indo-Pacific Species of *Parkia* (Leguminosae-Mimosoidae). *Kew Bull.* **1994,** *49*, 181–234.

Irvine, F. R. In *Wood Plants of Ghana*. Oxford University Press: Oxford, 1961; p 104.

Jena, A. K.; Sharma, P.; Deuri, R.; Bhuyan, T.; Singh, S. P. Tree Bean: A Miracle Vegetable of Arunachal Pradesh. *Pharma Innov. J.* **2018,** *7*, 541–543.

Kaur, N.; Singh, J.; Kamboj, S. S.; Agrewala, J. N.; Kaur, M. Two Novel Lectins from *Parkia biglandulosa* and *Parkia roxburghii*: Isolation, Physicochemical Characterization, Mitogenicity and Anti-proliferative Activity. *Protein Pept. Lett.* **2005,** *12*, 589–595.

Khangembam, V. C.; Srivastava, S. K.; Leishangthem, G. D.; Kataria, M.; Thakuria, D. Evaluation of Apoptosis Inducing Ability of *Parkia javanica*seed Extract in Cancer Cells. *Indian J. Pharm. Sci.* **2018,** *80*, 1069–1077.

Khumbongmayum, A. D.; Khan, M. L.; Tripathi, R. S. Ethnomedicinal Plants in the Sacred Groves of Manipur. *Indian J. Tradit. Knowl.* **2005,** *4*, 21–32.

Kumar, R.; Tapwal, A.; Borah, R. K. Identification and Controlling Verticillium Wilt Infecting *Parkia roxburgii* Seedlings in Manipur India. *Res. J. For.* **2012**.

Laha, R. Diversity of Antipyretic Plants in Mizoram, Northeast India. *Int. J. Sci. Res.* **2015,** *4*, 2251–2253.

Longvah. T.; Deosthale. Y. G. Nutrient Composition and Food Potential of *Parkia roxburghii*, A Less Known Tree Legume from Northeast India. *Food Chem.* **1998,** *62*, 477–481.

Luckow, M.; Hopkins, H. C. F. A Cladistic Analysis of *Parkia* (Leguminosae: Mimosoideae). *Am. J. Bot.* **1995,** *82*, 1300–1320.

Mullick, J. B.; Majumdar, T.; Reddy, K. V. R.; Mukherjee, S.; Sil, S. K. Activity of the Medicinal Plant *Parkia javanica* Against Multidrug-Resistant *Neisseria gonorrhoeae* and Other Clinical Isolates. *Asian J. Pharm. Clin. Res.* **2019,** *12*, 83–86.

Neill, D. A. *Parkia* nana (Leguminosae, Mimosoideae), A New Species from the Sub-Andean Sandstone Cordilleras of Peru. *Novon A J. Bot. Nomencl.* **2009,** *19*, 204–208.

Ong, H. C.; Ahmad, N.; Milow, P. Traditional Medicinal Plants used by the Temuan Villagers in Kampung Tering, Negeri Sembilan, Malaysia. *Stud. Ethno-Med.* **2011,** *5*, 169–173.

Patra, K.; Jana, S.; Sarkar, A.; Karmakar, S.; Jana, J.; Gupta, M.; Mukherjee, G.; Chandra De, U.; Mandal, D. P.; Bhattacharjee, S. *Parkia* Javanica Extract Induces Apoptosis in S-180 Cells via the Intrinsic Pathway of Apoptosis. *Nutr. Cancer* **2016,** *68*, 689–707.

Rathi, R. S.; Misra, A. K.; Roy, S.; Verma, S. K.; Singh, S. K. Potential of a Lesser Known Tree Species *Parkia roxburghii* G. Don of North East India. *Indian For.* **2012,** *138*, 476–479.

Saha, S.; Dinda, M.; Karmakar, P.; Sil, S. K. Immunomodulatory Effect of *Parkia javanica* Extract on Intracellular Expression OF IL-6, IL-8, IL-12 AND TNF-α. *J. Drug Deliv. Ther.* **2018,** *8*, 58–63.

Saha, S.; Karmakar, P.; Sil, S. K. Antibacterial Activities of *Parkia javanica* Extract Against Multidrug Resistant Gram-negative Bacteria Predominantly Found in Skin Wound. *Int. J. Pharm. Biol. Sci.* **2018,** *8*, 96–102.

Saha, S.; Mullick, J. B.; Choudhury, P. R.; Saha, P.; Chakraborty, D.; Sil, S. K. Anti-Vibrio Activity of *Parkia javanica*: Studies on MIC, MBC, Growth Curve Analysis and ROS Generation on Four *Vibrio cholarae* Strains. *Int. J. Curr. Microbiol. App. Sci.* **2016,** *5*, 538–544.

Sakunpak, A.; Panichayupakaranant, P. Antibacterial Activity of Thai Edible Plants Against Gastrointestinal Pathogenic Bacteria and Isolation of a New Broad Spectrum Antibacterial Polyisoprenylated Benzophenone, Chamuangone. *Food Chem.* **2012,** *130*, 826–831.

Samuel, A. J. S. J.; Kalusalingam, A.; Chellappan, D. K. Ethnomedical Survey of Plants used by the Orang Asli in Kampung Bawong, Perak, West Malaysia. *J. Ethnobiol. Ethnomed.* **2010,** *6*, 5.

Sheikh, Y.; Maibam, B. C.; Talukdar, N. C.; Deka, D. C.; Borah, J. C. In Vitro and In Vivo Anti-diabetic and Hepatoprotective Effects of Edible Pods of *Parkia roxburghii* and Quantification of the Active Constituent by HPLC-PDA. *J. Ethnopharmacol.* **2016,** *191*, 21–28.

Sil, S. K.; Saha, S.; Dinda, M.; Karmakar, P.; Jana, K. Cellular Basis of Wound Healing Activity of *Parkia javanica* (Lamk) Merr: In Vitro and In Vivo Studies. *Clin. Pediatr. Dermatol.* **2019,** *5*.

Singh, A. R.; Dutta, S. K.; Singh, S. B.; Boopathi, T.; Lungmuana, S.; Saha, S.; Dayal, V.; Akoijam, R.; Sharma, S. K; Singh, N. H. Occurrence, Severity and Association of Fungal Pathogen, *Botrydiplodia theobromae* with Sudden Death or Decline of Tree Bean (*Parkia timoriana*, (DC.) Merr) in North Eastern India. *Curr. Sci.* **2018,** *115*, 1133–1142.

Thangjam, R. *Cadra cautella* Walker (Lepidoptera: Crambidae: Phycitinae) – A Pest on *Parkia timoriana* (DC.) Merr. in Manipur. *Curr. Sci.* **2003,** *85*, 725–726.

CHAPTER 6

King Chili: A Future Potential Source of Pharmacological Applications

WAHENGBAM KABITA CHANU and
CHINGAKHAM BRAJAKISHOR SINGH

Institute of Bioresources and Sustainable Development, Takyelpat, Imphal West, Manipur, India

ABSTRACT

King Chili or Naga King Chili or Naga Chili or *Capsicum chinense* Jacq. is one of the hottest chilies in the world. This chili is native to the north-eastern region of India. It has attained a great economic potential for its food as well as medicinal value throughout the country. Traditionally, it has been used in the treatment of headache, gastritis, rheumatism, arthritis, muscle pain, and chronic indigestion by the local people of the North-Eastern states of India. Due to its high potential commercial value, many studies have been carried out in this crop including scientific cultivation, in vitro regeneration, pharmacological studies and its diseases, and their management, etc. This chapter attempts to bring into account the pharmacological properties so far which would be a source of treating many diseases.

6.1 INTRODUCTION

King chili is a herbaceous plant belonging to the family Solanaceae and the genus *Capsicum* belonging to more than 30 species (Bosland, 1994; Verma et al., 2013; Ahn et al., 2014). It is believed that chilies are originated from tropical South America. Chili was first introduced in India by the Portuguese

High-Value Plants: Novel Insights and Biotechnological Advances. Thangjam Anand Singh, Prakash Kumar Sarangi, and Ch. Brajakishore Singh (Eds.)
© 2025 Apple Academic Press, Inc. Co-published with CRC Press (Taylor & Francis)

in the 15th Century. It is widely cultivated in Manipur, Nagaland, Assam, and some other parts of India. A number of different varieties of this chili were noted in India with different local names such as Naga chili in Nagaland, Bhut Jolokia in Assam, and U-Morok in Manipur (Verma et al., 2013; Sanatombi et al., 2010). However, the King chili which is native to North-Eastern India has received the attention of the world scientific community due to its extremely high unique pungency aroma (Bhagowati and Changkija, 2009).

King Chili exhibits a considerable amount of diversity among species and their landraces, especially for traits like plant growth habit, pungency fruit color, size, and shape. However, many of the phenotypic traits are known to be influenced by environmental factors. Thus, it necessitates the use of molecular markers for ascertaining diversity between and within species. They have been widely used in marker-assisted selection in breeding programs, mapping of important genes, construction of linkage maps, a study of genetic diversity, and evolution in many organisms because of their locus-specific, co-dominant, PCR-based, multi-allelic nature, high-level polymorphism between different individuals, and accessions, significant abundance in the genome and high rates of transferability across the species (Ahn et al., 2014). Earlier King chili was identified as *Capsicum frutescens* Linn. However, its morphological characters resemble *Capsicum chinense* Jacq species which is known to produce the hottest chili peppers in past. They used Random Amplified Polymorphic DNA (RAPD) markers for species identification and concluded the dilemma on the authenticity of King Chili as genetic introgression occurred from *Capsicum frutescens* Linn into *Capsicum chinense* Jacq and placed it in a taxonomic position between *C. chinense* Jacq and *C. frutescens*, clustering more closely with the *C. chinense* Jacq group. Therefore, King Chili is considered as *C. chinense* Jacq (Ahn et al., 2014; Bosland and Baral, 2007). The five common cultivated species over 22 wild species belonging in Capsicum genus are *C. annum* L, *C. frutescens* Linn, *C. chinense* Jacq, *C. baccatum* L, and *C. pubescens* L (Bosland, 1994).

The King Chili is called by different names which change region to region. In Assam, it is called Bhut Jolokia, Ghost Chili, and Bih Jolokia. In Nagaland, it is called Naga Jolokia, Naga Chili, Naga King Chili, and King Chili. In Manipur, it is called Umorok and Go-morok., in western India, it is called ghost pepper (Verma et al., 2013; Baruah et al., 2014; Meghvansi et al., 2010). Subsequently, the geographical indication (GI) of the goods tag for this chili has been obtained by the Nagaland State Government. The people of North-eastern India claim that the King Chili pepper in this region is the hottest test in the world (Meetei et al., 2016). It was recorded in Guinness book of world record to be the hottest chili in the world in 2006 with a Scoville heat unit (SHU) rating of 1,001,304 (Bosland and Baral, 2007).

According to Yao, 1994 Jalapeno (*Capsicum annumm*) is a mildly hot pepper, having its value of heating rate 5000 SHU is found in Mexico, Jalapa, and mostly consumed in America. Nowadays due to races in landforms, geographical area, and climatic conditions, peppers are different in color, shape, size, and heat level. It is also found in Bangladesh, Sri Lanka, and now also grown in USA.

The pungency of *Capsicum* fruits is caused by lipophilic alkaloid capsaicin and its analogues, called capsaicinoids. The capsaicinoids content are mainly responsible for the functionality of chilies. Pungency is the most important quality attribute of the chili fruit and is due to the presence of alkaloid capsaicin and its analogue, collectively known as capsaicinoids (Chakradhar et al., 2013; Steward et al., 2005; Liu et al., 2013). It is used worldwide as a culinary condiment for its flavor, aroma, and color, and it is also commercialized as a fresh product, dry crushed pepper, paprika oleoresin, or pepper paste (Ananthan et al., 2018; Penagos-Calvete et al., 2019), chutneys, pickles, and other food preparations (Amruthraj et al., 2014) .

6.2 PLANT DESCRIPTION

FIGURE 6.1 King Chili plant bearing fruits, fresh King Chili fruits, dry King Chili fruits.

King Chili is a semi-perennial herb and self-pollinated plant (Gayathri et al., 2015; Kehie et al., 2016). The plant considerably cross-pollinated up to 10% when the insect population is high. In the North-Eastern states of India, the height of the plant ranges from 50–100 cm. The plant may grow even taller under semi-perennial situation. The leaf is ovate in shape and the surface looks crinkle. Flowers grow with the creamy white corolla, anthers are blue, and filaments are purple in color, having 2–3 flowers each node but at maturity, there are rarely more than two fruits in each node. The length of the elongated fruit is 5–7 cm, 2.5–3.0 cm in diameter. Fruits' color ranges from light green to green to bright red/bright orange. Fruits are elongated, sub-conical to conical in shape with rough wrinkle and spikes surface, possess 4–5 fruits having 25–35 slightly wrinkle seeds, weighing around 4–10 g. The seed color is light brown (Ahn et al., 2014; Meetei et al., 2016).

6.3 ETHNOMEDICINAL USES

The people of North-eastern India used the fruits of King Chili in different food formulations like flavoring curries due to its high-quality fragrance and pungency and also for various medicinal treatments like headache, rheumatism, night blindness, ankylosing spondylitis, arthritis, and digestive diseases. Hot infusions of King Chili are used for treating toothache and muscle pain and to reduce chronic congestion (Bhagowati and Changkija et al., 2009; Sarwa et al., 2012). It is used in pain relief of peripheral neuropathy, symptomatic treatment of arthritis, muscle and joint pains, and neurogenic pains such as Herpes zoster (Amruthraj et al., 2014; Caterina et al., 2000; Seth et al., 2018). Chilies detoxify the blood by removing waste products from our body and increase nutrient supply to the tissues. It also acts as gastrointestinal detoxicates, thereby helping indigestion (Purkayastha et al., 2012). Traditionally, the Maya people use Capsicum to cure asthma cough, and sore throat. The tender leaves of the plant as a fine paste over boils help for the removal of pus (Bhagowati and Changkija et al., 2009). In the Amazon region, the fruit is used as a medication for stomach ache and to cure hangover or hallucination. In Eastern Nicaragua, the decoction of the fruit and leave is also applied in fever, respiratory and pulmonary disorder, postpartum abdominal and back pain, skin rashes, and sores (Coe, 2008; Coe and Anderson, 1997).

6.4 CULTIVATION PRACTICES

According to Baruah et al. (2014) and Bhagowari et al. (2009), in India, King Chili is predominantly cultivated in the states of Nagaland, Manipur, and Assam. The crop is practiced in two planting seasons, that is, Kharif and Rabi. In hilly areas the crop is grown during February–March as a

Kharif Crop while in plains it is grown as rabi crops, that is, in the month of September–October.

The productivity and pungency of rabi grown crops are generally higher than Kharif grown crops. Sowing of seeds in September is more pungent than in February. In Nagaland, the farmer practices in paddy fields as a sporadic intercrop with summer paddy. In home gardens, people prefer to grow in shade areas rather than sunny hot areas as it yields fruits and enhances pungency. But the size of the fruits gradually reduced after 3 years (Baruah et al., 2014; Borgohain and Devi, 2007).

Direct seeding is practiced in Jhum cultivated paddy fields from February to March and peak harvested in between August and September. Crops are sown during September, that is, rabi crop exhibit more pungent smell than sown in the month of February (Moirangthem et al., 2014).

6.5 PHYTOCHEMICAL CONSTITUENTS

The fruit is a good source of various natural colorants like anthocyanins, carotenoids. The chili pepper contains vitamin A, B, C, and minerals along with iron, magnesium, ascorbic acid, and potassium (Bosland and Baral, 2007). The red color of chili is due to the presence of capsaicinoids, carotenoids, tocopherols, and ascorbic acid. Thus, the change in color indicates the declination in these alkaloid contents in chili (Adams et al., 2009).

Capsaicinoids, pungent principles of chili are the most dominant and important phytoconstituents of *Capsicum chinense* Jacq fruit. Capsaicinoids are the nonvolatile alkaloids, the acid amide of vanillylamide and C9-C11 branched fatty acids (Barbero et al., 2016). The placenta of chili is the main site of capsaicinoid biosynthesis (Sarwa et al., 2012; Wang et al., 2011). Till now, it has been reported that *pun*1 encoding putative aminotransferase (pamt) is the only gene for the qualitative role in determining pungency in fruits of chili (Lang et al., 2006). Capsaicinoids content was higher in the placenta than in other parts (Thapa et al., 2009).

Capsaicin consists of (trans-8-methyl-N-vanillyl-6-nonenamide) and dihydrocapsaicin (8-methyl-N-vanillylnonanamide), nordihydrocapsaicin, homocapsaicin, and homo-hydrocapsaicin. The main compounds are capsaicin and dihydrocapsaicin which generally represent around 77–98% of the total capsaicinoid content (Barbero et al., 2016; Howard and Wildman, 2007; De Aguiar et al., 2016).

The capsaicin is a vanilloid, the heaviest of this class of compounds in nature and the amount of these two ranges in ratio 1:1–2:1. The capsaicin content in King Chili pepper is higher than any other Indian chili (Rahman and Inden, 2012). In recent times, capsinoids another catalog of capsaicin was found in CH-19 sweet peppers. It is very similar to capsaicinoids bearing aliphatic hydroxyl groups in vanillyl alcohol with fatty acid only differing in central linkage. It is an amide moiety in capsaicinoids; however in capsinoids it is an ester linkage. Capsinoids are less pungent than capsaicinoids. The major components of capsinoids are capsiate, dihydrocapsiate, and nor-dihydrocapsiate (Luo et al., 2011).

6.6 PHARMACOLOGICAL EFFECTS

Capsaicinoids display many pharmacological properties like analgesic, anticancer, anti-inflammatory, antioxidant, and antiobesity properties. It also exhibits the benefit on cardiovascular and gastrointestinal systems. Therefore, this compound has the potential value in clinics for relieving pain, preventing cancer, and losing weight.

6.6.1 PAIN THERAPY

Capsaicin is the major component of capsaicinoids and the most studied component among capsaicinoids for relieving pain. It was used orally or locally to reduce the inflammation heat and noxious chemically hyperalgesia or to reduce pain from rheumatoid arthritis or fibromyalgia. It is the main ingredient of the drug Adlea which is used in the Phase 2 trial as a long-acting analgesic for treating postsurgical and osteoarthritis pain (Ramadevi et al., 2008). Being a pain reliever, it is added to many counter creams at a concentration of 0.075% or lower considering that creams containing capsaicin have a moderate or poor efficacy or chronic pain (Knotkova et al., 2008).

The mechanism underlying the effect of capsaicin in pain-relieving is through capsaicin receptor also known as vanilloid receptor 1 (VR1) or transient receptor potential vanilloid subfamily member 1 (TRPV1) (Bhave et al., 2002; Caterina et al., 1999; Zygmunt et al., 1999). This receptor is a nonselective cation channel, activated by a wide range of stimuli such as chemical substances, chemical factors like proton, noxious heat, and vanilloids. The role of TRPV1 in pain sensation was confirmed by the experiment done by Caterina et al. (1997, 2000), Christopher et al. (2008) using knock-out mice as a deletion of the TRPV1 gene of knockdown of TPRV1 gene using RNA interference mechanism.

Capsinoids (includes capsiate and dihydrocapsiate) also activated TRPV1 and induce a nociceptive response in mice models when capsinoids is injected subcutaneously in the hind paw by desensitizing the sensory nervous (Lida et al., 2003).

6.6.2 *CANCER PREVENTION*

Capsaicin and dihydrocapsaicin displayed the antitumor activity in "in vitro" as well as in "in vivo" studies. In cell culture, capsaicin was able to block the migration of breast cancer cells and also kill prostate cancer cells. However, it was reported that dihydrocapsaicin induces autophagy in human colon cancer cells HCT116. In an animal model, oral consumption of capsaicin was able to reduce the size of MDAMB231 breast cancer tumors by 50% in mice or potently inhibit the development of pre-neoplastic breast lesions by up to 80%, and direct injection of capsaicin led to 80% reduction in the size of the tumors (Oh et al., 2008; Thoennissen et al., 2010; Yang et al., 2010).

Capsaicin along with dihydrocapsaicin has been reported to suppress the growth of various immortalized or malignant cell lines via induction of cell cycle arrest, apoptosis, autophagy, and/or via the inhibition of cellular metabolic activation (Oh et al., 2008; Thoennissen et al., 2010; Choi et al 2010a; Ghosh and Basu 2010; Choi et al., 2010b; Zhang et al., 2003). It has been reported that these two compounds may inhibit an isoform of cytochrome P450, an enzyme involved in metabolic activation as well as detoxification of many low-molecular-weight carcinogens (Shing et al., 2001).

Capsaicin and dihydrocapsaicin induce cellular apoptosis by the direct pathway (receptor-independent one) and the indirect pathway (receptor-dependent one) (Ziglioli et al., 2009). In the direct pathway, apoptosis is induced by caspase (particularly 1 and 3) interacting with capsaicin. However, in the indirect pathway, there is needed for the interaction of capsaicin with TRPV1, leading to intracellular calcium increase and consequently to the evidence of precocious and late elements of apoptosis (Meetei et al., 2016).

From the epidemiologist studies of Baez et al. (2010), Pyun et al. (2008) revealed that consumption of a large number of chili peppers is at high risk of gastric cancer than those who do not consume. Further studies on capsaicin disclose the metabolites of capsaicin (such as the reactive phenoxy radicals) may attack the DNA and trigger the mutagenicity and malignant transformation. Hence, it is thought of like a double-edge sword, that is, having both carcinogenic and chemopreventive properties.

6.6.3 WEIGHT MANAGEMENT

Overweight or obesity is a rapidly growing threat to the health of many societies. It is widely accepted that increasing energy expenditure and reducing energy in-take form the basis for the management of this problem. Reductions in energy intakes were observed after participants consumed capsaicin containing (Westerterp-Plantenga et al., 2005). Capsaicin containing breakfast significantly reduced participants' rating of their hunger and their desire to eat before lunch (Yoshioka et al., 2004). Human trials and animal model intervention studies using capsaicinoids over the past decades have been found to increase energy expenditure. There is a piece of evidence that capsaicinoids play an important role in weight management (Cui and Himms-Hagen, 1992; Joo et al., 2010; Leung, 2008).

Intestinal mucosal afferent nerves between visceral and subcutaneous sites appear to play a role in controlling adipose tissue distribution. Capsaicin prevents adipogenesis by activating transient receptor potential channels vanilloid-1. Dietary capsaicin intake reduces metabolic dysfunction in obese mice by enhancing the expression of adiponectin along with its receptor (Leung, 2014). To understand the molecular mechanisms underlying the antiobesity effect of capsaicin, we have recently analyzed the protein alterations in white adipose tissue of rats in response to capsaicin treatment. The experiment markedly revealed lipid metabolism and thermogenesis-related proteins were altered upon capsaicin treatment, upholding the significant role of capsaicin in regulating energy metabolism (Joo et al., 2010).

Another study by Kang et al. (2007) demonstrated that dietary capsaicin intake may reduce obesity-induced glucose intolerance by not only suppressing the inflammatory responses but also enhancing the fatty acid oxidation in adipose tissue and/or liver, both of which are important peripheral tissues affecting insulin resistance. Several other mechanisms have been proposed for the antiobesity effect of capsaicin. The effects of capsaicin in adipose tissue and the liver are related to its dual action on peroxisome proliferator-activated receptor (PPAR) alpha and TRPV-1 expression/activation.

Saito et al. (2015) and Joo et al. (2010) suggested that brown adipose tissue (BAT) is prominent in the regulation of energy expenditure and body fat. TRPV1 activation by capsaicin can activate sympathetically mediated BAT thermogenesis and reduces body fat.

Red pepper consumption was reported to increase energy expenditure, lipid oxidation as well as decrease appetite. Although capsaicin has anti-obesity effects, clinically, there is limited application (Belza et al., 2007; Yoshioka et al., 1998).

6.6.4 ANTIOXIDANT AGENT

Chilies are a good source of dietary antioxidants due to the presence of pungent capsaicinoids (Rosa et al., 2002).

From the study of Dairam et al. (2008), the compound capsaicin significantly curtails iron- ($Fe2+$) and quinolinic acid (QA)-induced lipid peroxidation and potently scavenges the superoxide anion generated by 1 mM cyanide in rat brain homogenate. Later, Henderson et al. (1999) proved that the antioxidant behavior for capsaicin was primarily due to the phenolic moiety in the molecule, and that the main product of capsaicin oxidation is its dimer-dicapsaicin. However, according to Kogure et al. (2002) the C7-benzyl carbon, but not the phenolic OH group in capsaicin is the important functional group responsible for the potent antioxidant and free radical scavenging properties. Antioxidant properties of the capsaicinoids present in the King Chili may have a beneficial effect in the prevention of several oxidative stress-mediated diseases (Yoshioka et al., 1998).

6.6.5 ANTIMICROBIAL ACTIVITY

According to the report of Alcorn and Janis (1984) and Roys and Ralph Loveland (1931), *Capsicum* has been used to cure a variety of complicated ailments caused by microbial infection by the Mayan inhabitants of Meso-america. From the report of Cichewicz and Thorpe (1996), the extracts of *Capsicum* possess inhibitory effect against *Clostridium tetani, Bacillus cereus, Clostridium sporogenes, Bacillus subtilis*, and *Streptococcus pyogenes*. Soetarno et al. (1997) observed that the ethanolic extracts of the fruits of three varieties of *Capsicum* exhibited potencies in their anti-microbial activities against Gram (+) and Gram (–) bacteria, and fungi. It was reported from Bioautographic tests observed that capsaicin was the main antimicrobial component that disrupts the membrane of microorganisms (Alcorn and Janis, 1984; Cowan, 1999). Molina-Torres et al. (1999) reported that a high dose of capsaicin reduced the growth of Escherichia coli and *Pseudomonas solanacearum*, whereas growth of *Bacillus subtilis* was strongly inhibited. Zeyrek et al. (2005) observed capsaicin showed bactericidal effect against *Helicobacter pylori*, the major causal organism of gastroduodenal disease, gastric cancer, peptic, and gastric ulcer. Further, reports on antifungal activity on phytopathogens were published that pure capsaicin concentrations above 50 mg L-1 totally inhibited germination of conidial (Kraikruan et al., 2008). According to Kurita et al. (2002) based on DNA microarray analysis of yeast cells, it is suggested that the antimicrobial activity of capsaicin is attributed to the induction of an osmotic stress element, the key genes for membrane biosynthesis, and the pleiotropic drug resistance (PDR) network. In addition, capsaicin also represses the genes for ribosomal components and cellular growth.

6.6.6 CARDIOVASCULAR DISEASE TREATMENT

The main cardiovascular diseases that pose a threat potentially to human health are coronary heart disease, hypertension, myocardial infection, and atherosclerosis. The cardiovascular system is rich in capsaicin-sensitive sensory nerves. Capsaicin-sensitive sensory nerves play a considerable role in regulating the function of the cardiovascular system through the release of neurotransmitter substances like CGRP (Calcitonin gene-related peptide) substance P. It was shown that CGRP possesses cardiovascular protection (Peng and Li, 2010; Zvara et al., 2006; Chai at al., 2006).

It has been shown that the release of CGRP through activating TRPV1 is stimulated by capsaicin and therefore it has the potential benefits on cardiovascular function (Talukdar et al., 2015). Indeed, acute application of capsaicin can mimic the ischemic pre-conditioning-like cardiac protection, which was blocked by the selective CGRP receptor antagonist CGRP-(8-37) (Banerjee et al., 2014). However, the high dose of capsaicin treatment gave determinate effects by exhausting CGRP in the nerve endings and display the detrimental effects (Choi et al., 2011; Henning et al., 2011).

Recent reports suggested that capsaicin and dihydrocapsaicin inhibit platelet aggregation and activity of clotting factors VIII and IX. This evidence contributes to the benefits of prevention of the onset on the incidences of cardiovascular diseases (Talukdar et al., 2015).

6.6.7 GASTROINTESTINAL BENEFITS

The system of the gastro intestine is rich in capsaicin-sensitive sensory nerves. These nerves help in maintaining the gastrointestinal mucosal integrity against injurious interventions (Peng et al., 2010).

An epidemiological study conducted by Satyanarayan (2006) reported that Chinese people who consume more chilies are three times more prone to gastric ulcers than Indians and Malaysians.

There is also a report on the benefits of capsaicinoids as a protectant in gastrointestinal systems. The researchers demonstrated using different gastric injured animal models caused by hydrochloric acid, ethanol, ammonia, and aspirin, etc., which were attenuated in the presence of CGPR or TRPV1 agonist (Wang et al., 2005; Mozsik et al., 2007; Szolcsanyi and Bartho, 2001). In fact, depending on the dose/duration of capsaicin treatment, the effects may either be beneficial or determinate the gastrointestinal mucosa. Higher the dose of treatment, there is damage of capsaicin-sensitive sensory nerve cells. Hence, there are determinate effects on a high dose and vice versa (Nishihara et al., 2002).

6.7 CLINICAL USES

Chilies are rich in Vitamins A, B (thiamine), and C. Clinically, it has already been proved that capsaicin has the ability to dilate blood vessels, thus giving relief in chronic congestions and wheezing (Baruah et al., 2014).

It also helps to treat cough and sore throat. The leaves are reported to cure boils, headaches, and night blindness. It also controls animal diseases, Ranikhet disease by giving orally in the form of a paste combining with garlic (Kumart and Reghunath, 2005). There is also a recent report that a combination of capsaicin and an anesthetic drug, QX- 314, may help rebate trauma of labor pains, surgery, and treat chronic itching (Sharma, 2007). Although capsaicin can cause neurogenic inflammation under certain physiologic conditions, it also has analgesic and anti-inflammatory activities and is used currently in topical creams and gels (e.g., Axsain and Zostrix) to mitigate neurogenic pain (Caterina et al., 2000; Caterina et al., 1997). Chilies have antioxidant properties that can destroy cholesterol levels which could cause major diseases like atherosclerosis and other heart diseases. It also acts as gastrointestinal detoxicates helping indigestion (Wang et al., 2005; Moxsil et al., 2007; Szolcsanyi and Bartho, 2001).

Clinical trials suggest that capsaicinoids have anticancer (Yang et al., 2010), anti-inflammatory (Choi et al., 2011), and antioxidant properties (Henning et al., 2011).

In the clinical studies, the natural capsaicin showed the potentials to inhibit the growth of leukemic cells. Capsaicin suppressed the development of leukemic cells, but not mononuclear cells and normal bone marrow cells via induction of G0-G1 phase cell cycle arrest and apoptosis. With the elevation of intracellular reactive oxygen species production, Capsaicin-induced apoptosis occurred. NB4 and Kasumi-1 are leukemia cells expressing wild-type p53 which are sensitive to capsaicin. However, p53-defective cells including HL-60 and UF-1 cells were less sensitive to capsaicin. The study of Ito et al. (2004) analyzed the association between the sensitivity to capsaicin and the status of freshly isolated p53 cells from eight patients. It was found that the expression of wild-type p53 mRNA contributed to the sensitivity to capsaicin-induced apoptosis in the leukemic cells. Besides that, they also analyzed the levels of intracellular ROS generation during the treatment with capsaicin in capsaicin-sensitive and capsaicin-less sensitive cells. Further study showed that treatment with capsaicin GSH in capsaicin-sensitive cells like NB4 and leukemic cells caused a significant ROS generation with decreased intracellular. In contrast, in capsaicin-less sensitive cells such as HL-60, UF-1 there was less modulation in the intracellular ROS generation and GSH levels (Ito et al., 2004).

6.8 DISEASES OF KING CHILI

King Chili is susceptible to various biotic stresses caused by viruses, fungus, and bacteria. The most common diseases are "die-back," "anthracnose," and "leaf curl." The causal organism of Die-back is *Colletotrichum gloeosporioides*, Anthracnose/Fruit rot is caused by *Colletotrichum capsici* , Collar Rot by *Rhizoctonia solani*, stem rot and Wilt by *Sclerotinia sclerotiorum*, and Leaf spot by *Corynespora cassiicola* .

Besides these, the plant is also susceptible to other diseases caused by other plants. Some of the other diseases are given along with their causal organism in Table 6.1.

6.9 DISEASE MANAGEMENT STUDIES

Viral diseases of King Chili can be controlled by an integrated approach like treatments in a nursery and main field condition. The seeds treated

TABLE 6.1 Diseases of King Chili and Causal Organisms.

Disease type	Name of the disease	Causal organism	References
Viral diseases	Mosaic virus	*Cucumber mosaic virus, Potato virus Y*	Talukdar et al., 2015; Baruah et al., 2016
	Bud necrosis disease	*Groundnut bud necrosis virus*	Baruah et al., 2016
	Leaf curl disease	*Chili leaf curl virus*	Borgohain and Devi, 2007; Talukdar et al., 2015; Baruah et al., 2016
	Veinal motle disease	*Chili veinal motle virus*	Banerjee et al., 2014
	Leaf and stem necrosis disease	*Tomato spoted wilt virus*	Talukdar et al., 2015
Fungal diseases	Anthacnose/Fruit rot disease	*Colletotrichum capsica*	Borgohain and Devi, 2007; Talukdar et al., 2015
	Die-back disease	*Colletotrichum gloeosporoides*	Borgohain and Devi, 2007; Talukdar et al., 2015
	Stem rot and wilt disease	*Sclerotinia sclerotiorum*	Talukdar et al., 2015
	Collar rot disease	*Rhizoctonia solani*	Talukdar et al., 2015
	Leaf spot disease	*Corynespora cassicola*	Talukdar et al., 2015
	Damping-off disease of seedling; Root and stem rot in young transplants	*Rhizoctonia solani*	Ngullie et al., 2010
Bacterial disease	Bacterial wilt disease	*Ralstonia solanacearum*	Talukdar et al., 2015

imidacloprid@0.25 mL/l + nursery net+ foliar spray with imidacloprid @ 2mL/l at 15, 30, 45, 60 DAT is the most effective way to combat the infection of the plants (Baruah et al., 2016).

The fruit rot disease caused by *C. gloeosporioides* can be controlled by using a combination of *Trichoderma viride* and *Pseudomonas fluorescens*, both are biocontrol agents, which inhibit mycelial growth of the fungi and reduce the prevalence of the disease. Plant extracts of *Allium sativum* (10%) and *Azadirachta indica* (10%) also inhibit the mycelial growth of C. gloeosporioides. This helps in managing fruit rot diseases. However, the best result is obtained by the use of Bavistin (0.1%) with 80.84% disease reduction (Ngullie et al., 2010).

Aphids and mealybugs are also infected with the plants. To prevent such infection, very fine wood ash is sprinkled over the plants and the near soil. The sprinkling of fine wood ash makes unfavorable of insects for feeding the plant and also affects the reproduction of insects. On the other hand, ash contains potash, which may confer some sort of resistance against such infection (Talukdar et al., 2015; Ngullie et al., 2010; Marinus et al., 2013).

6.10 STATUS OF KING CHILI IN MANIPUR

In Manipur, harvesting of King Chili starts from the middle of July and continues till the end of April–May. The locally produced chilies are sold in the domestic market. The fresh chili is highly perishable and proper care has to be taken while harvesting and transporting. There is still a traditional marketing system prevailing in the state. The locally produced crops go through different supply chains to different markets in the state. The major markets are Imphal and Jiribam. There is no scientific storage facility or transport facility which can prevent it from spoilage. The major share of the crop is sold afresh to the local traders and traders from outside the village. Traditionally, King Chili is mainly grown in hill districts of Manipur of burned Jhum cultivated areas and also in kitchen gardens with some organic manures such as FYM, poultry droppings, etc. The crop is in high demand in both domestic markets and abroad. In September 2013, the Chief Secretary of Commerce and Industries, Government of Manipur in his Japan-China tour revealed that 1 kg of fresh King Chili fetched 50,000 INR in Japan (Malangmeih, 2016).

First organic King chili was also harvested in Shirarakhong, Ukhrul district, Manipur in August 2017 as a part of an initiative to boost organic farming, thereby marking the state's first venture into organic farming. The Department of Horticulture and Soil Conservation, Director as well as Project Officer of Manipur Organic Mission Agency (MOMA), S. Bhabichand, attended the initiative events as one of the dignitaries and stated that the inhabitants of Manipur will experience the taste of organic King Chili of Shirarakhong (http://e-pao.net/GP.asp?src=10..100817.aug17).

The grievances and hardship problems faced by the youths for getting State's government job would be overcome and they can be employed themselves by cultivating King chili as a lucrative business. This was shown by one farmer from the Bishnupur district of Manipur named Oinam Nilamani Singh, providing himself with self-employment by cultivating King Chili.

According to him, he sold King Chili at around Rs. 450 per kg. He told that he faced many problems and incurred losses in his initial stages. But later on, he was getting advice from the concerned department and put all his effort into taking care of the plants. He conveyed that a sum of Rs 10,000 as assets was spent in starting the business for buying the required equipment and fertilizers. He stated that he was able to earn a sum of Rs. 50,000 from that business. He further stated that "One can prevent unemployment and Generate money by providing a job to oneself" (http://kanglaonline.com/2017/09/King-chili-as-a-lucrative-business-for-state-farmer).

Lateron,inNovember23–24,2018,thefirstKingChilifestivalwascelebrated in Chaengdai (Akhui) village in the Tamenglong district of Manipur (https://easternmirrornagaland.com/manipur-king-chilli-festival-concludes/).

6.11 CONCLUSION

There is no doubt that the King Chili, being one of the hottest chili in the world, has enormous potential for commercialization given the significance and huge potentiality in the national and international stages. Since time immemorial, the indigenous people of North-East India is practising cultivation of King Chili. Apart from being used as a spices and vegetables, the fruit is used in treating various maladies. The great potential of this high-value crop is magnified not only by its pungency and capsaicinoids contents but also by its high medicinal value. It is known to possess a diverse range of medicinal, pharmaceutical properties, which created interest among local people for commercial production of this valuable plant resource, particularly in the North-East region of the country. The scientific literature on the role of capsaicinoids in pain therapy and its potential as an antioxidant molecule and antimicrobial, anti-inflammatory, anticancer, agent clearly supports the traditional application of King Chili in treating various ailments. Capsaicin has been shown also to have potential application in antiobesity treatments, chemo preventive and chemotherapeutic effects. Nevertheless, the researchers are still divided in their opinion regarding the role of capsaicin in carcinogenic processes, which also requires further exploration. Although capsaicin is used as part of the human diet without causing any known adverse health effects, the recent toxicological studies have revealed a large number of physiological and pharmacological effects on the biological system and hence a safe exposure level of capsaicin for human consumption needs to be established. Further efforts like phytochemical and proper pharmacological screening are required to be adopted for the exploration of some more new bioactive components from this plant. Moreover, biological methods are needed to produce stress-tolerant King Chili to explore new compounds and enhance the productivity.

KEYWORDS

- **capsicum**
- **chinense Jacq**
- **capsaicinoids**
- **pharmacological effects**
- **scoville heat unit (SHU)**
- **North-East India**
- **pungent**
- **genetic introgression**

REFERENCES

Adams, M. J.; Ahuja, K. D. K.; Geraghty, D. P. Effect of Capsaicin and Dihydrocapsaicin on In Vitro Blood Coagulation and Platelet Aggregation. *Thromb. Res.* **2009,** *124* (6), 721–723.

Ahn, Y. K.; Tripathi, S.; Kim, J. H.; Cho, Y. I.; Lee, H. E.; Kim, D. S.; Woo, J. G.; Yoon, M. K. Microsatellite Marker Information from High-Throughput Next-Generation Sequence Data of *Capsicum annuum* Varieties Mandarin and Blackcluster. *Sci. Hortic.* **2014,** *170*, 123–130.

Akinboro, A.; Bakare, A. A. Cytotoxic and Genotoxic Effects of Aqueous Extracts of Five Medicinal Plants on *Allium cepa* Linn. *J. Ethnopharmacol.* **2007,** *112* (3), 470–475.

Alcorn, J. B. In *Huastec Mayan Ethnobotany*, 1984.

Ananthan, R.; Subhash, K.; Longvah, T. Capsaicinoids, Amino Acid and Fatty Acid Profiles in Different Fruit Components of the World Hottest Naga King Chili (*Capsicum chinense* Jacq*)*. *Food Chem.* **2018,** *238*, 51–57.

Báez, S.; Tsuchiya, Y.; Calvo, A.; Pruyas, M,; Nakamura, K,; Kiyohara, C.; Oyama, M.; Yamamoto, M. Genetic Variants Involved in Gallstone Formation and Capsaicin Metabolism, and the Risk of Gallbladder Cancer in Chilean Women. *World J. Gastroenterol.* **2010,** *16*, 372.

Banerjee, A.; Dutta, R.; Roy, S.; Ngachan, S. V. First Report of Chilli Veinal Mottle Virus in Naga Chilli (*Capsicum chinense*) in Meghalaya, India. *VirusDisease* **2014,** *25* (1), 142–143.

Barbero, G. F.; Ali, L.; Latifa, A.; Miguel, P.; Carmelo, G. B. Capsaicinoid Contents in Peppers and Pepper-Related Spicy Foods. *Int. J Food Prop.* **2016,** *19* (3), 485–493.

Baruah, B. R.; Kashyap, A.; Nath, P. D. Incidence, Detection and Integrated Management of Viral Disease Complex in Bhut Jolokia, a Chili Cultivar in Assam. *Ann. Plant Prot. Sci.* **2016,** *24* (1), 136–141.

Baruah, S.; Zaman, Md. K.; Rajbongshi, P.; Das, S. A Review on Recent Researches on Bhut Jolokia and Pharmacological Activity of Capsaicin. *Int. J. Pharm. Sci. Rev. Res.* **2014,** *24* (2), 89–94.

Belza, A.; Frandsen, E.; Kondrup, J. Body Fat Loss Achieved by Stimulation of Thermogenesis by a Combination of Bioactive Food Ingredients: A Placebo-Controlled, Double-Blind 8-Week Intervention in Obese Subjects. *Int. J. Obes.* **2007,** *31* (1), 121–130.

Bhagowati, R. R.; Changkija, S. Genetic Variability and Traditional Practices in Naga King Chili Landraces of Nagaland. *Asian Agri-Hist.* **2009,** *13* (3), 171–180.

Bhave, G.; Zhu, W.; Wang, H.; Brasier, D. J.; Oxford, G. S.; Gereau 4th, W. G. cAMP-dependent Protein Kinase Regulates Desensitization of the Capsaicin Receptor (VR1) by Direct Phosphorylation. *Neuron* **2002,** *35* (4), 721–731.

Borgohain, R.; Devi, J. The hottest chilli: A new horizon in agri-entrepreneurship. *Sci. Tech. Entrepreneur.* **2007,** *15*.

Bosland, P. W. Chiles: History, Cultivation, and Uses. *Dev. Food Sci.* **1994**.

Bosland, P. W. *Capsicums:* Innovative Uses of an Ancient Crop. Progress in New Crops. *ASHS Press,* Arlington, VA, 1996; pp 479–487.

Bosland, P. W.; Baral, J. B. 'Bhut Jolokia'—The world's Hottest Known Chile Pepper is a Putative Naturally Occurring Interspecific Hybrid. *HortScience* **2007,** *42* (2), 222–224.

Caterina, M. J.; Leffler, A.; Malmberg, A. B.; Martin, W. J.; Trafton, J.; Petersen-Zeitz, K. R.; Julius, D. Impaired Nociception and Pain Sensation in Mice Lacking the Capsaicin Receptor. *Science* **2000,** *288* (5464), 306–313.

Caterina, M. J.; Rosen, T. A.; Tominaga, M.; Brake, A. J.; Julius, D. A Capsaicin-Receptor Homologue with a High Threshold for Noxious Heat. *Nature* **1999,** *398* (6726), 436–441.

Caterina, M. J.; Schumacher, M. A.; Tominaga, M.; Rosen, T. A.; Levine, J. D.; Julius, D. The Capsaicin Receptor: A Heat-Activated Ion Channel in the Pain Pathway. *Nature* **1997,** *389* (6653), 816–824.

Chai, W.; Mehrotra, S.; Danser, A. J.; Schoemaker, R. G. The Role of Calcitonin Gene-related Peptide (CGRP) in Ischemic Preconditioning in Isolated Rat Hearts. *Eur. J. Pharmacol.* **2006,** *531* (1-3), 246–253.

Chakradhar, T.; Reddy, B. P.; Srinivas, N.; Reddy, P. C. O. Identification and Validation of an Allele Specific Marker Associated with Pungency in *Capsicum* spp. *Curr. Trends Biotechnol. Pharm.* **2013,** *7* (1), 544–551.

Choi, C. H.; Jung, Y. K..; Oh, S. H. Autophagy Induction by Capsaicin in Malignant Human Breast Cells is Modulated by p38 and Extracellular Signal-Regulated Mitogen-Activated Protein Kinases and Retards Cell Death by Suppressing Endoplasmic Reticulum Stress-Mediated Apoptosis. *Mol. Pharmacol.* **2010a,** *78* (1), 114–125.

Choi, C. H., Jung, Y. K..; Oh, S. H. Selective Induction of Catalase-Mediated Autophagy by Dihydrocapsaicin in Lung Cell Lines. *Free Radic. Biol. Med.* **2010b,** *49* (2), 245–257.

Choi, S. E.; Kim, T. H.; Yi, S. A.; Hwang, Y. C.; Hwang, W. S.; Choe, S. J.; Lee, K. W. Capsaicin Attenuates Palmitate-Induced Expression of Macrophage Inflammatory Protein 1 and Interleukin 8 by Increasing Palmitate Oxidation and Reducing c-Jun Activation in THP-1 (Human Acute Monocytic Leukemia Cell) Cells. *Nutr. Res.* **2011,** *31*(6), 468–478.

Christoph, T.; Bahrenberg, G.; De Vry, J., Englberger, W.; Erdmann, V. A.; Frech, M., Kurreck, J. Investigation of TRPV1 Loss-of-Function Phenotypes in Transgenic shRNA Expressing and Knockout Mice. *Mol. Cell. Neurosci.* **2008,** *37* (3), 579–589.

Cichewicz, R. H.; Thorpe, P. A. The Antimicrobial Properties of Chile Peppers (*Capsicum* species) and Their uses in Mayan Medicine. *J. Ethnopharmacol.* **1996,** *52* (2), 61–70.

Coe, F. G. Rama Midwifery in Eastern Nicaragua. *J. Ethnopharmacol.* **2008,** *117* (1), 136–157.

Coe, F. G.; Anderson, G. J. Ethnobotany of the Miskitu of Eastern Nicaragua. *J. Ethnobiol.* **1997,** *17*, 171–214.

Cowan, M. M. Plant Products as Antimicrobial Agents. *Clin. Microbiol. Rev.* **1999,** *12* (4), 564–582.

Cui, J. Y.; Himms-Hagen, J. Long-Term Decrease in Body Fat and in Brown Adipose Tissue in Capsaicin-Desensitized Rats. *Am. J. Physiol. Regul. Integr. Comp. Physiol.* **1992,** *262* (4), R568–R573.

Daira, A.; Fogel, R.; Daya, S.; Limson, J. L. Antioxidant and Iron-Binding Properties of Curcumin, Capsaicin, and S-allylcysteine Reduce Oxidative Stress in Rat Brain Homogenate. *J. Agric. Food Chem.* **2008,** *56* (9), 3350–3356.

De Aguiar, A. C.; Silva, L. P. S.; de Rezende, C. A.; Barbero, G. F.; Martínez, J. Encapsulation of Pepper Oleoresin by Supercritical Fluid Extraction of Emulsions. *J. Supercrit. Fluids* **2016,** *112*, 37–43.

Gayathri, N.; Gopalakrishnan, M.; Sekar, T. In Vitro Micropropagation of Capsicum Chinense Jacq.(Naga King chili). *Asian J. Plant Sci. Res.* **2015,** *5*, 13–18.

Ghosh, A. K.; Basu, S. Fas-associated Factor 1 is a Negative Regulator in Capsaicin Induced Cancer Cell Apoptosis. *Cancer Lett.* **2010,** *287* (2), 142–149.

Henderson, D. E.; Slickman, A. M.; Henderson, S. K. Quantitative HPLC Determination of the Antioxidant Activity of Capsaicin on the Formation of Lipid Hydroperoxides of Linoleic Acid: A Comparative Study Against BHT and Melatonin. *J. Agric. Food Chem.* **1999,** *47* (7), 2563–2570.

Henning, S. M.; Zhang, Y.; Seeram, N. P.; Lee, R. P.; Wang, P.; Bowerman, S.; Heber, D. Antioxidant Capacity and Phytochemical Content of Herbs and Spices in Dry, Fresh and Blended Herb Paste Form. *Int. J. Food Sci. Nutr.* **2011,** *62* (3), 219–225.

Howard, L. R.; Wildman, R. E. Antioxidant Vitamin and Phytochemical Content of Fresh and Processed Pepper Fruit (*Capsicum annuum). In Handbook of Nutraceuticals and Functional Foods 2nd ed.;* CRC Press: Boca Raton, 2007; pp 165–191.

http://e-pao.net/GP.asp?src=10..100817.aug17

http://kanglaonline.com/2017/09/King-chili-as-a-lucrative-business-for-state-farmer

https://easternmirrornagaland.com/manipur-king-chilli-festival-concludes/

Iida, T.; Moriyama, T.; Kobata, K.; Morita, A.; Murayama, N.; Hashizume, S.; Tominaga, M. TRPV1 Activation and Induction of Nociceptive Response by a Non-pungent Capsaicin-like Compound, Capsiate. *Neuropharmacology* **2003,** *44* (7), 958–967.

Ito, K.; Nakazato, T.; Yamato, K.; Miyakawa, Y.; Yamada, T.; Hozumi, N.; Kizaki, M. Induction of Apoptosis in Leukemic Cells by Homovanillic Acid Derivative, Capsaicin, Through Oxidative Stress: Implication of Phosphorylation of p53 at Ser-15 Residue by Reactive Oxygen Species. *Cancer Res.* **2004,** *64* (3), 1071–1078.

Joo, J. I.; Kim, D. H.; Choi, J. W.; Yun, J. W. Proteomic Analysis for Antiobesity Potential of Capsaicin on White Adipose Tissue in Rats Fed with a High Fat Diet. *J. Proteome Res.* **2010,** *9* (6), 2977–2987.

Kang, J. H.; Tsuyoshi, G.; Han, I. S.; Kawada, T.; Kim, Y. M.; Yu, R. Dietary Capsaicin Reduces Obesity-Induced Insulin Resistance and Hepatic Steatosis in Obese Mice Fed a High-Fat Diet. *Obesity* **2010,** *18* (4), 780–787.

Kehie, M.; Kumaria, S.; Devi, K. S.; Tandon, P. Genetic Diversity and Molecular Evolution of Naga King Chili Inferred from Internal Transcribed Spacer Sequence of Nuclear Ribosomal DNA. *Meta Gene* **2016,** *7*, 56–63.

Knotkova, H.; Pappagallo, M.; Szallasi, A. Capsaicin (TRPV1 Agonist) Therapy for Pain Relief: Farewell or Revival. *Clin. J. Pain* **2008,** *24* (2), 142–154.

Kogure, K.; Goto, S.; Nishimura, M.; Yasumoto, M.; Abe, K.; Ohiwa, C.; Terada, H. Mechanism of Potent Antiperoxidative Effect of Capsaicin. *Biochim. Biophys. Acta Gen. Sub.* **2002,** *1573* (1), 84–92.

Kraikruan, W.; Sangchote, S.; Sukprakarn, S. Effect of Capsaicin on Germination of *Colletotrichum capsici* Conidia. *Agric. Natl. Res.* **2008,** *42* (3), 417–422.

Kumary, S.; Reghunath, B. R. Chilies can Cure. *Indian J. Arecanut Spices Med. Plants* **2005,** *7*, 104–105.

Kurita, S.; Kitagawa, E.; Kim, C. H.; Momose, Y.; Iwahashi, H. Studies on the Antimicrobial Mechanisms of Capsaicin using Yeast DNA Microarray. *Biosci. Biotechnol. Biochem.* **2002,** *66* (3), 532–536.

Lang, Y.; Yanagawa, S.; Sasanuma, T.; Sasakuma, T. A Gene Encoding a Putative Acyl-transferase Involved in Pungency of *Capsicum*. *Breed. Sci.* **2006,** *56* (1), 55–62.

Leung, F. W. Capsaicin-Sensitive Intestinal Mucosal Afferent Mechanism and Body Fat Distribution. *Life Sci.* **2008,** *83* (1-2), 1–5.

Leung, F. W. Capsaicin as an anti-obesity drug. In *Capsaicin as a Therapeutic Molecule*, **2014;** pp 171–179.

Liu, S.; Li, W.; Wu, Y.; Chen, C.; Lei, J. De Novo Transcriptome Assembly in Chili Pepper (*Capsicum frutescens*) to Identify Genes Involved in the Biosynthesis of Capsaicinoids. *PloS One* **2013,** *8* (1), e48156.

Luo, X. J.; Peng, J.; Li, Y. J. Recent Advances in the Study on Capsaicinoids and Capsinoids. *Eur. J. Pharmacol.* **2011,** *650* (1), 1–7.

Malangmeih, L.; Rahaman, S. M. Economics of Fresh Naga King Chili in Manipur, India-a Case Study. *Int. J. Environ. Ecol.* **2016,** *6* (1), 151–162.

Marinus, N.; Loli, D. Efficacy of Biocontrol Agents in Controlling Rhizoctonia Solani on Naga King chili (*Capsicum chinense* Jacq.). *J. Exp. Biol. Agric. Sci.* **2013,** *1* (3), 197–201.

Meetei, N. T.; Singh, A. K.; Singh, B. K.; Mandal, N. Recent Advances in Naga King chili (capsicum chinense JACQ.) Research. *Int. J. Agric. Environ. Biotechnol.* **2016,** *9* (3), 421.

Meghvansi, M. K.; Siddiqui, S.; Khan, M. H.; Gupta, V. K.; Vairale, M. G.; Gogoi, H. K.; Singh, L. Naga Chili: A Potential Source of Capsaicinoids with Broad-spectrum Ethnopharmacological Applications. *J. Ethnopharmacol.* **2010,** *132* (1), 1–14.

Moirangthem, S. S.; Gogoi, S.; Thongbam, P. D.; Ramya, K. T; Fiyaz, R. A.; Pandey, D. S. Effect of Sowing Time and Crop Geometry on the Capsaicinoid Content in Bhoot Jolokia (*Capsicum chinense* Jacq.). *J. Food Sci. Technol.* **2014,** *51* (9), 1974–1981.

Molina-Torres, J.; García-Chávez, A.; Ramírez-Chávez, E. Antimicrobial Properties of Alkamides Present in Flavouring Plants Traditionally used in Mesoamerica: Affinin and Capsaicin. *J. Ethnopharmacol.* **1999,** *64* (3), 241–248.

Mózsik, G.; Szolcsányi, J.; Dömötör, A. Capsaicin Research as a New Tool to Approach of the Human Gastrointestinal Physiology, Pathology and Pharmacology. *Inflammopharmacology* **2007,** *15* (6), 232–245.

Nagoth, J. A.; Amruthraj, J. P.; Preetam Raj, J. P.; Label, L. A. Effect of Vegetable Oil in the Solubility of Capsaicinoids Extracted from *Capsicum chinense* Bhut Jolokia. *Asian J. Pharm. Clin. Res.* **2014,** 48–51.

Ngullie, M.; Daiho, L.; Upadhyay, D. N. Biological Management of Fruit Rot in the World's Hottest Chili (*Capsicum chinense* Jacq.). *J. Plant Prot. Res.* **2010**.

Nishihara, K.; Nozawa, Y.; Nakano, M.; Ajioka, H.; Matsuura, N. Sensitizing Effects of Lafutidine on CGRP-Containing Afferent Nerves in the Rat Stomach. *Br. J. Pharmacol.* **2002,** *135* (6), 1487–1494.

Oh, S. H.; Kim, Y. S.; Lim, S. C.; Hou, Y. F.; Chang, I. Y.; You, H. J. Dihydrocapsaicin (DHC), A Saturated Structural Analog of Capsaicin, Induces Autophagy in Human Cancer Cells in a Catalase-Regulated Manner. *Autophagy* **2008,** *4* (8), 1009–1019.

Penagos-Calvete, D.; Guauque-Medina, J.; Villegas-Torres, M. F.; Montoya, G. Analysis of Triacylglycerides, Carotenoids and Capsaicinoids as Disposable Molecules from Capsicum Agroindustry. *Hortic. Environ. Biotechnol.* **2019,** *60* (2), 227–238.

Peng, J.; Li, Y. J. The Vanilloid Receptor TRPV1: Role in Cardiovascular and Gastrointestinal Protection. *Eur. J. Pharmacol.* **2010,** *627* (1-3)

Purkayastha, J.; Alam, S. I.; Gogoi, H. K.; Singh, L. Capsicum assamicum sp. Nov (Solanaceae), from Assam, Northeastern India. *Ozean J. Appl. Sci.* **2012,** *5*, 55–66.

Pyun, B. J.; Choi, S.; Lee, Y.; Kim, T. W.; Min, J. K.; Kim, Y.; Kwon, Y. G. Capsiate, A Nonpungent Capsaicin-Like Compound, Inhibits Angiogenesis and Vascular Permeability Via a Direct Inhibition of Src Kinase Activity. *Cancer Res.* **2008,** *68* (1), 227–235.

Rahman, M. J.; Inden, H. Effect of Nutrient Solution and Temperature on Capsaicin Content and Yield Contributing Characteristics in Six Sweet Pepper (Capsicum annuum L.) Cultivars. *J. Food Agric. Environ.* **2012,** *10*, 524–529.

Remadevi, R.; Szallisi, A. Adlea (ALGRX-4975), an Injectable Capsaicin (TRPV1 Receptor Agonist) Formulation for Longlasting Pain Relief. *IDrugs***2008,** *11* (2), 120–132.

Rosa, A.; Deiana, M.; Casu, V.; Paccagnini, S.; Appendino, G.; Ballero, M.; Dessí, M. A. Antioxidant Activity of Capsinoids. *J. Agric. Food Chem.* **2002,** *50* (25), 7396–7401.

Roys, R. L. The Ethno-Botany of the Maya (No. F 1435.3. M4. R69), 1931.

Saito, M.; Yoneshiro, T.; Matsushita, M. Food Ingredients as Anti-obesity Agents. *Trends Endocrinol. Metab.* **2015,** *26* (11), 585–587.

Sanatombi, K.; Sen-Mandi, S.; Sharma, G. J. DNA Profiling of *Capsicum* Landraces of Manipur. *Sci. Hortic.* **2010,** *124* (3), 405–408.

Sarwa, K. K.; Kira, J.; Sahu, J.; Rudrapal, M.; Debnath, M. A Short Review on *Capsicum chinense* Jacq. *J. Herb. Med. Toxicol.* **2012,** *6* (2), 7–10.

Satyanarayana, M. N. Capsaicin and Gastric Ulcers. *Crit. Rev. Food Sci. Nutr.* **2006,** *46* (4), 275–328.

Seth, T.; Lyngdoh, Y. A.; Chattopadhyay, A.; Sharangi, A. B.; Mishra, G. P. Export of Onion, Garlic and Chili: Three Essential Spices in Daily Kitchen. In *Indian Spices*; Springer: Cham, 2018; pp 359–378.

Sharma, S. Chilies Made you Cry, now theirs Soothe You. *Hindustan Times, New Delhi Edition,* Oct 5, 2000.

Singh, S.; Asad, S. F.; Ahmad, A.; Khan, N. U.; Hadi, S. M. Oxidative DNA Damage by Capsaicin and Dihydrocapsaicin in the Presence of Cu (II). *Cancer Lett.* **2001,** *169* (2), 139–146.

Soetarno, S.; Sukrasno, S.; Yulinah, E.; Sylvia, S. Antimicrobial Activities of the Ethanol Extracts of Capsicum Fruits with Different Levels of Pungency. *Jurnal Matematika dan Sains.* **1997,** *2* (2), 57–63.

Stewart Jr, C.; Kang, B. C.; Liu, K.; Mazourek, M.; Moore, S. L.; Yoo, E. Y.; Jahn, M. M. The Pun1 Gene for Pungency in Pepper Encodes a Putative Acyltransferase. *Plant J.* **2005,** *42* (5), 675–688.

Szolcsanyi, J.; Bartho, L. Capsaicin-Sensitive Afferents and Their Role in Gastroprotection: An Update. *J. Physiol.Paris* **2001,** *95* (1-6), 181–188.

Talukdar, J.; Saikia, A. K.; Borah, P. Survey and Detection of the Diseases of Bhut Jolokia (Capsicum chinense Jacq.) in Assam. *J. Crop Weed* **2015,** *11*, 186–192.

Thapa, B.; Skalko-Basnet, N.; Takano, A.; Masuda, K.; Basnet, P. High-Performance Liquid Chromatography Analysis of Capsaicin Content in 16 *Capsicum F*ruits from Nepal. *J. Med. Food* **2009,** *12* (4), 908–913.

Thoennissen, N. H.; O'kelly, J.; Lu, D.; Iwanski, G. B.; La, D. T.; Abbassi, S.; Koeffler, H. P. Capsaicin Causes Cell-cycle Arrest and Apoptosis in ER-Positive and-Negative Breast Cancer Cells by Modulating the EGFR/HER-2 Pathway. *Oncogene* **2010,** *29* (2), 285–296.

Verma, P. K.; Rawat, K. K.; Das, N.; Pradhan, B. A Botanical Enigma of India's Hottest Chili Bhoot Jolokia (*Capsicum chinense* Jacq). *N. Y. Sci. J.* **2013,** *6* (11), 49–51.

Wang, L.; Hu, C. P.; Deng, P. Y.; Shen, S. S.; Zhu, H. Q.; Ding, J. S.; Li, Y. J. The Protective Effects of Rutaecarpine on Gastric Mucosa Injury in Rats. *Planta Med.* **2005,** 71(05), 416-419.

Wang, J.; Peng, Z.; Zhou, S.; Zhang, J.; Zhang, S.; Zhou, X.; Peng, B. A Study of Pungency of Capsaicinoid as Affected by Their Molecular Structure Alteration. *Pharmacol. Pharm.* **2011,** *2* (3), 109.

Westerterp-Plantenga, M. S.; Smeets, A.; Lejeune, M. P. G. Sensory and Gastrointestinal Satiety Effects of Capsaicin on Food Intake. *Int. J. Obes.* **2005,** *29* (6), 682–688.

Yang, Z. H.; Wang, X. H.; Wang, H. P.; Hu, L. Q.; Zheng, X. M.; Li, S. W. Capsaicin Mediates Cell Death in Bladder Cancer T24 Cells Through Reactive Oxygen Species Production and Mitochondrial Depolarization. *Urology* **2010,** *75* (3), 735–741.

Yao, J.; Nair, M. G.; Chandra, A. Supercritical Carbon Dioxide Extraction of Scotch Bonnet *(Capsicum annuum*) and Quantification of Capsaicin and Dihydrocapsaicin. *J. Agric. Food Chem.* **1994,** *42* (6), 1303–1305.

Yoshioka, M.; Imanaga, M.; Ueyama, H.; Yamane, M.; Kubo, Y.; Boivin, A.; Kiyonaga, A. Maximum Tolerable Dose of Red Pepper Decreases Fat Intake Independently of Spicy Sensation in the Mouth. *Br. J. Nutr.* **2004,** *91* (6), 991–995.

Yoshioka, M.; St-Pierre, S.; Suzuki, M.; Tremblay, A. Effects of Red Pepper Added to High-Fat and High-Carbohydrate Meals on Energy Metabolism and Substrate Utilization in Japanese Women. *Br. J. Nutr.* **1998,** *80* (6), 503–510.

Zeyrek, F. Y.; Oguz, E. In Vitro Activity of Capsaicin Against *Helicobacter pylori. Ann. Microbiol.* **2005,** *55* (2), 125–127.

Zhang, J.; Nagasaki, M.; Tanaka, Y.; Morikawa, S. Capsaicin Inhibits Growth of Adult T-cell Leukemia Cells. *Leukemia Res.* **2003,** *27* (3), 275–283.

Ziglioli, F.; Frattini, A.; Maestroni, U.; Dinale, F.; Ciufifeda, M.; Cortellini, P. Vanilloid-Mediated Apoptosis in Prostate Cancer Cells Through a TRPV-1 Dependent and a TRPV-1-independent Mechanism. *Acta Bio-medica: Atenei Parmensis* **2009,** *80* (1), 13–20.

Zvara, Á.; Bencsik, P.; Fodor, G.; Csont, T.; Hackler Jr, L.; Dux, M.; Ferdinandy, P. Capsaicin-Sensitive Sensory Neurons Regulate Myocardial Function and Gene Expression Pattern of Rat Hearts: A DNA Microarray Study. *FASEB J.* **2006,** *20* (1), 160–162.
Zygmunt, P. M.; Petersson, J.; Andersson, D. A.; Chuang, H. H.; Sørgård, M.; Di Marzo, V.; Högestätt, E. D. Vanilloid Receptors on Sensory Nerves Mediate the Vasodilator Action of Anandamide. *Nature* **1999,** *400* (6743), 452–457.

CHAPTER 7

Chenopodium: Future Smart Crop from Indian Himalaya

RAJKUMARI JASHMI DEVI,[1] NIKHIL KUMAR CHRUNGOO,[2] and BISESHWORI THONGAM[1]

[1]*Plant Systematics and Conservation Laboratory, Institute of Bioresources and Sustainable Development, Takyelpat, Imphal, Manipur*

[2]*Department of Botany, North-Eastern Hill University, Shillong, Meghalaya*

ABSTRACT

The Himalayan ranges of India are extremely rich in floristic wealth and are home to a large variety of traditional crops that could form an important component of human diet in times to come. The genus Chenopodium of family Amaranthaceae is one of the important but underutilized nutraceutical crops of this region. It comprises about 150 species, of which Chenopodium quinoa and C. album are important for their nutritional value. This chapter discussed the morphological variations, SDS-PAGE profiling of soluble seed protein, DNA-based molecular profiling of C. quinoa and C. album, and also complete chloroplast genome of C. album. Evolutionary ages of C. quinoa and C. album were also discussed. This chapter further discussed Single Nucleotide Polymorphism (SNP) based C. quinoa specific primer, R1RQ-AFR, designed from rbcL sequences that could differentiate C. quinoa from 64 genera including 13 species of the genus Chenopodium.

7.1 INTRODUCTION

The current world population is estimated to hit 9 billion by 2050, and global crop production needs to increase at twice the current rate to meet

High-Value Plants: Novel Insights and Biotechnological Advances. Thangjam Anand Singh, Prakash Kumar Sarangi, and Ch. Brajakishore Singh (Eds.)
© 2025 Apple Academic Press, Inc. Co-published with CRC Press (Taylor & Francis)

global food demand. The increase in demands for cereals and homogeneity of food preferences driven by food companies and consumers stifle the efforts made to meet global food and nutritional security (Ahmadian et al. 2001; Anderson, 1999). Agricultural research has focused on major crops, specifically, maize, wheat, and rice, with relatively little attention in orphan crops. These major cereal crops are now showing signs of slowing down in the annual yield despite the phenomenal improvement of agricultural productivity over the last half-century. There is a need to consume more of the less-popular foods or underutilized crops in order to sustain food security (Baar, 1913; Bazile, 2015). Considering ever-increasing demand for food materials, it is not only necessary to use the available rich diversity and wide genetic resources and to improve the existing conventional cultivars but also to look for nonconventional lesser known and underutilized food crops which have remained inadequately exploited for their characteristics and neglected by research and conservation, which can effectively serve as sources of food and nutritional security for people. The Himalayan ranges of India are extremely rich in floristic wealth and are home to a large variety of traditional crops that could form an important component of human diet in times to come. The region is home to a rich diversity of several plant species, many of which are underutilized. Among the existing known plant resources, the International plant Genetic Resources Institute (IPGRI) and Consultative Group on International Agriculture (CGIAR) have identified Chenopodium as one of the important but underutilized nutraceutical crops of the region. The genus belongs to the family Amaranthaceae and comprises about 250 herbaceous, suffrutescent, and arborescent perennials species (Bera and Mukherjee, 1992). With a basic chromosome number of x = 9, this genus includes numerous wild species with chromosome numbers of 2n = 18, 36, 54, and shows disomic inheritance (Bhargava, 2005b; Bhargava, 2005; Bhargava et al., 2007a). While Wilson (1980, 1990) has described the genus Chenopodium as a "taxonomic receptacle" because of the extreme morphological heterogeneity, Rahiminejad and Gornall (2004) have described it as a complex group emphasizing that not only did the group lack good morphological characteristics to distinguish between species, it also lacked informative sites that can be used to differentiate between species at the level of a Biotype.

The economic importance of the genus arises due to the domesticated species C. quinoa, C. berlandieri subsp. nuttalliae, C. pallidicaule, C. album, and C. giganteum, which have been used as grain, as leafy vegetable, or as a forage crop (Brinegar and Goundan, 1993). While the leaves of C. album are known to be a good source of vitamins and micronutrients (Brinegar et al., 1996), C. quinoa has gained importance because of the gluten-free flour of its grains, high-quality protein of its grains with abundance of sulfur-rich amino acids (Bundock et al., 2005; Chrungoo et al., 2019) and high content of minerals and vitamins (Brinegar and Goundan, 1993; Cole, 1962). Both the species, however, show great morphological heterogeneity, especially C. album which has been reported as a complex of diploid (2n = 18), tetraploid (2n = 36), and hexaploid (2n = 54) cytotypes (Devi and Chrungoo, 2017a, 2013). The domesticated species of Chenopodium are divided into two subsections on the basis of pericarp and perianth morphology and crossing relationships (Drzewiecki et al., 2003). The first subsection Cellulata contains diploid allotetraploids (2n = 4x = 36) like C. quinoa and C. berlandieri subsp. nuttaliae. The second subsection Leiosperma includes domesticated and semi-domesticated forms like C. pallidicaule (2n = 18) and C. album (2n = 18, 36, 54) (Bhargava et al., 2006; Duke and Crawford, 1979). Most of the work on genetic diversity and phylogeny in Chenopodium has focused on domesticated species like C. quinoa and C. berlandieri subsp. Nuttalliae (Fairbanks et al., 1990; Fuentes-Bazan et al., 2012b). And very few studies have been undertaken on important weed species like C. album, C. berlandieri, C. ficifolium, C. glaucum, C. murale, and C. strictum. This taxonomic complex was studied on the basis of cytology (Devi and Chrungoo, 2013; Gangopadhyay et al., 2002), karyotypic analysis (Devi and Chrungoo, 2017a; Devi and Thongam, 2017b), flavonoids (Bhargava et al., 2003), RAPD profiles (Bhargava et al., 2005; Duke and Crawford, 1979; Fuentes-Bazan et al., 2012), ISSR markers (Giusti, 1970) directed amplification of minisatellite DNA (DAMDA) (Fuentes-Bazan et al., 2012), microsatellite markers (Fuentes-Bazan et al., 2012a), ribosomal DNA (Fuentes-Bazan et al., 2012b) and clearly indicated the existence of C. album as the most polymorphic plant species of genus Chenopodium. Ruas et al. (1999), used RAPD markers to distinguish C. album from C. berlandieri and found 90% similarity between the two which perhaps gives us a reason why these species are often misidentified.

Chenopodium quinoa Willd., belonging to the family amaranthaceae, is known as Goosefoot or Pigweed, or Inca wheat. Quinoa (C. quinoa) was domesticated in the Bolivian and Peruvian Andean region 7000 years ago and continues to be cultivated in the region today (Gomez-Pando, 2015; Iwaki et al., 2002). The in situ conservation systems for quinoa germplasm and its wild relatives are reported to be still found in the traditional systems of Peruvian farming communities, encompassing both farmer's' fields and nonfarmed areas (Hong et al., 2017; Jarvis et al., 2017; Ji et al., 2013). Quinoa is allotetraploid species with 2n = 4x = 36 chromosomes. It was reported to be derived from hybridization of a female American Chenopodium diploid (A genome) with a male Old World diploid (B genome), dating back 3.3–6.3 million years (Joshi, 1999; Karcz et al., 2005) Quinoa belongs to the subfamily Chenopodioideae, which encompasses new genera such as Chenopodiastrum, Oxybasis, Lipandra, and Dysphania, in addition to the previously recognized genera Chenopodium, Atriplex, Blitum, and Spinacia (Khoury et al., 2014; Kim et al., 2005).

Chenopodium album L. is commonly known as lambsquarters or fat hen. In India, it is known by various vernacular names, viz. Monsaobi (Manipuri), Bathua sag (Hindi), Chandan betu (Bengali), Parupukkirai (Tamil), Pappukura (Telgu), and Katu ayamoddakam (Malyalam). C. album L., of subg Chenopodium, sect. Chenopodium, subsect. Chenopodium (Kolano et al., 2008), is a cosmopolitan weed of family Amaranthaceae. Some forms are grown for their seed and as leafy vegetables in the Himalayas (Brinegar et al., 1996; Kolano et al., 2006; Kolano et al., 2012). The semi-domesticated Himalayan forms whether ascribed to C. album or C. giganteum D. Don are invariably hexaploid (2n = 6x = 54) (Devi and Chrungoo, 2017a) and not tetraploid as wrongly identified earlier (Brinegar and Goundan, 1993; Kolano et al., 2012; Koziol, 1992). The C. album complex has traditionally functioned as a convenient taxonomical receptacle to include diverse entities not assignable to any other species (Bhargava et al., 2006). Furthermore, the taxa included in this complex are ascribed to three ploidy levels, diploid (2n = 2x = 18), tetraploid (2n = 4x = 36), hexaploid (2n = 6x = 54) (Cole, 1962). The 2x and 6x are found as winter weeds throughout North Indian Plains appearing in mid-November and maturing in April, while the 4x cytotype is restricted to North western part where it appears in the middle of June and matures in August. The existence of wide morphological variation (Brinegar et al., 1996; Lohse et al., 2013; Mason et al., 2005) and different ploidy levels may be the result of frequent natural hybridization (Bhargava et al., 2003; Duke and Crawford, 1979; Matilla et al., 2005), high level of autogamy (Maughan et al., 2006) which maintains distinct forms, and parallel evolution (Maughan et al., 2009). Unfortunately this confusion has persisted because very scarce studies existed till recently which explain interrelationships between various components of this complex or between the constituent genomes of the polyploidy forms.

7.1.1 *CULTIVARS OF HIMALAYAN CHENOPODS*

The cultivars of Himalayan Chenopods (tetraploid, 2n = 4x = 36) are designated as C. album; however, they are morphologically very distinct from the pantropic C. album which is hexaploid (2n = 6x = 56) (Maughan et al., 2019). Wilson (1980, 1990) has ascribed this discrepancy to the fact that C. album has been used as a "convenient taxonomic receptacle" for materials not readily assigned to other species of the genus. The domesticated chenopods of the Himalayas have been classified into four cultivars on the basis of seed color. While three of these viz. black, brown, and red cultivars appear to be correctly assigned to the C. album complex, the fourth cultivar, bearing white color seed, has been suggested to be morphologically similar to C. quinoa (Maughan et al., 2019), thereby raising a question mark on the taxonomic status of this cultivar (Kolano et al., 2012). Joshi (1991) had suggested the Himalayan chenopod (C. album) to be an assemblage of more than one species, Gangopadhyay (2002) and Bhargava16 have proposed it to be an assemblage of heteromorphic and heterocytotic (2x, 4x, 6x) forms.

This chapter focuses on effective discrimination of C. quinoa and C. album using advanced biotechnological approaches and development of marker for effective discrimination of C. quinoa from other species of genus Chenopodium.

7.2 MORPHOLOGICAL CHARACTERS

The qualitative and quantitative morphological characters of Chenopodium quinoa Willd. and Chenopodium album L. were studied by Devi and Chrungoo (2015, 2017a, 2013). Their study was the primary investigation conducted to develop marker for effective discrimination of C. quinoa from other species of genus. List of accessions of Chenopodium that were studied are listed in Table 7.1.

Devi and Chrungoo (2015, 2017a, 2013) raised these accessions to full maturity in the experimental garden of the North-Eastern Hill University, Shillong. Accessions belonging to C. quinoa and C. album were evaluated in a randomized block design with three replications for two successive years at the experimental garden of the University. 25 diagnostic morphological traits were studied which included 15 qualitative and 10 quantitative characters. The qualitative traits assessed included flower color, leaf color, leaf arrangement, leaf shape, leaf apex, leaf base, leaf margin, seed color, seed

TABLE 7.1 List of Accessions of Chenopodium quinoa and Chenopodium album.

Sl. No.	Accession no.	Species	Sourcea
1	IC341704	*C. album*	NBPGR, Shimla
2	NIC22517	*C. album*	NBPGR, Shimla
3	IC341700	*C. album*	NBPGR, Shimla
4	IC447575	*C. album*	Chaura, Kinnaur, Himachal Pradesh, India
5	EC359447	*C. album*	NBPGR, Shimla
6	EC359451	*C. album*	NBPGR, Shimla
7	EC507744	*C. quinoa*	NBPGR, Shimla
8	EC507742	*C. quinoa*	Chile
9	IC411825	*C. album*	Thakma, Leh, Jammu and Kashmir, India.
10	IC411824	*C. album*	Thakma , Leh, Jammu and Kashmir, India.
11	EC507738	*C. quinoa*	Peru
12	EC507739	*C. quinoa*	Ecuador
13	EC5077391	*C. quinoa*	NBPGR, Shimla
14	EC507740	*C. quinoa*	USA
15	EC5077401	*C. quinoa*	USA
16	EC5077402	*C. quinoa*	USA
17	EC507741	*C. quinoa*	USA
18	EC507747	*C. quinoa*	USA
19	EC507748	*C. quinoa*	Argentina

[a]National Bureau of Plant Genetic Resources, Shimla station, India.

Source: Reprinted from Devi & Chrungoo. © 2013. NECEER, Imphal.

shape, seed texture, seed position, seed coat, seed edge, pollen aperture, and pollen type. Plants were reported to have yellow flowers, alternate leaf arrangement, leaves with an obtuse base, seeds with a lenticular shape positioned vertically in the flower, and sunken type of pollen (Table 7.2). Variations were, however, reported in leaf color, shape and margins, type of pollen grains, seed color, seed coat texture, and morphology of seed edges (Figs. 7.1a, 7.1b, 7.1c) (Mujica and Jacobsen, 2006). While Bhargava12 has reported a high degree of intraspecific variation in 16 phenotypic traits including plant height, days to flowering, days to maturity, leaf area, seed size, inflorescence length, and dry weight per plant in C. quinoa. They did not reveal any variations in the 15 qualitative morphological characters among different accessions of C. quinoa. However, marked variations were reported in the morphology of seeds in different accessions of C. album (Fig.

TABLE 7.2 Variations Observed in the Qualitative Morphological Parameters Studied in the Present Investigation.

Sl. No.	Accessions	A	B	C	D	E	F	G	H	I	J	K	L	M	N	O
1	IC341704	Yellow	Green	Alternate	Lanceolate	Obtuse	Obtuse	Entire	Black	Lenticular	Shiny	Vertical	Reticulate	Pattern	Sunken	Perforate
2	NIC22517	Yellow	Red	Alternate	Rhombic	Acute	Obtuse	Dentate	Black	Lenticular	Shiny	Vertical	Smooth	Pattern	Sunken	Perforate
3	IC341700	Yellow	Red	Alternate	Rhombic	Acute	Obtuse	Dentate	Black	Lenticular	Shiny	Vertical	Reticulate	Pattern	Sunken	Perforate
4	IC447575	Yellow	Red	Alternate	Rhombic	Acute	Obtuse	Dentate	Brown	Lenticular	Shiny	Vertical	Smooth	Smooth	Sunken	Perforate
5	EC359447	Yellow	Red	Alternate	Rhombic	Acute	Obtuse	Dentate	Black	Lenticular	Shiny	Vertical	Smooth	Pattern	Sunken	Perforate
6	EC359451	Yellow	Red	Alternate	Rhombic	Acute	Obtuse	Dentate	Black	Lenticular	Shiny	Vertical	Smooth	Pattern	Sunken	Perforate
7	IC411824	Yellow	Green	Alternate	Rhombic	Acute	Obtuse	Dentate	White	Lenticular	Dull	Vertical	Smooth	Smooth	Sunken	Foveate
8	IC411825	Yellow	Green	Alternate	Rhombic	Acute	Obtuse	Dentate	White	Lenticular	Dull	Vertical	Smooth	Smooth	Sunken	Foveate
9	EC507738	Yellow	Green	Alternate	Rhombic	Acute	Obtuse	Dentate	White	Lenticular	Dull	Vertical	Smooth	Smooth	Sunken	Foveate
10	EC507739	Yellow	Green	Alternate	Rhombic	Acute	Obtuse	Dentate	White	Lenticular	Dull	Vertical	Smooth	Smooth	Sunken	Foveate
11	EC5077391	Yellow	Green	Alternate	Rhombic	Acute	Obtuse	Dentate	White	Lenticular	Dull	Vertical	Smooth	Smooth	Sunken	Foveate
12	EC507740	Yellow	Green	Alternate	Rhombic	Acute	Obtuse	Dentate	White	Lenticular	Dull	Vertical	Smooth	Smooth	Sunken	Foveate
13	EC5077401	Yellow	Green	Alternate	Rhombic	Acute	Obtuse	Dentate	White	Lenticular	Dull	Vertical	Smooth	Smooth	Sunken	Foveate
14	EC5077402	Yellow	Green	Alternate	Rhombic	Acute	Obtuse	Dentate	White	Lenticular	Dull	Vertical	Smooth	Smooth	Sunken	Foveate
15	EC507741	Yellow	Green	Alternate	Rhombic	Acute	Obtuse	Dentate	White	Lenticular	Dull	Vertical	Smooth	Smooth	Sunken	Foveate
16	EC507742	Yellow	Green	Alternate	Rhombic	Acute	Obtuse	Dentate	White	Lenticular	Dull	Vertical	Smooth	Smooth	Sunken	Foveate
17	EC507744	Yellow	Green	Alternate	Rhombic	Acute	Obtuse	Dentate	White	Lenticular	Dull	Vertical	Smooth	Smooth	Sunken	Foveate
18	EC507747	Yellow	Green	Alternate	Rhombic	Acute	Obtuse	Entire	White	Lenticular	Dull	Vertical	Smooth	Smooth	Sunken	Foveate
19	EC507748	Yellow	Green	Alternate	Rhombic	Acute	Obtuse	Dentate	White	Lenticular	Dull	Vertical	Smooth	Smooth	Sunken	Foveate

Parameter codes: A-Flower color, B-leaf color, C-Leaf arrangement, D-Leaf shape, E-leaf apex, F-Leaf base, G-Leaf margin, H- seed color, I-Seed shape, J-Seed texture, K- Seed position, L-Seed coat, M-Seed edge, N- Pollen aperture, O-pollen type. Accessions of Chenopodium album 1-8, Chenopodium quinoa 9–19.

Source: Reprinted from Mujica and Jacobsen (2006). © 2015 Rajkumari Jashmi Devi and Nikhil K Chrungoo.

7.1c). Baar (1913), who made the first attempt to describe seed heterogeneity in C. album, reported the presence of both black and brown seeds within the same plant. On the contrary, Williams and Harper (1965), Roberts and Neilson (1980), Kamayeva (1987), and Matilla et al. (2005) have postulated the existence of cryptic heterospermy manifested by the presence of black seeds of various sizes and of their capability for rapid or delayed germination in C. album. Devi and Chrungoo (2015, 2017a, 2013) reported that seeds of C. album varied from black seed coat with a shiny texture and patterned edges, brown seed coat with a shiny texture and smooth edges, white seed coat with a dull texture and smooth edges, reticulate seed coat to smooth seed coat (Fig. 7.1a). Pollen was perforate type with a pollen aperture diameter of =/< 1.0 mm in C. album and foveate type in C. quinoa with a pollen aperture diameter of >1.0 mm (Fig. 7.1b). No significant differences

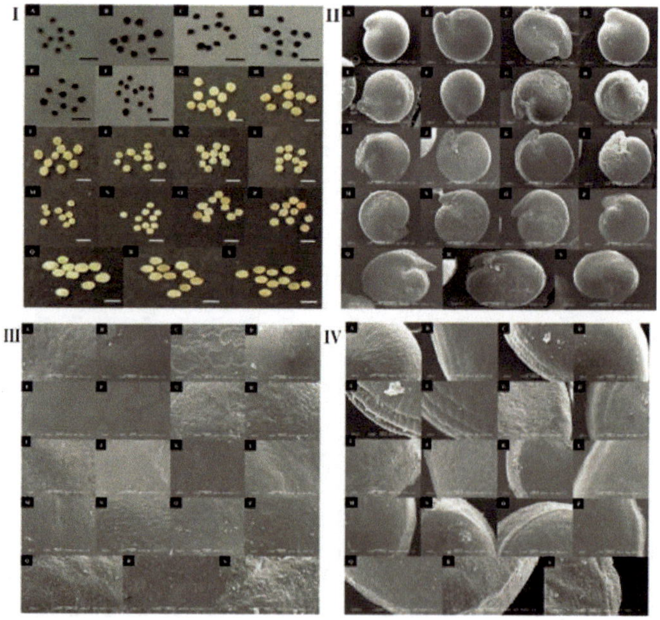

FIGURE 7.1(a) Photographs showing variations in (I) seed color, (II) seed shape, (III) seed coat surface morphology, and (IV) seed edges in different accessions of Chenopodium album: (A) IC-341704, (B) NIC-22517, (C) IC-341700, (D) IC-447575, (E) EC-359447, (F) EC-359451, (G) IC-411824, (H) IC-411825, and Chenopodium quinoa: (I) EC-507744, (J) EC-507742, (K) EC-507738, (L) EC-507739, (M) EC-5077402, (N) EC-5077391, (O) EC-5077401, (P) EC-507741, (Q) EC-507747, (R) EC-507748, (S) EC-507740. Horizontal bars in "I" represent scale at 5 mm.

FIGURE 7.1(b) Photographs showing variations in Leaf margin (A) dentate margin in NIC-22517 of C. album, (B) entire margin in IC-341704 of C. album, (C) dentate margin in EC-507744 of C. quinoa, (D) entire margin in EC507747 of C. quinoa).

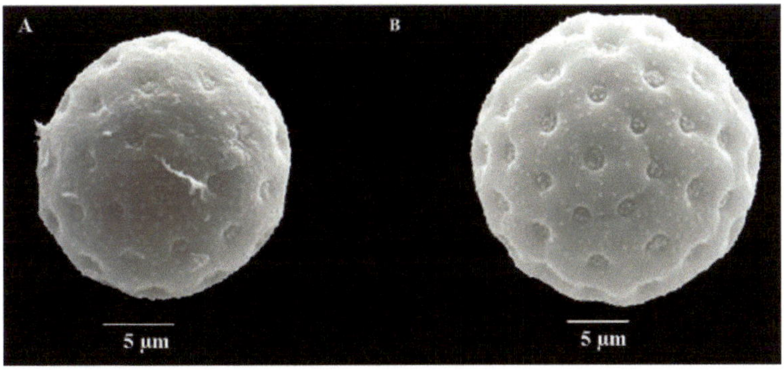

FIGURE 7.1(c) Photographs showing (A) Perforated type of pollen in C. album and (B) Foveate type of pollen in C. quinoa.

in other pollen characters were reported between accessions of C. quinoa and C. album such as in pollen diameter (D), chord (C), C/D ratio, number of apertures on equatorial plane, number of tectal conical tuberculates per unit area. While the average length of chord (C) in accessions of C. album and C. quinoa ranged between 3.5 and 3.7 µm, the pollen diameter for accessions of both the species ranged from 19.6 to 20.4 µm. Pollen grains of accessions belonging to both the species showed a C/D ratio of 0.2 (Mujica and Jacobsen, 2006). These reports on pollen diameter (D), distance between two pores on pollen wall (Chord, C) and C/D ratio for the pollen of accessions of

C. album having seeds with either black or brown seed coat are in agreement with those of Hamdi et al. (2009) who reported a pollen diameter (D) of 23.36 μm, chord length (C) of 4.8 μm, and C/D ratio of 0.2 in Iranian accessions of C. album. While the number of apertures in the equatorial plane of the pollen grains belonging to C. quinoa and C. album ranged from 34–36, the number of tectal conical tuberculates per area of 25 μm2 ranged from 26 in C. quinoa to 33 in C. album. However, there were no differences in the number of tectal conical tuberculates per aperture and the depth of sunken aperture among the accessions.

The seed weight for different accessions of Chenopodium ranged from 0.4 mg to 3.9 mg. Differences were reported in the seed weight between accessions belonging to C. quinoa and C. album. While the weight of per seed for C. album ranged between 0.4 and 0.6 gm, weight of per seed for C. quinoa ranged between 2.1 and 3.9 mg. This showed conformity with the observations of Spehar et al. (2003), who reported an average seed weight of 2.42 gm 1000-1 for seeds of C. quinoa and 0.52 gm 1000-1 gm for C. album. While Karcz et al. (2005) associated tuberculate and smooth seed coats with C. quinoa and C. album, respectively, Devi and Chrungoo (2015, 2017a, 2013) reported smooth seed coat for the seeds of C. quinoa, and smooth as well as reticulate seed coats for the seeds of C. album. Besides the black and brown seeds observed in different accessions of C. album, they also observed white seeds in two accessions viz. IC-411824 and IC-411825, which have been identified by National Bureau of Plant Genetic Resources, India as C. album but have foveate type pollen grains, dull texture of seeds with a smooth surface which is typical of C. quinoa. Except for the accessions IC-341704 and IC-341700, which had a reticulate seed coat, all other accessions belonging to C. album had a smooth seed coat. All accessions of C. quinoa had white seeds with a smooth seed coat having dull texture and smooth edge (Fig. 7.1a). The dendrogram generated on the basis of Jaccard's similarity coefficient for the 15 qualitative traits clustered the accessions into two major clusters. While cluster I comprised six accessions of C. album viz. IC-341704, NIC-22517, IC-341700, IC-447575, EC-359447, and EC-359451 with the similarity coefficient between the accessions ranging from 0.45 to 1.0, cluster II comprised 10 accessions of C. quinoa viz. EC-507738, EC-507739, EC-507739-1, EC-507740, EC-507740-1, EC-507740-2, EC-507741, EC-507742, EC-507744, EC-507748, and two accessions viz. IC-411824 and IC-411825 which have been identified by National Bureau of Plant Genetic Resources (India) as C. album.

These observations on seed surface morphology of different accessions of the genus Chenopodium, however, indicated that it may not be possible to distinguish between C. album and C. quinoa on the basis of their carpological features. Even though seed coat morphology has been used as a phenotypic character in species identification in the genus Chenopodium (Pal and Shukla, 1990; Rahiminejad and Gornall, 2004), the existence of heteromorphs in different species of the genus necessitates the use of other characters including molecular markers in species identification in this genus.

7.3 PROFILES OF SOLUBLE SEED PROTEINS

A total of 22 protein bands ranging in molecular mass from 26–68 kDa were recorded in C. album and C. quinoa in SDS PAGE profile of total soluble seed proteins. While 6 protein bands having molecular masses of 65.50, 62.44, 51.58, 44.69, 39.34, and 26.42 kDa were monomorphic, 16 bands representing proteins with molecular masses 68.70, 67.62, 60.48, 58.59, 54.97, 55.85, 50.76, 48.40, 46.14, 41.27, 36.91, 37.51, 35.76, 33.55, 30.98, and 29.54 kDa were polymorphic. Six protein bands having molecular weights 68.7, 58.59, 55.85, 48.40, 41.27, and 37.51 kDa were detected in C. quinoa and two accessions of C. album which seed color is white (IC-411824 and IC-411825). On the other hand, five protein bands of molecular weights 67.62, 60.48, 54.97, 50.76, and 36.91 kDa were detected in C. album (Mujica and Jacobsen, 2006). An important feature reported in the profiles was the detection of a duplex of 33.55 and 29.59 kDa protein bands in C. quinoa (Fig. 7.1). The average polymorphic Information Content (PIC) calculated from the seed protein profiles was 0.42. The Jaccard's similarity coefficient derived from the similarity matrix of binary data of SDS-PAGE profiles of seed proteins of different accessions of Chenopodium ranged from 0.34 to 1.00 (Fig. 7.2). The Dendrogram generated on the basis of UPGMA (Unweighted Pair-Group Method using Arithmetic averages) analysis resolved the accessions into two broad clusters of C. album and C. quinoa where each cluster further resolved into two subgroups (Mosyakin and Clements, 1996; Mujica and Jacobsen, 2006)., These results are in conformity with the observations of Bhargava67 who has also reported the segregation of C. quinoa and C. album into two clusters on the basis of their seed protein electrophoretic profiles.

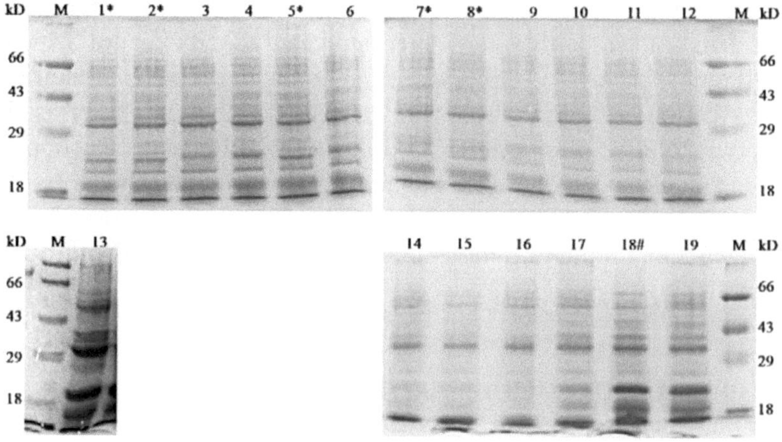

FIGURE 7.2 SDS-PAGE profile of soluble seed proteins extracted from different accessions of Chenopodium quinoa and C. album.
(1) EC-507744, (2) EC-507742, (3) IC-411825, (4) IC-411824, (5) EC-507738, (6) EC-507739, (7) EC-5077391, (8) EC-507741, (9) EC-5077401, (10) EC-5077402, (11) EC-507740, (12) EC-507748, (13) EC-507747, (14) EC-359447, (15) IC-447575, (16) IC-341700, (17) NIC-22517, (18) IC341704. (19) EC359451. While "*" C. quinoa detected with the duplex of 33.55 and 29.59 kDa protein bands, "#"indicates the accession IC-341704 of C. album with one protein band of molecular mass 35.76 kDa. M: standard molecular weight marker.

Source: Reprinted from Mujica and Jacobsen (2006). © 2015 Rajkumari Jashmi Devi and Nikhil K Chrungoo.

While cluster I comprised of the accessions IC-341704, NIC-22517, IC-341700, IC-447575, EC-359447, and EC-359451, all of which belong to C. album and showed a similarity coefficient ranging from 0.84 to 1.0, cluster II comprised of all accessions of C. quinoa and the accessions IC-411825, IC-411824, which have been identified by NBPGR as C. album. Fairbanks et al. (1990) have reported the presence of three polymorphic polypeptides having molecular masses of 34.3, 35.6, and 36.2 kD from the globulin fraction of seed proteins of C. quinoa. This is in conformity with the observations reported by Devi and Chrungoo (2015, 2017a, 2013).

Drzewiecki et al. (2003) reported 41 protein bands in SDS-PAGE profiles of seed proteins of C. quinoa. While Drzewiecki et al. (2003) focused on total seed protein profiling, investigation of Devi and Chrungoo (2015, 2017a, 2013) focused only on the soluble seed proteins. Quinoa protein consists of two major protein fractions viz. an 11S globulin type seed storage protein which accounts for 37% of the total protein content (Risi and Galwey, 1984) and a 2S albumin type protein with a molecular mass of 9 kDa that accounts

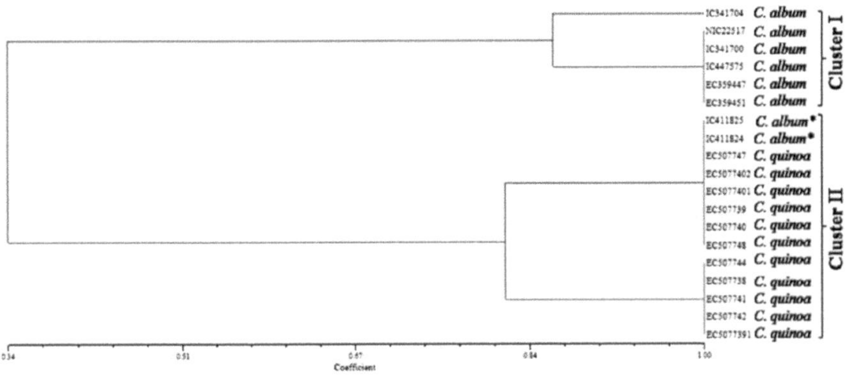

FIGURE 7.3 Dendrogram generated on the basis of UPGMA analysis of the similarity matrix of SDS PAGE profiles of soluble seed proteins of different accessions of Chenopodium. *Two accessions identified as C. album by NBPGR.

Source: Reprinted from Mujica and Jacobsen (2006). © 2015 Rajkumari Jashmi Devi and Nikhil K Chrungoo.

for 35% of the total protein in quinoa (Roberts and Neilson, 1980). The 11S globulin is reported to comprise of two heterogeneous sets of polypeptides viz. the 30–40 kDa (acidic) subunits and the 20–25 kDa (basic) subunits which are linked by disulfide bridges. It is possible that the duplex of 33.55 and 29.59 kDa or the 30.98 kDa protein could be the acidic subunit(s) and the 26.42 kDa protein could be the basic subunit of Chenopodin, the 11S seed storage globulin of Chenopodium. This suggestion is based on the observations of Rout and Chrungoo (1999) on the 26 kDa basic subunit of 13S globulin in buckwheat (Fagopyrum esculentum).

The similarity matrices developed on the basis of both morphology as well as SDS-PAGE profiles of seed proteins placed the accessions IC-411824 and IC-411825 closer to C. quinoa than C. album., identified by National Bureau of Plant Genetic resources (India) has identified the two accessions as C. album. Thus it suggested a relook at the identification of accessions IC-411824 and IC-411825 as C. album.

7.4 DNA-BASED MOLECULAR PROFILING

7.4.1 AMPLIFIED FRAGMENT-LENGTH POLYMORPHISM (AFLP)

Chrungoo et al.73 studied accessions of C. quinoa and C. album procured from NBPGR, Shimla for AFLP profiles following the procedure described by Vos et al. (1995) with slight modifications. Twenty four primer pairs were screened, out of which eight primer combinations were reported to generate the best polymorphic banding patterns. A total of 373 bands ranging in size from 79 bp for the primer pair E-ACG+ M-CAG to 620 bp for the primer pair E-ACG+ M-CTGC were reported. Out of the 373 bands scored, 291 bands representing 78.01% of the total number of bands were polymorphic and 82 bands representing 21.99% of the total number of bands were monomorphic. The number of fragments scored per primer combination ranged from 34 for the primer pair E-ACT+ M-CTAG to 51 for the primer pair E-AGG+ M-GAC. The highest number of polymorphic bands was recorded for the primer pair E-ACG + M-CAG (Fig. 7.3). While the lowest PIC value of 0.24 was recorded from the electrophoretic band profiles generated with primer pair E-ACT + M-CTAG, the highest PIC value of 0.34 was recorded for the amplicons generated with primer pair E-ACG + M-CTGC. The values for resolving power (Rp) for the amplicons generated with the eight primer pairs ranged from 0.39 for the primer pair E-ACT + M-CTAG to 0.48 for primer pairs E-ACT + M-CTAG and E-ACT + M-GAC. The values for effective multiplex ratio (EMR) for the amplicons generated with different primer combinations ranged from 1.59 for the primer combination E-ACT + M-CTAG to 2.83 for the primer combination E-ACG + M-CAG. The primer pair E-ACG + M-CAG also generated the highest number of polymorphic bands. The marker index (MI) value ranged from 0.48 for the primer combination E-ACT+ M-CTAG to 0.86 for the primer combination E-ACT+ M-GAC. Heterozygosity, a measure of allelic diversity was reported for all primer combinations and the accessions were found highly polymorphic with the value ranged between 0.79 of primer combination E-ACG + M-CTGC to 0.89 of primer combination E-ACT+M-CTAG.

In the UPGMA dendrogram generated on the basis of scoring matrix of banding profiles obtained from AFLP analysis of different accessions of Chenopodium using primer pairs AGC + MCAA, EACT + MGAC, EAGG + MGAC, EACG + MCAG, EAGG + MCAT, EACT + MCTG, EACG + MCTGC, and EACT + MCGAT, the lowest similarity coefficient of 0.12 was reported between accessions belonging to C. quinoa and C. album, the highest coefficient of 1.0 was reported between the accessions EC-507742 which belongs to C. quinoa and IC-411825 which has been identified as C. album (Rout and Chrungoo, 1999). This observation further strengthens the argument that IC-411825 belongs to C. quinoa rather than C. album. The AFLP clustering pattern clearly placed the accessions IC-411824 and IC-411825, which were identified by NBPGR as C. album, in the cluster comprised of

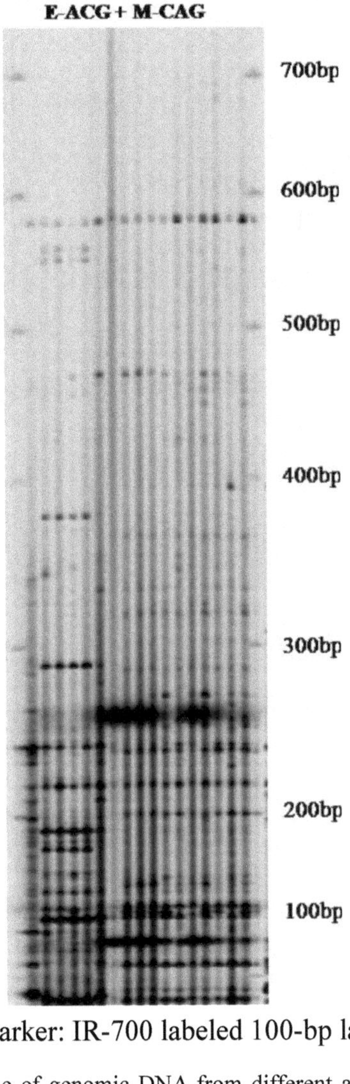

Marker: IR-700 labeled 100-bp ladder.

FIGURE 7.4 AFLP profile of genomic DNA from different accessions of Chenopodium showing highest number of polymorphic bands amplified using primer pairs E-ACG+ M-CAG.

accessions belonging to C. quinoa. Since the data on morphology, SDS-PAGE profile of seed proteins and AFLP clearly point toward the accessions IC-411824 and IC-411825 being C. quinoa AFLP results too re-suggested a relook at the taxonomic identity of these two accessions.

7.4.2 NUCLEAR INTERNAL TRANSCRIBED SPACER REGION (ITS)

In the nucleotide sequence of nuclear internal transcribed spacer region (ITS) region, 24 single-nucleotide polymorphic sites were reported in ITS1 and 18 single-nucleotide polymorphic sites in ITS2 region discriminating between accessions of C. quinoa and C. album (Rout and Chrungoo, 1999). PCR amplification of ITS region from genomic DNA of each accession was carried out with the forward (ITS5:5 GGAAGTAAAAGTCG-TAACAAGG 3) and reverse (ITS4: 5 TCCTCCGCTTATTGATATGC 3) primers designed by White et al. (1990). The length of ITS1, 5.8S and ITS2 were ranged from 263–270, 169–172, and 229–248 bases, respectively. No significant single-base change was reported in 5.8S region. Presence of "CAAGGAA" motif as a part of the conserved angiosperm motif GGCR-(4–7n)-GYGYCAAGGAA was reported at P'204-210, EcoRV site represented by the sequence "GATATC" was also reported at P'301-306 of 5.8 S region and the 14 bp angiospermic motif "GAATTGCAGAATCC" was reported at P'355-368 which is 20 bases pairs downstream of the 16 bp motif "CGATGAAGAACGTAGC" located at P'320-335'(Ruas et al., 1999).76

The secondary structure of the ITS2 region was also reported for C. quinoa and C. album. The secondary structures of ITS2 showed four conserved helices comprising of helix III, a U–U mismatch in helix II, AAA between helix II and III, and an UGGGU/UGGU/GGU near the apex of helix III. The predicted secondary structure of consensus ITS2 sequence of C. album and C. quinoa revealed additional loops between helix II–III and helix IV–I; the sequences from C. quinoa showed an additional loop between helix III–IV (Fig. 7.4). A significant feature of the analysis was the similarity in the secondary structure of ITS2 region between the accessions of C. album which seed color was white and C. quinoa. Their results on sequence variations in the ITS region indicated that ITS1 is more conserved than ITS2. A significant feature of ITS1 and ITS2 spacer region was the ability to discriminate accessions belonging to C. quinoa and C. album. Chrungoo et al.73 suggested that the accession IC-411824 belongs to C. quinoa rather than C. album based on secondary structure of ITS2 region.

The four common helices of conserved structure of ITS2 are numbered I–IV.

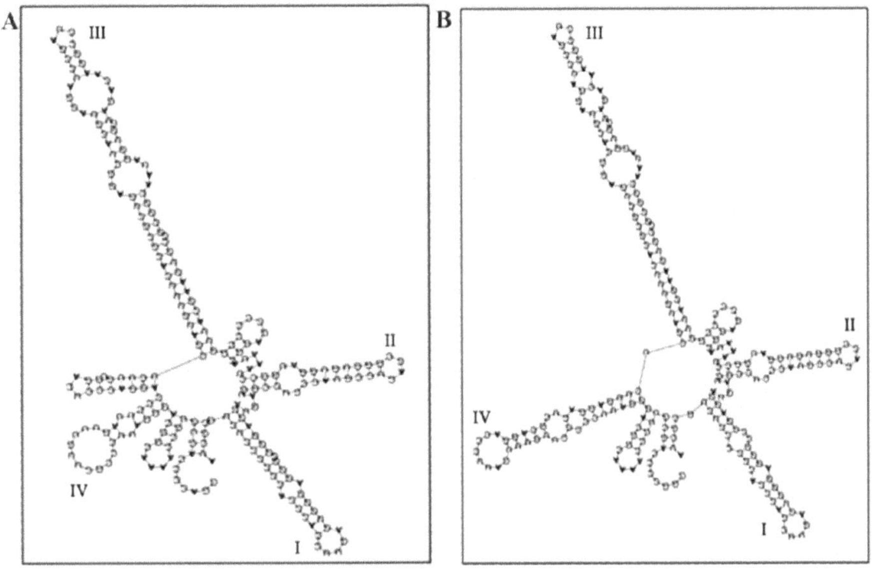

FIGURE 7.5 Graphical visualization (2D-plot) of the secondary structure of sequences representing ITS2 spacer region of (A) C. quinoa and (B) C. album. A B

7.4.3 CHLOROPLAST RBCL, MATK REGIONS, AND ESTIMATION OF EVOLUTIONARY AGE

22 indels and 4 single-base transversions with 95.79% conserved sites, 1.26% parsimony-informative sites, and 0.68% interspecific sequence diversity with a transition/transversion ratio of 0.57 were reported in the nucleotide sequence of chloroplast rbcL region of C. quinoa and C. album (Devi and Chrungoo, 2015, 2017a, 2013). And 32 indels and 9 single-base changes were reported in nucleotide sequences of matK region with 17.27% conserved sites, 4.97% parsimony-informative sites, and 2.81% interspecific sequence diversity. Chloroplast rbcL and matK regions from DNA of each accession was amplified with primer pairs jrF (5'-ATTATACTCCTGAGTATGA-3')-jrR (5'-ACTCCATTTGCTAGCTTC-3') (Mosyakin and Clements, 1996) and AF (5'-CTATATCCACTTATCTTTCAGGAGT-3')-8R (5'-AAAGTTCTAGCA-CAAGAAAGTCGA-3') (Simmonds, 1976), respectively.

The evolutionary ages of C. quinoa and C. album were reported from 87 sequences of rbcL and matK genes representing 62 species from 19 families of the order Caryophyllales including nine species of the genus Chenopodium, 4 species of the family Ranunculaceae, and one species of the family Amborellaceae (Mosyakin and Clements, 1996). They used the reported ages of the Polygonaceae, Cactaceae, Ranunculaceae, and Amborellaceae families as calibration nodes. Accessions of C. quinoa and C. album were reported to resolve into two different clusters in maximum clade credibility phylogenetic tree generated with BEAST 1.8.0 from nucleotide sequence data of the rbcL and matK regions. The cluster consisted of all eleven accessions of C. quinoa and the accessions IC-411824 and IC-411825 which resolved as a single clade with a posterior probability value of 0.97. The accessions IC-411824 and IC-411825, which have white seeds, were reported to show closeness toward C. quinoa rather than C. album (Fig. 7.5). This observation was reported to be supported by the time tree that indicated IC-411824 and IC-411825 to be a sister clade to accessions of C. quinoa with an estimated age of 1.2 Mya. In the same context, the six accessions of C. album clustered together as a single clade with a posterior probability value of 0.98. Of these, four Himalayan accessions showed an estimated age of 10.5–4.1 million years (Myr) and the other two exotic accessions showed an estimated age of 13.1–7.2 Myr (Mosyakin and Clements, 1996). Their results established a lower age for Himalayan C. album than for the Andean chenopods. But the estimated crown age of C. quinoa was reported to be 13 Myr with 95% higher probability distribution and that of C. album was reported as 17 Myr. The mean divergence time separating C. quinoa and C. album was reported as 21 Myr during Miocene epoch of the Cenozoic era (recent life). The genus Chenopodium was reported to originate during the Eocene epoch of the Cenozoic era. Their observations on the evolutionary lineage of different species of the genus Chenopodium were in conformity with those of Fuentes-Bazan et al. (2012, 2012a, 202b), and Kadereit et al. (2003, 2005), thereby establishing a paraphyletic origin of the genus Chenopodium. But Kadereit et al. (2003, 2005) reported the arrival of Chenopodium subgenera Chenopodium/Rhagodia in Australia during the late Miocene to Pliocene epochs when aridification and increasing salinity changed the landscape of many parts of the continent, Devi and Chrungoo (2015, 2017a, 2013) indicated the origin of the genus Chenopodium during the Eocene to Pliocene epochs.

Branch lengths in the tree correspond to age of the species. Time scale is given at the bottom of the tree in terms of millions years ago (Mya). The tree is rooted to basal angiosperm Amoborella trichopoda of amborellaceae (estimated as 154 Myr), whose crown age was indicated to be 179–158 Myr (Spehar et al., 2003).

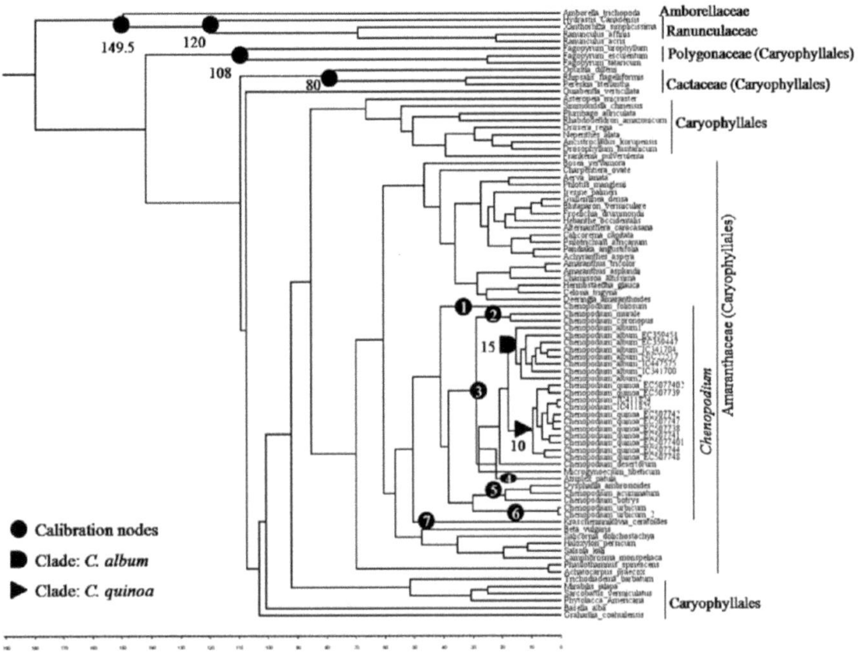

FIGURE 7.6 Maximum clade credibility phylogenetic tree generated from nucleotide sequence data of the rbcL and matK regions of 46 genera belonging to 18 families of caryophyllales including 9 species of Chenopodium.

Source: Reprinted from Mosyakin and Clements (1996). © 2015 Rajkumari Jashmi Devi and Nikhil K Chrungoo.

7.5 SNP-BASED SPECIES SPECIFIC MARKER

7.5.1 GENOTYPING OF SNPS BY ALLELE-SPECIFIC PRIMER EXTENSION

The first C. quinoa-specific SNP-based primer, R1RQ-AFR, designed from rbcL sequences was reported by Devi and Chrungoo (2015, 2017a, 2013) that could differentiate quinoa from 64 genera including 13 species of the genus Chenopodium. They validated SNPs in the cprbcL and matK regions of 36 accessions belonging to C. quinoa and C. album using allele-specific PCR and primers carrying a single base change at the 3' end.

A total of thirteen single-base changes discriminating accessions of C. quinoa and C. album were reported from the sequence alignment of rbcL and matK genes. Out of the 13 identified SNPs, three each viz. G/C, G/T,

T/G at Position 607, 820, and 1070 from the rbcL gene sequences and C/A, G/C, G/A at Position 57, 220, and 517 from the matK gene were selected for further genotyping to develop C. quinoa specific markers (Mosyakin and Clements, 1996). These six SNPs were selected based on the consistency of species-specific SNPs among the accessions and the type of substitution which discriminated C. quinoa and C. album. Two forward primers, with a single base mismatch for the template DNA at the 3' end, were used for each SNP, wherein one primer was specific for each of the two species. Two reverse primers (1 for each set of 6 forward primers for matK and rbcL) were used to discriminate C. quinoa from C. album. All the primer pairs were reported to test individually for the amplification of the target DNA. They confirmed specificity of the primers for the target species by performing PCR with DNA from different accessions of Chenopods and taxonomically closely related genera viz. Beta vulgaris and Amaranthus cordatus of the family amaranthaceae and Fagopyrum esculentum of the family polygonaceae. The primer pairs R1RQ-AFR (5' CAT TAC TTG AAT GCT ACT GCC3'- 5'GTC CCT CAT TAC GAG CTT GTC C3') of rbcL gene and R2MQ-AFM (5'GAT AAT CGA TTG ATA TAG AC3'- 5'CAA ACT CTT CGC TAC TGG TTG3') for matK region were reported to be the most effective. The primer pairs amplified the target DNA from all the accessions belonging to C. quinoa as well as IC411824 and IC411825 but did not amplify the target DNA from IC-447575, IC-341704, EC-359451, NIC-22517, IC-341700, and EC-359447 which belong to C. album (Fig. 7.6). The present chapter would conclude that C. quinoa specific single-allele primer pairs, R1RQ-AFR and R2MQ-AFM, could widely discriminate Chenopodium quinoa from other 19 species of Chenopodium viz. C. botrys, C. frutescens, C. cristatum, C. auricomum, C. desertorum, C. urbicum, C. coronopus, C. foliosum, C. glaucum, C. bonus-henricus, C. simplex, C. ficifolium, C. polysperum, C. rubrum, C. vulvaria, C. murale, C. acuminatum, C. capitatum, and C. album, thereby suggesting these SNP-based allele-specific primer as C. quinoa species specific.

Among the various SNP genotyping methods, the cleaved amplification polymorphic site or polymerase chain reaction-restriction fragment length polymorphism (CAPS or PCR-RFLP) and the derived CAPS (Walsh et al., 2015; Vos et al., 1995) are widely applied (Ward, 2000; White et al., 1990; Wikström, 2001). However, these methods are time consuming and not applicable to all SNPs because of the high cost of some restriction enzymes (Williams and Harper, 1965). While the single-base change at the

3' end of the primer is often not sufficient to ensure reliable discrimination between the two SNP alleles (Wilson, 1980), and two reactions are needed, one for each allele (Wilson, 1990). Devi and Chrungoo (2015, 2017a, 2013) highlight a simple yet reliable PCR protocol with a single-base substitution at the 3' end based on SNPs in the chloroplastic rbcL and matK genes to identify Chenopodium quinoa from other species of the genus. They also suggested the accessions IC411824 and IC411825 to be C. quinoa rather than C. album. Compared to other PCR-based techniques the SNP validation method applied in their study was simple, easy, accurate, inexpensive, reliable, cost-effective, and time-efficient.

7.6 CHLOROPLAST GENOME

The first complete chloroplast genome sequence of C. album from north-eastern India was reported by Devi and Thongam (2017b) bearing NCBI GenBank accession number MF418659. Plant material of C. album was collected from purul (25°22'30.14"N, 94°13'43.05"E), Senapati district of Manipur, India. The complete chloroplast genome sequence was determined from the high-quality 2 × 250 bp pair end library using Illumina Hiseq2500 platform. The complete chloroplast genome sequence of C. album is 150,272 base pair in length. It has quadripartite structure with large single copy (LSC) of 82,417 bp, small single copy (SSC) of 23,271 bp, and two inverted repeats of 22,292 bp (IRA and IRB). The genome has an overall GC content of 36.97%. It has a total of 113 genes comprising of 80 protein-coding gene, 29 tRNAs coding genes, and 4 rRNAs coding genes. The two inverted repeats have 17 duplicated genes viz. rrn5, rrn4.5, rrn23, trnA-UGU, trnI-GAU, trnV-GAC, trnL-CAA, trnI-CAU, rpl2, rpl23, rps7, rps12, ycf2, ycf15, ycf68, and ndhB (Fig. 7.7). Till present, no report has been found on complete chloroplast genome of C. quinoa from the Indian Himalayan region. However, Hong et al. (2017) reported comparative analysis of chloroplast genome of C. quinoa and C. album obtained from the National Agrobiodiversity Center of the Rural Development Administration, Korea.

7.7 CONCLUSION

This chapter discussed similarities/dissimilarities between accession between C. quinoa and C. album and suggested a look at the taxonomic identity of

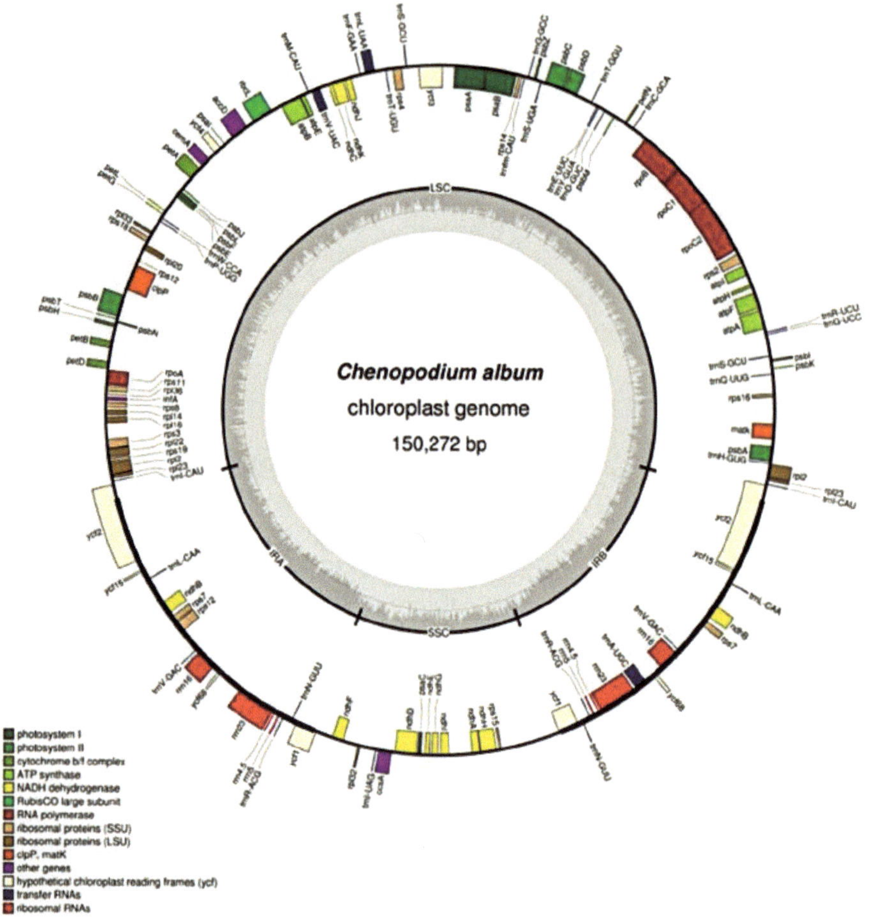

FIGURE 7.7 Circular map of chloroplast genome of Chenopodium album (NCBI GenBank accession number MF418659 (Yamanaka et al., 2004)) drawn using OGDraw v1.2 (Yao et al., 2010).

the two accessions of Himalayan cultivar of C. album which seeds color is white. The development of simple yet efficient strategy of SNP genotyping method discussed here is basically based on the single allele specific primer amplification where utility and informativeness are evaluated by DNA resequencing. The present chapter will add extensive information on genetics and evolutionary relationship analysis of genus Chenopodium.

KEYWORDS

- **Indian Himalayas**
- **Chenopodium**
- ***C. quinoa***
- ***C. album***

REFERENCES

Ahmadian, A.; Gharizadeh, B.; O'Meara, D.; Odeberg, J.; Lundeberg, J. Genotyping by Apyrase-Mediated Allele-Specific Extension. *Nucleic Acids Res.* **2001,** *29,* e121.

Anderson, W. R. Genetic Relationships Among 19 Accessions of Six Species of *Chenopodium* L. by Random Amplified Polymorphic DNA Fragments (RAPD). *Euphytica* **1999,** *105,* 25–32.

Baar, H. Zur Anatomie und Keimungsphysiologie Heteromorpher Samen von *Chenopodium album* und *Atriplex nitens*. *Sitzungsber Kaiserl Akad Wiss Wien* **1913,** *121,* 21–40.

Bazile, D. Le quinoa, les enjeux d'une conquête. Editions Quae, 2015.

Bera, B.; Mukherjee, K. K. Electrophoretic Studies on Diploid and Polyploidy Cytotypes of *Chenopodium album* Linn. In *Perspectives in Cytology and Genetics*, 1992**;** pp 635–643.

Bhargava, A.; Rana, T. S.; Shukla, S.; Ohri, D. Seed Protein Electrophoresis of Some Cultivated and Wild Species of *Chenopodium* (Chenopodiaceae). *Biol. Plant* **2005b,** *49* (*4*), 505–511.

Bhargava, A.; Shukla, S.; Ohri, D. Genetic Variability and Heritability of Selected Traits During Different Cuttings of Vegetable *Chenopodium*. *Indian J. Genet.* **2003**, *63*, 359–360.

Bhargava, A.; Shukla, S.; Ohri, D. Determination of Genetic Diversity in *Chenopodium* spp. *Indian J. Genet*. **2005**, *65*, 202–206.

Bhargava, A.; Shukla, S.; Ohri, D. Karyotypic Studies on Some Cultivated and Wild Species of *Chenopodium* (Chenopodiaceae). *Genet. Resour. Crop Evol.* **2006,** *53,* 1309–1320.

Bhargava, A.; Shukla, S.; Ohri, D. Evaluation of Foliage Yield and Leaf Quality Traits in *Chenopodium* spp. in Multiyear Trials. *Euphytica* **2007a**, *153,* 199–213.

Brinegar, C.; Goundan, S. Isolation and Characterization of Chenopodin, the 11 S Seed Storage Protein of *quinoa* (*Chenopodium quinoa*). *J. Agric. Food Chem.* **1993,** *4*, 182–185.

Brinegar, C.; Sine, B.; Nwokocha, L. High-cysteine 2S Seed Storage Proteins from *Quinoa* (*Chenopodium quinoa*). *J. Agric. Food Chem.* **1996,** *44,* 1621–1623.

Bundock, P. C.; Cross, M. J.; Shapter, F. M.; Henry, R. J. Robust Allele-Specific Polymerase Chain Reaction Markers Developed for Single Nucleotide Polymorphisms in Expressed Barley Sequences. *Theor. Appl. Genet.* **2005,** *112,* 358–365.

Chrungoo, N. K.; Devi, R. J.; Goel, S.; Das, K. Deciphering Species Relationships and Evolution in *Chenopodium* Through Sequence Variations in Nuclear Internal Transcribed Spacer Region and Amplified Fragment-Length Polymorphism in Nuclear DNA. *J. Genet.* **2019,** *98*, 1–11.

Cole, M. J. Interspecific Relationships and Interspecific Variation of *Chenopodium* Album L. in Britain. II The Chromosome Numbers of *Chenopodium* Album and Other Species. *Watsonia* **1962**, *5,*117–122.

Devi, R. J. Development of Molecular Profiles for Characterization of Accessions of *Chenopodium Quinoa* Willd., an Important Underutilized Crop from Indian Himalayas. Doctoral Dissertation, North Eastern Hill University, Shillong, India, 2015. http://hdl.handle.net/10603/169713

Devi, R. J.; Chrungoo, N. K. Diversity in Seed and Pollen Morphology of *Chenopodium quinoa* and *Chenopodium* Album: Potential Underutilized Pseudo-cereal Crops of Genus *Chenopodium* Complex. *NeBIO* **2013**, *4*, 36–41.

Devi, R. J.; Chrungoo, N. K. Species Relationships in *Chenopodium quinoa* and *Chenopodium* Album on the Basis of Morphology and SDS-PAGE Profiles of Soluble Seed Proteins. *J. Appl. Biol. Biotechnol.* **2015,** *3*, 029–033.

Devi, R. J.; Chrungoo, N. K. Evolutionary Divergence in *Chenopodium* and Validation of SNPs in Chloroplast *rbc*L and *mat*k Genes by Allele-Specific PCR for Development of *Chenopodium quinoa*-Specific Markers. *Crop J.* **2017a,** *5*, 32–42.

Devi, R. J.; Thongam, B. Complete Chloroplast Genome Sequence of *Chenopodium* Album from Northeastern India. *Genome Announc.* **2017b,** *5* (47).

Drzewiecki, J.; Delgado-Licon, E.; Haruenkit, A. Identification and Differences of Total Proteins and Their Soluble Fractions in Some Pseudocereals Based on Electrophoretic Patterns. *J. Agric. Food Chem.* **2003,** *51*, 7798–7804.

Duke, J. L.; Crawford, D. J. Character Compatibility and Phyletic Relationships in Several Closely Related Species of *Chenopodium* of Western United States. *Taxon* **1979,** *28*, 307–314.

Fairbanks, D. J.; Burgener, K. W.; Robison, L. R.; Andersen, W. R.; Ballon, E. Electrophoretic Characterization of Quinoa Seed Proteins. *Plant Breed.* **1990,** *104*, 190–195.

Fuentes-Bazan, S.; Mansion, G.; Borsch, T. Towards A Species Level Tree of the Globally Diverse Genus *Chenopodium* (Chenopodiaceae). *Mol. Phylogenet. Evol.* **2012,** *62*, 359–374.

Fuentes-Bazan, S.; Mansion, G.; Borsch, T. Towards a Species Level Tree of the Globally Diverse Genus *Chenopodium* (Chenopodiaceae). *Mol. Phylogenetics Evol.* **2012a,** *62*, 359–374.

Fuentes-Bazan, S.; Uotila, P.; Borsch, T. A Novel Phylogenybased Generic Classification for *Chenopodium* sensu lato, Dand A Tribal Rearrangement of Chenopodioideae (Chenopodiaceae). *Willdenowia* **2012b**, *42*, 5–24.

Gangopadhyay, G.; Das, S.; Mukherjee, K. K. Speciation in *Chenopodium* in West Bengal. *Genet. Resour. Crop Evol.* **2002,** *49*, 503–510.

Giusti, L. El género *Chenopodium*en Argentina 1: Nú-meros de cromosomas. *Darwiniana* **1970,** *16*, 98–105.

Gomez-Pando, L. Quinoa Breeding. In *Quinoa: Improvement and Sustainable Production*; Murphy, K., Matanguihan, J., Eds.; Wiley: Hoboken, 2015; pp 87–107.

Hamdi, S. M. M.; Malekloo, M.; Assadi, M.; Nejadsatari, T. Pollen Micromorphological Studies of the Genus *Chenopodium* (Chenopodiaceae) in Iran. *Asian J. Plant Sci.* **2009,** *8*, 129–137.

Hong, S. Y.; Cheon, K. S.; Yoo, K. O.; Lee, H. O.; Cho, K. S.; Suh, J. T.; Kim, Y. H. Complete Chloroplast Genome Sequences and Comparative Analysis of *Chenopodium quinoa* and C. album. *Front. Plant Sci.* **2017,** *8*, 1696.

Iwaki, K.; Nishida, J.; Yanagisawa, T.; Yoshida, H.; Kato, K. Genetic Analysis of Vrn-B1 for Vernalization Requirement by Using Linked dCAPS Markers in Bread Wheat (*Triticum aestivum* L.). *Theor. Appl. Genet.* **2002**, *104*, 571–576.

Jarvis, D. E.; Ho, Y. S.; Lightfoot, D. J.; et al. The Genome of *Chenopodium quinoa*. *Nature* **2017,** *542*, 307–312.

Jellen, E. N.; Kolano, B. A.; Sederberg, M. C.; Bonifacio, A.; Maughan, P. J. *Chenopodium*. In *Wild Crop Relatives: Genomic and Breeding Resources, Legume Crops and Forages*; Kole, C., Ed.; Springer: Berlin, 2011; pp. 35–61.

Jellen, E. N.; Maughan, P. J.; Fuentes, F. F.; Kolano, B. Botany, Phylogeny, and Evolution. In *State of the Art Report on Quinoa Around the World in 2013;* Bazile, D., Ed.; Santiago de Chile, y CIRAD, FAO: Montpellier, 2015; pp 12–25.

Ji, Q.; Xu, X.; Wang, K. Genetic Transformation of Major Cereal Crops. *Int. J. Dev. Biol.* **2013,** *57*, 495–508.

Joshi, B. D. Genetic Resources of Leaf and Grain Amaranthus and Chenopod. In *Biodiversity*; Swaminathan, M. S., Jana, S., Eds.; Macmillan India Ltd: Madras, 1991; pp 121–134.

Kadereit, G.; Borsch, T.; Weising, K.; Freitag, H. Phylogeny of Amaranthaceae and Chenopodiaceae and the Evolution of C4 Photosynthesis. *Int. J. Plant Sci.* **2003,** *164,* 959–986.

Kadereit, G.; Gotzek, D.; Jacobs, S.; Freitag, H. Origin and Age of Australian Chenopodiaceae. *Org. Divers. Evol.* **2005,** *5,* 59–80.

Kamayeva, G. M. Heterocarpy and Heterospermy of the Weedy *Chenopodium* and *Atriplex* Occurring in Voronezh Province. In *Vegetation of the Black Earth Region and its Protection*; Khmelyov, K. F., Ed.; Univ. Press: Voronezh, 1987; pp 57–62.

Karcz, J.; Kolano, B.; Maluszynska, J. In *SEM Studies on Fruit and Seed of Some Chenopodium L. Species (Chenopodiaceae)*, XII International Conference on Plant Embryology; Cracow: Poland, 2005.

Khoury, C. K.; Bjorkman, A. D.; Dempewolf, H.; Ramirez-Villegas, J.; Jarvis, A.; Rieseberg, L. H.; Struik, P. C.; Guarino, L. Increasing Homogeneity in Global Food Supplies and the Implications for Food Security. *PNAS* **2014,** *111*, 4001–4006.

Kim, M. Y.; Van, K.; Lestari, P.; Moon, J. K.; Lee, S. H. SNP Identification and SNAP Marker Development for a GmNARK Gene Controlling Supernodulation in Soybean. *Theor. Appl. Genet.* **2005,** *110,* 1003–1010.

Kolano, B.; Plucienniczak, A.; Kwasniewski, M.; Maluszynska, J. Chromosomal Localization of a Novel Repetitive Sequence in the *Chenopodium quinoa* genome. *J. Appl. Genet.* **2008,** *49*, 313–320.

Kolano, B.; Siwinska, D.; Maluszynska, J. Molecular Cytogenetic Analysis of Genome Structure in *Chenopodium album* Complex. In *Variability and Evolution-New Perspectives*, **2006**; vol *72*, pp 507–512.

Kolano, B.; Tomczak, H.; Molewska, R.; Jellen, E. N.; Maluszynska, J. Distribution of 5S and 35S rRNA Gene Sites in 34 *Chenopodium*species (Amaranthaceae). *Bot. J. Linn. Soc.* **2012,** *170*, 220–231.

Koziol, M. J. Chemical Composition and Nutritional Value of *quinoa* (*Chenopodium quinoa* Willd.). *J. Food Compos. Anal.* **1992,** *5,* 35–68.

Lohse, M.; Drechsel, O.; Kahlau, S.; Bock, R. OrganellarGenomeDRAW—A Suite of Tools for Generating Physical Maps of Plastid and Mitochondrial Genomes and Visualizing Expression Data Sets. *Nucleic Acids Res.* **2013**, *41(W1)*, W575–W581.

Mason, S. L.; Stevens, M. R.; Jellen, E. N.; Bonifacio, A.; Fairbanks, D. J.; Coleman, C. E.; Maughan, P. J. Development and Use of Microsatellite Markers for Germplasm Characterization in *quinoa* (*Chenopodium quinoa* Willd.). *Crop Sci.* **2005,** *45*, 1618–1630.

Matilla, A.; Gallardo, M.; Puga-Hermida, M. I. Structural, Physiological and Molecular Aspects of Heterogeneity in Seeds: A Review. *Seed Sci. Res.* **2005,** *15*, 63–76.

Maughan, P. J.; Kolano, B. A.; Maluszynska, J.; Coles, N. D.; Bonifacio, A.; Rojas, J.; Jellen, E. N. Molecular and Cytological Characterization of Ribosomal RNA Genes in *Chenopodium quinoa* and *Chenopodium berlandieri*. *Genome* **2006,** *49*, 825–839

Maughan, P. J.; Chaney, L.; Lightfoot, D. J.; Cox, B. J.; Tester, M.; Jellen, E. N.; Jarvis, D. E. Mitochondrial and Chloroplast Genomes Provide Insights into the Evolutionary Origins of *quinoa* (*Chenopodium quinoa*. Willd.). *Sci. Rep.* **2019,** *9*, 185.

Maughan, P. J.; Turner, T. B.; Coleman, C. E.; Elzinga, D. B.; Jellen, E. N.; Morales, J. A.; Udall, J. A.; Fairbanks, D.; Bonifacio, A. Characterization of salt overly sensitive (SOS1) gene homoeologs in *quinoa* (*Chenopodium quinoa* Willd). *Genome* **2009**, *52*, 647–657.

Mehra, P.; Malik, C. Cytology of Some Indian Chenopodiaceae. *Caryologia* **1963,** *16,* 67–84.

Michaels, S. D.; Amisino, R. M. A Robust Method for the Detecting Single-Nucleotide Changes as Polymorphic Markers by PCR. *Plant J.* **1998,** *14*, 381–385.

Mosyakin, S. L.; Clements, S. E. New Infrageneric Taxa and Combinations in *Chenopodium* L. (Chenopodiaceae). *Novon* **1996,** *6*, 398–403.

Mujica, A.; Jacobsen, S. E. La quinua *Chenopodium quinoa* Willd. sus parientes silvestres. La Paz: Botánica Económica Los Andes Cent., Universidad Mayor de San Andrés, 2006; pp 449–457.

Mukherjee, K. K. A Comparative Study of Two Cytotypes of *Chenopodium album* in West Bengal, India. *Can. J. Bot.* **1986,** *64*, 754–759.

Neff, M. M.; Neff, J. D.; Chory, J.; Pepper, A. E. dCAPS, A Simple Technique for the Genetic Analysis of Single Nucleotide Polymorphisms: Experimental Applications in *Arabidopsis thaliana* Genetics. *Plant J.* **1998,** *14*, 387–392.

Ooi, K.; Endo, Y.; Yokoyama, J.; Murata, J. Useful Primer Designs to Amplify DNA Fragments of Plastid gene *mat*K from Angiosperm Plants. *J. Jap. Bot.* **1995,** *70*, 328–331.

Ortiz, R.; Mujica, A.; Rossel, J.; Tapia, M.; Apaza, V.; Canahua, A. Los Parientes Silvestres De la Chenopodiaceas en Puno. Presented at the Parientes silvestres de los Cultivos Nativos en el Peru. In Situ Project and National Agrarian University La Molina, Universidad Nacional agrarian La Molina, Lima, Peru, 2002; pp 47–50.

Pal, M.; Shukla, S. A Hexaploid Grain Chenopod from Eastern Himalayas. *Newslett Himalayan Bot.* **1990,** *8*, 12–14.

Partap, T.; Joshi, B. D.; Galway, N. W. Chenopods: *Chenopodium* spp. Promoting the Conservation and Use of Underutilized Crops; Institute of Plant Genetics and Crop Science Research, Gatersleben/International Plant Genetic Resources Institute: IPGRI, 1998; pp 1–67.

Partap, T.; Joshi, B. D.; Galwey, N. W. Chenopods: *Chenopodium* spp. Promoting the Conservation and Use of Underutilized and Neglected Crops; IPGRI, Rome, Italy, 1998.

Partap, T.; Kapoor, P. The Himalayan Grain Chenopods II. Comparative Morphology. *Agric. Ecosyst. Environ.* **1985,** *14,* 201–220.

Rahiminejad, M. R.; Gornall, R. J. Flavonoid Evidence for Allopolyploidy in the *Chenopodium* Album Aggregate (Amaranthaceae). *Plant Syst. Evol.* **2004,** 246, 77–87.

Rana, T. S.; Narzary, D.; Ohri, D. Genetic Diversity and Relationships Among Some Wild and Cultivated Species of *Chenopodium* L. (Amaranthaceae) using RAPD and DAMD Methods. *Curr. Sci.* **2010,** *98*, 840–846.

Rana, T. S.; Narzary, D.; Ohri, D. Molecular Differentiation of *Chenopodium album* Complex and Some Related Species using ISSR Profiles and ITS Sequences. *Gene* **2011,** *495*, 29–35.

Ray, D. K.; Mueller, N. D.; West, P. C.; Foley, J. A. Yield Trends are Insufficient to Double Global Crop Production by 2050. *PLoS One* **2013,** *8*, e66428.

Risi, J.; Galwey, N. W. The *Chenopodium* Grains of the Andes: Inca Crops for Modern Agriculture. *Ann. Appl. Biol.* **1984**, *10*, 145–216.

Roberts, H. A.; Neilson, J. E. Seed Survival and Periodicity of Seedling Emergence in Some Species of *Atriplex, Chenopodium, Polygonum* and *Rumex*. *Ann. Appl. Biol.* **1980,** *94*, 111–120.

Rojas, W.; Pinto, M.; Alanoca, C.; Gomez Pando, P.; Leon-Lobos, L.G.; Alercia, A.; Quinoa Genetic Resources and Ex Situ Conservation. In *State of the Art Report on Quinoa Around the World in 2013*; Bazile, D., Ed.; CIRAD: Montpellier, 2015; pp 65–94.

Rout, M. K.; Chrungoo, N. K. The Lysine and Methionine Rich Basic Subunit of Buckwheat Grain Legumin: Some Results of a Structural Study. *Int. J. Biochem. Mol. Biol.* **1999,** *47,* 921–926.

Ruales, J.; Nair, B. M. Nutritional Quality of the Protein in *quinoa* (*Chenopodium quinoa,* willd) Seeds. *Plant Foods Hum. Nutr*. **1992,** *42,* 1–11.

Ruas, P. M.; Bonifacio, A.; Ruas, C. F.; Fairbanks, D. J.; Andersen, W. R. Genetic Relationships Among 19 Accessions of Six Species of *Chenopodium* L. by Random Amplified Polymorphic DNA Fragments (RAPD). *Euphytica* **1999,** *105*, 25–32.

Ruiz, K.; Biondi, S.; Oses, R.; Acuña, I. S.; Antognoni, F.; Martinez, E. A. Quinoa Biodiversity and Sustainability for Food Security Under Climate Change: A Review. *Agron. Sustain. Dev.* **2014,** *34*, 349–359. httpsf://doi.org/10.1007/s1359f3-013-0195-0.

Simmonds, N. W. Ed.; *Quinoa* and Relatives. In *Evolution of Crop Plants*; Longman: New York, 1976; pp 29–30.

Spehar, C. R.; Santos, R. L. B.; Nasser, L. C. B. Differences Between *quinoa* and the Weed *Chenopodium album. Weed* **2003,** *21*, 487–491.

Storchova, H.; Drabesova, J.; Chab, D.; Kolar, J.; Jellen, E. N. The Introns in FLOWERING LOCUS T-LIKE (FTL) Genes are Useful Markers for Tracking Paternity in Tetraploid *Chenopodium quinoa* Willd. *Genet. Resour. Crop Evol.* **2015**, *62*, 913–925.

Tapia, M.; Canahua, A.; Ignacio, S. Las razas de Quinuas del Perú, 1st ed.; Lima: ANPE Peru and CONCYTEC, 2014.

United Nations. World Population Prospects 2019. Highlights (ST/ESA/SER.A/423), 2019.

Vos, P.; Hogers, R.; Bleeker, R.; Reijans, M.; Van de Lee, T.; Hornes, M. AFLP: A New Technique for DNA Fngerprinting. *Nucleic Acids Res.* **1995,** *23*, 4407–4414.

Walsh, B. M.; Adhikary, D.; Maughan, P. J.; Emshwiller, E.; Jellen, E. N. *Chenopodium* Polyploidy Inferences from Salt Overly Sensitive 1 (SOS1) Data. *Am. J. Bot.* **2015**, *102*, 533–543.

Ward, S. M. Response to Selection for Reduced Grain Saponin Content in *Chenopodium quinoa* Willd. *Field Crops Res.* **2000,** *68,* 157–163.

White, T. J.; Bruns, T. D.; Lee, S.; Taylor, J. Analysis of Phylogenetic Relationships by Amplification and Direct Sequencing of Ribosomal RNA Genes. In *PCR Protocols: A Guide to Methods and Applications*; Innis, M. A., Gelfand, D. H., Sninsky, J. J., White, T. J., Ed.; Academic Press: New York, 1990; pp 315–322.

Wikström, N.; Savolainen, V.; Chase, M. W. Evolution of the Angiosperms: Calibrating the Family Tree. *Proc. R. Soc. Lond. B* **2001,** *268,* 2211–2220.

Williams, J. T.; Harper, J. L. Seed Dimorphism and Germination. I. The Influence of Nitrates and Low Temperatures on the Germination of *Chenopodium album*. *Weed Res.* **1965**, *5*, 141–150.

Wilson, H. D. Artificial Hybridization among Species of *Chenopodium* sect. *Chenopodium*. *Syst. Bot.* **1980,** *5*, 253–263.

Wilson, H. D. *Quinoa* and Relatives (*Chenopodium* sect. *Chenopodium* subsect. Cellulata). *Econ. Bot.* **1990**, *44,* 92–110.

Yamanaka, S.; Nakamura, I.; Watanabe, K. N.; Sato, Y. I. Identification of SNPs in the Waxy Gene Among Glutinous Rice Cultivars and their Evolutionary Significance During the Domestication Process of Rice. *Theor. Appl. Genet.* **2004,** *108*, 1200–1204.

Yanagisawa, T.; Kiribuchi-Otobe, C.; Hirano, H.; Suzuki, Y.; Fujita, M. Detection of Single Nucleotide Polymorphism (SNP) Controlling the Waxy Character in Wheat by using a Derived Cleaved Amplified Polymorphic Sequence (dCAPS) marker. *Theor. Appl. Genet.* **2003,** *107,* 84–88.

Yao, S.; Lan, H.; Zhang, F.; Variation of Seed Heteromorphism in *Chenopodium album* and the Effect of Salinity Stress on the Descendants. *Ann. Bot.* **2010,** *105,* 1015–1025.

CHAPTER 8

Unearthing of Sea Buckthorn (*Hippophae* spp) in India

THANGJAM ANAND SINGH, PRAKASH K SARANGI, and NG. JOYKUMAR SINGH

College of Agriculture, Iroisemba, Central Agricultural University, Imphal, Manipur, India

ABSTRACT

Sea buckthorn (*Hippophae* L.) is an economically important, multipurpose plant extensively grown in the Himalayan belts of Asia, Europe, and Canada. In India the plant is relatively unknown and the immense industrial potential is essentially untapped. This chapter deals with the nutritional, biochemical, and bioactive components of seabuckthorn. The traditional use of seabuckthorn by the natives is also highlighted along with probable modern applications. Healthy lipids and fatty acid of the berries and seed contribute to the overall nutritional benefits of sea buckthorn products to a consumer. It is also rich in vitamins particularly ascorbic acid, tocopherols, and minerals. Presence of both fat-soluble (tocopherols, carotenoids) and water-soluble antioxidant compounds (flavonoids, tannins, phenolics, ascorbic acids) in high quantities makes it a remarkable option in nutraceutical and functional food industries. Above this seabuckthorn has been known to possess many bioactive compounds which have potential to treat chronic diseases. But a sustainable and viable seabuckthorn based industry required some change in policy and R&D supports by the Indian government. If few issues were addressed the seabuckthorn industry in India could bloom and become one of the leading havens of seabuckthorn business.

High-Value Plants: Novel Insights and Biotechnological Advances. Thangjam Anand Singh, Prakash Kumar Sarangi, and Ch. Brajakishore Singh (Eds.)
© 2025 Apple Academic Press, Inc. Co-published with CRC Press (Taylor & Francis)

8.1 INTRODUCTION

Seabuckthorn (SB), a hardy shrub, known for its perceived and documented values is considered highly nutritious with medicinal properties. It is highly adaptive and grows in extreme climatic conditions and environment. SB is native of Asia and Europe (Rousi, 1971) and has been exploited by natives living in high lands of India, China, and other Himalayan regions for centuries. The ancient Greek scholars such as Dioscorides and Therophast have noted this unique plant for its various medicinal properties. This plant was also recorded in ayurvedic medicine in India, in Chinese folk medicine, and many oriental medicinal treaties like *rGyud bzi* of Tibetan medical classic (Li and Guo, 1989). Recently, SB has fascinated the modern pharmaceutical as well as functional and nutraceutical industries after being long considered as a thorny shrub without any industrial or economic values. This plant plays an important environmental role in the Himalayan highlands where it is the Keystone species. The seabuckthorn, since then, has been recognized as a multi-utility plant with high value unique functional food and environmentally important plant. This plant has recently been a good candidate for agroforestry in many Western countries including new player countries like Canada in North America. The immense potential of the plant has been subjected to various research in Asia, Europe, and more recently in North America. In India pioneering works on SB were carried out by DRDO, Leh where it is available in plenty and locally known as "Leh Berry." Sikkim is another state which has plenty of natural seabuckthorn belts particularly in Lachung and Lachen of North Sikkim in the foothills of Himalayas.

8.2 AREA AND PRODUCTION

It grows in cold regions of Asia, Europe, and North America. It is distributed in Himalayan regions including India, Nepal, Bhutan, Pakistan, Afghanistan, China, Mongolia, Russia, Kazakhstan, Hungary, Romania, Switzerland, Germany, France, Britain, and northwards to Scandinavian countries (Bernath and Foldesi, 1992; Rousi, 1971). The wide distribution of SB is reflected in its habited-related variation not only in morphology, yield, growth rhythms, and cold hardiness, but also in berry-related characters such as fresh weight, chemical, and sensory attributes (Alam, 2004). SB includes 5 or more species namely *H. rhamnoides*, *H. salicifolia*, *H. tibetana*, *H. neurocarpa*, and *H. goniocarpa*. Mainly three species of SB, viz. *H. rhamnoides* var.

turkestanica, H. salcifolia, and *H. tibetana* are found in India. Among these *H. rhamnoides* var. *turkestanica* is the most common in India (Ali and Kaul, 2011).

In India, SB is found in the Himalayan regions of Himachal Pradesh, Ladakh, Uttaranchal, Sikkim, and Arunachal Pradesh and usually concentrated in cold deserts of Trans-Himalayas at an altitude from 2500 msl to 4500 msl. Indian Himalayas has second or third largest area under SB (Ali and Kaul, 2011) in natural forest. Ladakh, which has the highest production in India, grows approximately 12,000 hectare area and covers major part of the forest area. Though Sikkim state has very high naturally occurring seabuckthorn forest (Fig. 8.1) in the North Sikkim district, the plant density

FIGURE 8.1 Seabuckthorn plant in Lachen, Sikkim. (A) Seabuckthorn in natural habitat. (B) A seabuckthorn plant. (C) A branch of seabuckthorn bearing berries. (D) Harvested seabuckthorn berries.

is very low (200 plants per ha) against the plant density of planted orchards (1600 plants/ha.). A natural SB plant can yield 750–1500 kg fruit/ha (Li and Schroeder, 1996). In a season a fruiting plant gives approximately 6–7 kg of berries that would produce about 4–5 L of seabuckthorn juice.

8.3 DESCRIPTION OF THE PLANT

SB (*Hippophae spp.*) is the most important economic plant belonging to Elaeagnaceae family. Two terms "Hippo" which literally means horse and "Phoas" meaning shine (Rousi, 1971) were given as native Tibetans believed giving seabuckthorn to horses makes the animal healthy and shiny. The shrub have narrow leaves, reaching 2–4 m in height and developing a tree-like appearance as it usually sprout in the upper buds and branch. It has dense and stiff branches and is very thorny. The leaves are pale silvery-green, lanceolate, 3–8 cm long and less than 7 mm broad. The male plants produce brownish flowers which produce wind-distributed pollen. The female plants produce orange berry-like fruit which is 6–9 mm in diameter, soft, juicy, and rich in oils. The roots of the plant rapidly distribute extensively and are known for non-leguminous nitrogen fixation role in surrounding soils.

TABLE 8.1 Biochemical and Nutritional Properties of Seabuckthorn.

Nutrients/attributes	g/100 g		References
	[a]DW	[b]FW	
Moisture	78–81		Jaroszewska et al., 2018
Crude protein	8.6–10.0		
Total carbohydrate	36.8–42.0		
Crude fiber	6.2–7.3		
Oil	36.5–44.3		
Ash	4.0–4.1		
Vitamin C		0.025–2.50	Bal et al., 2011
Vitamin E		0.003–0.021	
Vitamin K		0.11–0.23	
Flavonoids		0.35–1.00	
Phytosterols		1.30–2.00	
Carotenoids		0.002–0.04	
Phenolics		21.31–55.38 GAE/g	Maheshwari et al., 2011

[a]DW, dry weight basis; [b]FW, fresh weight basis.

8.4 CLIMATE AND SOIL

The SB-growing areas in the Trans-Himalaya have harsh geo climatic conditions like high altitude, extreme temperature variations (–30 to 30°C). These areas were mostly low in precipitation, low in oxygen with dry soil. Being a hardy plant, it survived in most soil types though sandy loam favored its growth. SB is well adapted to extreme cold, water stress, as well as saline and alkaline soils. SB also acts as a nitrogen fixer (Annonymous, 2000; Li and Schroeder, 1996) like many other plants of Elaeagnaceae family plants. The complex root system with nitrogen-fixing nodules could provide dense growth in very harsh environment, which makes the plant an ideal tool for water and soil conservation in eroded land of the cold desert.

8.5 BIOCHEMICAL AND NUTRITIONAL COMPOSITIONS OF SB BERRIES

SB berry is renowned for its nutritional values, being rich in amino acids, carbohydrates, organic acids, protein, vitamins, flavonoids, and phenolic compounds (Alam, 2004). Many studies have already shown that the biochemical composition of the fruit varies greatly according to the species and climatic and environmental conditions where the plant was grown (Zhao, 1991). The moisture content of SB fruit has been reported to range from 74.0 to 86.7%; carbohydrate content in the fresh berries of Chinese origin varies between 8.0% and 14.0% with glucose and fructose as the major carbohydrates (Table 8.1). Both the sugars are present in approximately equal quantities. The crude protein content of the berries varies with the variety and geographical location. It was reported in the range from 0.79 to 3.11% on a fresh weight basis (Alam, 2004). SB has reputation of being an admirable source of Ascorbic acid and is reported in the range (200–2500 mg/100 g juice) (Bal et al., 2011; Kallio, 2002). SB berries have acidity in the range of pH 2.7 to pH 3.5. Malic acid being the major organic acid along with significant amounts of citric and tartaric acid is reported. The whole berry oil content of Chinese SB was 2.1% and 1.7% of the seed and pulp/peel, respectively (Parimelazhagan, 2005), whereas the whole berry oil content of Finnish SB was reported to be 3.5% with an oil content of 11.3% of the seed and 2.8% of the pulp/peel (Table 8.3). Linoleic (18:2 n-6) and linolenic acids (18:3 n-3) are major components with about 70%

of seed oil Fatty acid. Palmitoleic acid (16:1 n-7) is another fatty acid with 12.1–39.0% of oil in pulp/peel being another important component of SB pulp oil. But palmitoleic acid was virtually absent in the seed oil. 61% Phospholipids and 39% Galactolipids were polar lipids which contained mainly 16:0, 16:1 (n-7), 18:1 (n-9), 18:1 (n-7), and 18:2 (n-9, 6) fatty acid. Galactolipids were shown to be rich in 18:1 (n-9) and 18:3 (n-9, n- 6, n-15) fatty acid, while phospholipids contained higher concentrations of 16:0 and 18:1 (n-9) fatty acids (Yang and Kallio, 2002). The total sterol content were reported in the range from 1200–1800, 240–400, and 340–520 mg/kg in the oils from seeds, fresh pulp/peel, and whole berries, respectively (Mironov, 1989). *Sitosterol*, the most abundant sterol in the SB oil, constituted 57–76% and 61–83% of the seed and pulp/peel oils, respectively. *D-Sitosterol*, 24-methylene-cycloartanol, *citrostadienol* and *uvaol* were detected in the unsaponifiable portion of a pentane extract of the fruit pulp (Beverridge, 1999). Among the Phytosterols in SB (*H. rhamnoides* L.) seed oil, *sitosterol* and 5-*avenasterol* were the most abundant and also most important (Oomah and Mazza, 1999). The carotenoid content of SB berries normally ranges from 30 to 40 mg/100 g fruit (Lu, 1993) with *D-carotene* as a major component with approximately 45% of the total carotenoids whereas phenolic content of SB fruit were in the range from 114 to 244 mg/100 g fruit (Yang and Kallio, 2002). Phenolic compounds such as *quercetin 3- O-glycosides, catechins,* and *hydroxybenzoic acids* with a catechol structure were known to exhibit good antioxidant activities though phenolic compounds accounted less than 5% of the total antioxidant activity of the filtered juice. The major antioxidant activities were derived from ascorbic acid which is significantly present in SB juice (Bal et al., 2011). The flavonoid content in the leaves and fruit of SB has been reported to range from 310 to 2100 mg/100 g dried leaf and from 120 to 1000 mg/100 g fresh fruit, respectively (Hardy, 2000). Phytopharmaceuticals in SB include *quercetin*, *kaempherol*, and *isorhamnetiz* (Lu, 1993).

8.6 USES OF SB BERRIES

SB berries, in recent times, were in great demand mainly for its medicinal value and great economic potential. All the parts of the plants including fruit, branches, seeds, and leaves have been in use for more centuries in traditional medicines in India, China, Mongolia, and Tibet (Bal et al., 2011). The fruit and seed extracts were used for the treatment of burns, cancer,

CVD, gastric ulcers, and oral inflammation (Bal et al., 2011; Bernath and Foldesi, 1992). SB oil is already approved for clinical uses in China and Russia and listed in Pharmacopoeia as well (Mironov, 1989). SB berries are becoming a billion dollar industry due to its perceived and explored bioactive properties of the plant and were being processed by many nutraceuticals and cosmaceutical industries in China and Russia for products (Table 8.2). Many drugs of varied applications have been developed from SB in many countries particularly China, Russia, and more recently India (Menrad, 2003).

TABLE 8.2 Fatty Acid Profiles of Seabuckthorn Pulp and Seed Oil.

Sl no	Fatty acids	Pulp oil (%)	Seed oil (%)	References
	Myristic acid $C_{14:0}$	0.14	0.15	Zheng et al., 2017
	Pentadecyclic acid $C_{15:0}$	0.04	0.14	
	Palmitic acid $C_{16:0}$	27.11	8.71	
	Palmitoleic acid $C_{16:1}$	0.81	37.07	
	Stearic acid $C_{18:0}$	1.37	2.60	
	Oleic acid $C_{18:1}$ (n9)	18.14	20.81	
	Vaccenic acid $C_{18:1}$ (n7)	8.7	1.93	
	Linoleic acid $C_{18:2}$	4.99	38.98	
	Linolenic acid $C_{18:3}$	2.59	25.78	
	Gadoleic acid $C_{20:1}$	0.07	0.25	

8.7 APPLICATIONS IN PHARMACOLOGICAL, FUNCTIONAL, AND NUTRACEUTICAL INDUSTRIES

In these last few decades SB has been widely used as functional food in nutraceuticals and cosmaceutical applications (Table 8.3). The physiological effects of SB berries and berry products were associated with high antioxidant, radio-protective, antitumor, inhibition of LDL cholesterol oxidation and platelet aggregation, antihypertensive, immunomodulation and cytoprotective effects, protection from gastric ulcer, reduction of atopic dermatitis, and wound healing (Pundir et al., 2021; Ciesarovaa et al., 2020; Wani et al., 2016).

TABLE 8.3 Bioactive Compounds and Its Pharmacological Effects.

Sl no	Part of SB used	Bioactive compounds	Treatment
1	Leaves	Kaempferol-3-O-β-D-(6″-O-coumaryl) glycoside, 1-feruloyl-β-D-glucopyranoside, isorhamnetin-3-O-glucoside, quercetin-3- O-β-D-glucopyranoside, quercetin-3-O-β-Dglucopyranosyl-7-O-α-L-rhamnopyranoside, and isorhamnetin-3-O-rutinoside	Rheumatoid arthritis, Cholesterol reduction, Inflammations , Cytotoxicity, Cold hypoxia
2	Branches	Carotenoids, flavonoids, and organic acids	Inflammations, diarrhoea, Colitis
3	Fruits	Catechin, epicatechin, gallocatechin, and epigallocatechin	Gastric, skin problems, Thrombosis, wound healing, cardiovascular diseases
4	Seed/berry oil	Carotenoids, β-sitosterol Tocopherols , tocotrienols, MUFA	Dermatitis, thrombosis, eczema, conjunctivitis, lupus erythematosus

Source: Adapted from Wani et al., 2016.

8.8 TRADITIONAL USES AND RECENT DEVELOPMENTS IN INDIA

Though the SB is regarded as highly versatile, multi-utility plant with huge economic potential, it is relatively unknown and unexplored in India till recently. No authentic scientific report on SB in Sikkim which is the second most production area in India after Ladakh is available as of now. In earlier studies it was usually mentioned as not so important plant and reported as animal feed and firewood for the poor and classified as non-timber forest produce (NTFP). In Sikkim and Ladakh, traditionally, the leaves of SB were used as fodder, woody parts as firewood and fruits were used as tonic. SB was also mentioned in traditional medicine preparations also particularly with gastric ulcer and bowel movement problems. The juice extracts were applied to minor burns. Even the local considered it as having magical properties if taken in a specific way and claimed that the juice could treat mentally unstable people with highly reliable accuracy. The juice was also given to feral horses as the locals believed it could tame the animal. Another common application of SB juice by the local people was dyeing cloths to give a fresh look of bright orange colour finish that would fetch them higher price in local markets. The juice was routinely put into the meat broth particularly pork meat. The juice acts as tenderising agent as well as it give a unique pleasant flavour to the soup. This is still very much in practice in the Lachung and Lachen area in Sikkim and Leh in Ladakh. It was also given to people who are recovering from high fever and other diseases as it could provide much needed vitamins and minerals.

The locals, particularly Bhutia's harvest the berries by chopping the branches and hand plucking it from the cut branches. They also used a globe made of thick cloth in plucking the berries from the twigs. They make juice from the berries by crushing but discarded the seeds after extracting juice of the fruit thus 6–10 MT of SB seed is thrown out as waste in a season. This *waste* can be a good resource as the oil extracted from it is still very rich source of essential Fatty acid and vitamins. There huge scope of expanding the SB area in Sikkim and Ladakh Highlands where no other plant can sustain particularly in the Himalayan foothills.

8.9 HARVEST AND POSTHARVEST MANAGEMENT

Harvesting of SB is one of the most important units of operation which has a substantial consequence on the products that will be developed later on. Maturity indices of SB are the color and texture of the berries. Harvesting was done manually either by shaking the tree or by cutting the branches bearing the berries and hand plucking it from the cut branches. In fruiting season, round yellow-orange berries cover the female plants. Sea buckthorn berries have a strong musky odor with rancid taste when overripe which is even detectable in the field. This odor could be reduced or changed by washing in running water (Beverridge, 1999) which leads to avoidable loss of berries. Correct harvesting of berries and quick transportation to cooling facilities at temperatures around 4°–6°C would considerably reduce the loss. Storing in low temperature would also appreciably retard the microbial activities. If longer period of storage were required, it must be frozen, if at all possible by individual quick freezing techniques. Later the berries were thawed and processed to products as required. Juice extracted by pressing or centrifugal techniques was stored under refrigeration after pasteurization or freezing for long-term storage. Alternatively fruits were processed, finished products were subjected to pasteurization or sterilization. This allowed storage of products in room temperature for longer period up to 3–6 months depending on the products. The shelf life of products was limited even after sterilization but was improved considerably in refrigerated storage.

FIGURE 8.2 Value chain of seabuckthorn.

8.10 PROCESSING OF SB BERRIES

Presently, largest producers and consumers of sea-buckthorn products are China, Russia, and Mongolia. They all have large-scale processing facilities. Processed products include: oil, juice, alcoholic beverages, candies, ice-cream, tea, jam, biscuits, vitamin C tablets, food colors, medicines, cosmetics, and shampoos (Bal et al., 2011; Lu, 1993; Oomah and Mazza, 1999). The SB berries were squeezed to get juice, pulp, and seed (Fig. 8.2). Juice, pulp oil, seed oil, cream, and pigments are the main commercial products from SB berries. Ideally processing and harvesting of berries should be seamless. The harvested berries were cleaned by removing diseased and damaged berries, stems, leaves, and other debris and washed. Washing with lukewarm water or adding little mild detergents or wetting agents increased the juice yield. Screw pressing, cloth pressing, and serpentine pressing, etc. were basically utilized for the separation of juice from berries. Juice obtained may be turbid as high content of suspended solids and pulp oil may be present (Lu, 1993). This might lead to the formation of an unwanted oily layer on the top during storage. The shortcoming may be overcome by centrifugation of unheated juice to separate the floating cream phase. The stalk centrifuge or a cream separator could be employed to separate the cream from the juice (Zhang, 1989) and fatty acid removal in low temperature was recommended (Beverridge, 1999). Solvent extraction has been the simplest method for oil recovery, but it is not advised for nutraceutical and pharmaceutical applications owing to the potential damage to the bioactive phytochemicals during desolventization. Fresh pressed juice separates into three phases when allowed to stand overnight in the refrigerator: an upper cream phase, juice in the middle portion, and sediment at the bottom. Enzymatic hydrolysis with commercial, broad spectrum carbohydrate hydrolyzing enzyme preparations reduces the juice viscosity, assists juice separation, and provides an opalescent juice (Beverridge, 1999). Super critical-CO_2 extraction produced superior quality, solvent free oils. But for this operation, the berries have to be dried before extraction. Again drying the berries will lead to loss of juice and some temperature-sensitive phytonutrients during drying.

8.11 STATUS AND POTENTIAL OF SEABUCKTHORN PRODUCTION AND PROCESSING IN INDIA

The species available in India has, however, never been studied scientifically and needs validation by traditional as well as modern molecular techniques. Leh, Ladakh, and Sikkim are the most important SB areas in India. The epicenter of SB in Lachung is in and around the Lachung bridge riverbed comprising long stretches of SB bushes (N-27?41.424' and E-88?44.772') at an altitude of 8680 ft msl and Temchi Zema area in Lachen (N-27?43.162' and E-88? 33.479') with an altitude of 8460 ft msl.

SB is a non-timber forest product and the product is handled by the forest department and the local people through Joint Forest Management Committees (JFMC). In a season a fruiting plant gives approximately 6–7 kg of berries (4–5 L of juice). Rough estimate shows that a hectare of seabuckthorn will give at least 5–6 MT of seabuckthorn berries. India has potential for producing 1200–7000 thousand liters of seabuckthorn juices currently.

Seabuckthorn has adapted well in Himalayan highland particularly at an altitude of 2500 m msl–4000 m msl (or 8000 ft–10,000 ft msl). As the plant itself is hardy and survives in most soil types, it can be expanded many folds. The inhabitable hilly terrain having favorable altitude could be explored for expanding the production areas of the plant. As of now cost of production of this plant in India is almost nil as no inputs such as irrigation, weeding fertilizers, manures, etc., are required. Hand harvesting is commonly performed by the collector/farmer themselves. There is huge scope of socioeconomic development of the people in Leh, North Sikkim in particular through Seabuckthorn production with the intervention of State and Central Government in promoting Seabuckthorn.

The advantages of establishing Seabuckthorn production and processing industry in Sikkim and Ladakh are favorable soil and climate, vast land which cannot be used for any other agricultural crops, relatively cheap labor, water, and power at reasonable price and enthusiastic people who already have utilized the plant and possess indigenous knowledge of the plant and its uses. Ready consumers will be available for the products developed from the plant.

Though highly plausible there are numerous constraints in establishing viable and sustainable seabuckthorn-based industry. Perhaps the foremost is lack of R&D support as well as the prevailing forest and other laws of the states. Also, there is no definite technology for the Indian varieties and the available technologies were not tested in Indian conditions. Many prior researches have documented the varietal and geographical differences in nutritional qualities of seabuckthorn; nobody is sure of the quality of the Indian Varieties. Also suitable seabuckthorn cultivar has not been identified for Indian conditions for multiplications. India could be one of the major havens for seabuckthorn if the issues were addressed. Based on the information, it is found that the SB needs some R&D for its commercialization. The various researchable issues are: establishing the species available in India; biochemical and phytonutrient profile of the berries; animal model experiments to establish the various medicinal properties of the berries; commercially successful value-added products and waste utilization.

8.12 CONCLUSION

Seabuckthorn is a high altitude plant with exceedingly nutritious, multifaceted plant. Various products range from specialized food supplements, cosmetics to nutraceuticals are widely consumed in parts of Europe and Asia. Its popularity has been increased manifold in recent times with the event of functional and nutraceutical industries. India is a huge untapped market for the products of seabuckthorn; there will be immense scope for processing industries. The nutritional qualities of seabuckthorn studied so far in nook and corner of the world promise healthy fatty acids and ascorbic acids and other healthy nutrients. The seabuckthorn berry with considerable antioxidant properties could be marketed as health drinks or sports drinks.

There is massive potential for developing seabuckthorn production and processing in India and this helps in upliftment of socioeconomic status of the natives. But ample R&D support and appropriate government policy will be essential for developing such industry in India. Immediate steps must be taken up for sustainable utilization of this unique gift of nature. Seabuckthorn could be a game changer in the cold deserts of Indian Himalayas which is still a very much untapped valuable resource.

KEYWORDS

- **bioactive**
- **nutraceuticals**
- **processing**
- **antioxidant**
- **hipophae**

REFERENCES

Alam, Z. Chemical and Nutritional Constituents of Seabuckthorn Juice. *Pakistan J. Nutr.* **2004,** *3* (2), 99–106.

Ali, A.; Kaul, V Seabuckthorn: A Valuable Resources of the Cold Dessert (Ladakh). ENVIS Bulletin: Himalayan Ecology, 2011; vol 19.

Annonymous. Seabuckthorn – A Promising Multi – Purpose Crop for Saskatchewan. *Agri. Agri. Food. Canada* **2000,** 1– 8.

Bal, L. M.; Meda, V.; Naik, S. N.; Satya, S. Seabuckthorn Berries: A Potential Source of Valuable Nutrients for Nutraceuticals and Commaceuticals. *Food Res. Int.* **2011,** *44,* 1718–1727

Bernath, J.; Foldesi, D. Seabuckthorn (*Hippophae rhamnoides* L.): A Promising New Medicinal and Food Crop. *J. Herbs Spices Med. Plants* **1992,***1* (1/2), 27–35.

Beverridge, T.; Li, T. S. C.; Oomah, B. D.; Smith, A. Seabuckthorn Products: Manufacture and Composition. *J. Agris. Food Chem.* **1999,** *50* (1), 113–116.

Ciesarovaa, Z.; Murkovicb, M.; Cejpekc, K.; Krepsd, F.; Tobolkovaa, B.; Koplikc, R.; et al. Why is Sea Buckthorn (*Hippophae rhamnoides* L.) so Exceptional? A Review. *Food Res. Int.* **2020,** *133,* 109170.

Gao, X.; Ohlander, M.; Jeppsson, N.; Bjork, L.; Trajkovski, V. Changes in Antioxidant Effects and their Relationship to Phytonutrients in Fruits of Seabuckthorn (*Hippophae rhamnoides*, L.) During Matiuration. *J. Agric. Food Chem.* **2000,** *48* (50), 1485–1490.

Hardy, G. Nutraceutical and Functional Foods: Introduction and Meaning. *Nutrition* **2000,** *16* (7 -8), 688–697.

Jaroszewska, A.; Biel, W.; Telesiński, A. Effect of Mycorrhization and Variety on the Chemical Composition and Antioxidant Activity of Sea Buckthorn Berries. *J. Elementol.* **2018,** *23* (2), 673–684. https://doi.org/10.5601/jelem.2017.22.3.1434.

Kallio, H.; Yang, B. R.; Peippo, P. Effects of Different Origins and Harvesting Time on Vitamin C, Tocopherols, and Tocotrienols in Seabuckthorn (*Hippophae rhamnoides*) Berries. *J. Agri. Food Chem.* **2002,** *50*, 6136–6142.

Lanev, E. The Effect of an Extract of Seabuckthorn on the Healing of Experimental Skin Wounds in Rats. *Dermatology* **1995,** *48* (3), 30–33.

Li, F.; Guo, T. In *Application of Hippophae rhamnoides L. in Tibetan Medicine*, Proceedings of International Symposium of Seabuckthorn (*H. rhamnoides)*; Xian, China, 1989; pp 409–412.

Li, T. S. C.; Schroeder, W. R. Seabuckthorn (*Hippophae rhamnoides* L.): A Multipurpose Plant. *Hort. Technol.* **1996,** *6*, 370–380.

Lu, R. In *The Chemical Composition of Hippophae Fruits in China*, Proceedings of International Symposium on Seabuckthorn (*Hippophae rhamnoides* L.); Barnaul, Russia, 1993.

Maheshwari, D. T.; Yogendra Kumar, M. S.; Verma, S. K.; Singh, V. K.; Singh, S. N. Antioxidant and Hepatoprotective Activities of Phenolic Rich Fraction of Seabuckthorn (Hippophae rhamnoides L.) Leaves. *Food Chem. Toxicol.* **2011,** *49*, 2422–2428. http://dx.doi.org/10.1016/j.fct.2011.06.061

Masoodia, K Z.; Wania, W.; Dara, Z. A.; Mansoora, S.; Anam-ul-Haqa, S.; Farooqb, I.; et al. Sea buckthorn (*Hippophae rhamnoides* L.) Inhibits Cellular Proliferation, Wound Healing and Decreases Expression of Prostate Specific Antigen in Prostate Cancer Cells *in vitro*. *J. Funct. Food* **2020,** *73*, 104102.

Menrad, K. Market and Marketting of Functional Food in Europe. *J. Food Eng.* **2003,** *56* (2-3), 181–188.

Mironov, V. A. In *Chemical Composition of Hippophae rhamnoides of Different Populations of the USSR*, Proceedings of International Symposium on Seabuckthorn (*H. rhamnoides* L.); Xian, China, 1989.

Oomah, B. D.; Mazza, G. Health Benefits of Phytochemicals from Selected Canadian Crops. *Trends Food Sci. Technol.* **1999,** *10* (6-7), 193–198.

Parimelazhagan, T.; Chaurasia, O. P.; Ahmed, Z. Seabuckthorn: Oil with Promising Medicinal Value. *Curr. Sci.* **2005,** *88* (1), 8–9.

Pundir, S.; Garg, P.; Dwivedi, A.; Ali, A.; et al. Ethnomedicinal Uses, Phytochemistry and Dermatological Effects of Hippophae Rhamnoides L.: A Review. *J. Ethnopharmacol.* **2021,** *266*, 113434.

Rousi, A. The Genus *Hippophae*, L. – A Taxonomic Study. *Ann. Bot. Fenn.* **1971,** *8*, 177–227.

Wani, T. A.; Wani, S. M.; Ahmad, M.; Gani, A.; Masoodi, F. A. Bioactive Profile, Health Benefits and Safety Evaluation of Seabuckthorn (Hippophae rhamnoides L.): A Review. *Cogent Food Agric.* **2016,** *2* (1), 1128519.

Xing, J. Effects of Seabuckthorn Seed and Pulp Oils on Experimental Models of Gastric Ulcer in Rats. *Fitoterapia* **2002,** *73* (7-8), 644–650.

Yang, B.; Kallio, H. Composition and Physiological Effects of Seabuckthorn (Hippophae) Lipids. *Trends Food Sci. Technol.* **2002,** *13* (5), 160–167.

Yao, Y. Effects of Temperature on Vitamin C Concentration and Yield of Seabuckthorn (*H. rhamnoides*) Fruit: Optimal Time of Fruit Harvest. *Agri. Food Sci.* **1993,** *2*, 497–505.

Zhang, P.; Yan, J.; Duo, B.; Ren, B.; Guo, J. In *Preliminary Study of Biochemical Constituents of Seabuckthorn Berries Growing in Shanxi Province and their Changing Trend*, Proceedings of International Symposium on Seabuckthorn (*H. rhamnoides* L.); Xian, China, 1989; pp 96–105.

Zhao, H.; Zhu, C.; Li, H.; Liu, H.; Sun, W. Geographic Variation of Fruits of Chinese Seabuckthorn and Selection Provseness for Fruit. *Hippophae* **1991,** *4*, 15–18.

Zheng, L.; Shi, L. K.; Zhao, C. W.; Jin, Q. Z.; Wang, X. G. Fatty Acid, Phytochemical, Oxidative Stability and In Vitro Antioxidant Property of Sea Buckthorn (*Hippophae rhamnoides* L.) Oils Extracted by Supercritical and Subcritical Technologies. *LWT Food Sci. Technol.* **2017,** *86*, 507–513. https://doi.org/10.1016/j.lwt. 2017.08.042.

Zhou Y. Study on the Effect of *Hippophae* Seed Oil Against Gastric Ulcer. Institute of Medicinal Plants Resource Development, The Chinese Academy of Medical Sciences, Beijing, 1998.

CHAPTER 9

Perspectives and Potential of Indian Organic Basmati Rice

ANKUR SAXENA and ASHUTOSH SINGH

College of Agribusiness Management, G. B. Pant University of Agriculture and Technology, Pantnagar, Uttrakhand, India

ABSTRACT

A vast majority of people in the country and abroad have developed liking for basmati rice. India is the world's largest producer of aromatic and premier variety basmati rice. The lack of technical knowhow, non-availability of technology, low demand of local market, and credit facility at reasonable rate of interest were found major constraints. The study suggest two pronged strategies, that is, (1) continue training programs by state department of horticulture and food processing, (2) government department should facilitate the sale of organic basmati. Apart from India, all the major rice growing countries have their own traditional and evolved basmati varieties. India being the big player in basmati export, the market competitiveness of the product has not been given its due importance so far. Export competitiveness is related to the country's ability to compete within export markets, maintain economic growth and employment. The demand for the increase in export competitiveness mostly is based on the necessity to consider the globalization challenges, affecting the competitiveness, and apply adequate decisions regarding the increase of export competitiveness. Uttarakhand Organic Commodity Board (UOCB), a government supporting institution provides backward and forward linkages though both state and central government should promote aggressively in global market.

High-Value Plants: Novel Insights and Biotechnological Advances. Thangjam Anand Singh, Prakash Kumar Sarangi, and Ch. Brajakishore Singh (Eds.)
© 2025 Apple Academic Press, Inc. Co-published with CRC Press (Taylor & Francis)

9.1 INTRODUCTION

In the International market, rice is traded under two main categories, such as fragrant and non-fragrant.

TABLE 9.1 Share of Organic Basmati in Total Domestic Organic Product Sale.

S. No.	Product	Sale (tons)	Percentage share
1	Fruits and vegetable	1801	28.51
2	Basmati rice	1178	18.65
3	Tea	945	14.96
4	Wheat	900	14.25
5	Pulses	678	10.73
6	Coffee	546	8.64
7	Spices	267	4.22
	Grand total	6315	

Source: ORG-Marg and rice India.

9.2 EXPORT COMPETITIVENESS, CONCEPT, AND IMPORTANCE

Export competitiveness can be measured in terms of cost of products, reliability of producers, and delivery schedules including other aspects like image of the country/company and brand equity. Collectively, these factors define the competitive aspect of product in free market.

The detailed analysis of export competitiveness requires systematic consideration of inter-related factors, which form an overall export competitiveness. Export competitiveness can be measured in several ways viz., analyzing one or various factors of the country's export, creating composite indices, factors which stimulate the international trade, etc. One of the most frequently applied competitiveness measurement methods involves measurement by index. This index is described as an instrument artificially created for the analysis of multi-aspect problem, consisting of sub- indicators and based on which the analyzed objects can be rated (Sneiska and Bruneckiene, 2019). In modern practice, the indices for the measurement of export specialization are applied. However, theoretical and practical interpretations on their applicability have neither been formed nor comprehensively discussed by researchers yet. The proposed study will attempt to identify and analyze the factors felt important for increasing the export competitiveness of Uttarakhand organic basmati rice.

9.3 BASMATI PRODUCTION

As is clear from Table 9.2 that Punjab contributes about 40% to total production out of 18% of total basmati area as compared with Haryana, which contributes about 42% to total production out of the 50% area. This is because Punjab has the highest productivity in India for basmati. Major basmati rice producers in India include the Union territory of Jammu and Kashmir, Himachal Pradesh, Punjab, Haryana, Delhi, Uttarakhand, and Uttar Pradesh. Out of these organic basmati initiative has been taken up in part of western UP, Punjab, Haryana, parts of Himachal Pradesh, Uttarakhand, and Jammu and Kashmir only.

TABLE 9.2 State-Wise Basmati Production (lakh tons).

States	2014–2015	2015–2016	2016–2017	2017–2018	2018–2019	2019–2020
Punjab	31.65	28.31	28.32	22.82	22.92	34.98
Haryana	24.32	27.52	26.77	22.61	28.98	37.01
Uttar Pradesh	9.86	14.43	20.66	14.28	12.70	12.60
Uttarakhand	0.58	0.97	0.80	0.53	0.54	0.66
J & K	0.79	0.79	0.94	0.96	0.92	2.40
Himachal Pradesh	0.00	0.10	0.43	0.05	0.03	0.02
Delhi	0.05	0.40	0.06	0.06	0.40	0.03
Total	67.27	72.17	77.98	61.33	66.16	87.73

Source: www.airea.net and BEDF.

9.4 EXPORT OF BASMATI RICE FROM INDIA

The UAE, Saudi Arabia, and other gulf countries are the major importers of Indian basmati rice contributing to about 70% of the total export. Organic food exports increased from to 78 m$ (2014–2015) to 100 m$ in (2015–2016). Out of which the export of organic basmati rice was 5630 MT (2015–2016) (Singh, 2016). During 2015–2016, India exported 86 items with total volume of 37,533 MT, valued around Rs. 498 crores worth and cotton (43% contribution) leads (16,503 MT) among the products exported followed by basmati rice which contributes for 15% of export volume, 13% of export value amounting to Rs. 59.2 crores. (APEDA, 2016).

9.5 STATUS OF ORGANIC BASMATI IN UTTARAKHAND

Diversified Agriculture Support Project (DASP) introduced the concept of organic farming in select regions of the state in 1998. Dehraduni Basmati

Project was started in May 2002 by DASP. It was a serious attempt to increase the production of Dehraduni Basmati by motivating the farmers to cultivate it and in turn helping them in realizing premium price for the produce. After completion of the DASP in March 2004, the organic basmati Production Program was adopted by UOCB and was renamed Organic Basmati Export Program (OBEP). A total of 3600 tons of organic basmati paddy were produced in both areas in 2016. In Uttarakhand, the area is reported to be 1400 ha and expanding (UOCB). In India, as a whole, about 44,335 MT of rice is produced out of 18,134 ha area under organic management. Table 9.3 depicts the year-wise total Basmati rice produced in the state.

TABLE 9.3 Area, Production, and Productivity in Uttarakhand.

Year	Area (ha)	Production (MT)	Productivity
2010–2011	273,662	540,128	19.74
2011–2012	259,805	523,203	20.14
2012–2013	275,711	528,524	19.17
2013–2014	270,608	551,716	20.39
2014–2015	270,416	495,083	18.31
2015–2016	261,319	544,497	20.84
2016–2017	255,037	561,339	22.01
2017–2018	246,959	551,758	22.34
2018–2019	247,698	561,856	22.68
2019–2020	25,0297	585,921	23.41

Source: Directorate of agriculture, Dehradun.

TABLE 9.4 District -Wise Production of Organic Basmati.

District	Area		Production		Productivity	
	2018–2019	2019–2020	2018–2019	2019–2020	2018–2019	2019–2020
Dehradun	9179	10,095	17,672	21,283	19.25	21.00
Haridwar	12,985	14,715	28,856	31,436	22.22	21.36
U.S. Nagar	101,482	102,280	330,419	347,132	32.55	33.93
Nainital	11,597	10,165	32,544	28,793	28.06	28.00

9.6 REGIONAL ISSUES FACED BY BASMATI RICE EXPORTERS

The basmati paddy raw material in Uttarakhand is not sufficient for fulfilling round the year demand of the rice export industry. Exporters from Uttarakhand have to source from the neighboring mother state Uttar Pradesh to fulfill their requirement. As per the export policy of Uttar Pradesh, rice exporters in Uttar Pradesh are exempted from the mandi fee. But if an exporter in Uttarakhand procures the same paddy in Uttar Pradesh, he has to pay a mandi fee of 2.5%, even if the Basmati is for export. This double taxation is hampering competitiveness of basmati exporters of Uttarakhand.

Uttarakhand has a Basmati Export Zone in U.S. Nagar and the contiguous districts of Uttar Pradesh (Bareilly, Pilibhit, and Rampur), from which zone paddy is procured. It also falls under the Basmati Export Zone.

For exports, exports need phytosanitary certificate and the cost of obtaining this certificate is Rs. 100. The certifying official must be called from Delhi for inspection and the export company must meet all expenses of calling him. the expenses would come to approximately Rs. 3000–4000. Moreover, exporters require this certificate every time a consignment needs to be shipped. Exporters need independent and private quality testing lab to be setup in nearby area to settle disputes arising between farmers and exporters regarding the quality of rice. If exporters are making direct purchases from mandi, they are required to pay the mandi fee of 2.5%, which makes them less competitive. Exporters have to pay a sale tax of 4% of inter-state selling of basmati rice, even if the purchased basmati is finally meant for exports. Even the Uttarakhand to get exemption from mandi fee for basmati rice meant for export, exporters have to get prior permission for deciding on quotas for a year by submitting copies of the export orders in advance. Moreover, permission procedure takes 3–4 weeks for approval making this a great problem for the exporters. Interrupted power supply results in additional operating cost of Rs. 10,000 per hour. Moreover the bigger problem is that, the power breakdown occurs without prior information due to which sudden shut down occurs, which causes damage to the expensive machines. If the prior information of power breakdown is given by the electricity department, then the systematic shut down of machines can be done and damage to machines and raw materials can be minimized. For instance, due to frequent interrupted power supply, the percentage of broken rice increases, thereby reducing the quality of rice. Also on restarting, the efficiency of machines decreases. Exporters also face a lot of problem of DNA testing of rice as presently, this facility is available only in Hyderabad, and it becomes a cumbersome and time-consuming process to send samples for testing to Hyderabad and wait for the result.

Inland container depot (ICD) is located in Rudrapur, which is so far from Dehradun and Haridwar, so the exporters must send consignments to Moradabad in Uttar Pradesh, which costs Rs. 3000–4000 per container additionally. If this container facility will be provided in Dehradun, then this expenditure can be saved. Also, there shall be single window contact, information and facilitation for providing all information and escort services to the entrepreneurs in the state.

9.7 CURRENT STATUS AND CONCERNS IN UTTARAKHAND

9.7.1 SHRINKAGE IN AREA

In the Uttarakhand districts of Dehradun, Haridwar, Pauri, Tehri, and Almora, most of the organic rice growing farmers are few or marginal food security is the foremost concern for them. They have now better technological options like improved high-yielding varieties, both non-aromatic and mild aromatic types. Alternate and more profitable cropping systems have also evolved. Sugarcane-based cropping system, for example, has become prevalent in Dehradun and Haridwar belts often without rice in the sequence. Many areas close to cities are now used for housing and other non-agricultural uses.

specifically speaking, U. S. Nagar, which is having more ideal environment for basmati cultivation, Tarai Development Corporation (TDC) has made a contract type arrangement with farmers for seed production which occupies about half of the rice area for seed production only.

9.7.2 INCREASING DEMAND FOR LONGER GRAIN TYPE

Since late seventies and early eighties, the international rice market has undergone a sea change. The demand for long-grain non-aromatic and aromatic rice, even with a mild scent, but extra-long slender grain, has increased. Thus, there is now a less demand for once well-known traditional basmati varieties. In northern India, Basmati 370, Type 3, and Hansraj are the most popular and widely grown varieties. At present, Pusa Basmati-1 and Taraori Basmati are fast replacing these traditional aromatic varieties, especially by virtue of their extra-long slender grains, although, except for the grain length, other quality parameters are almost same or even inferior to Basmati 370 or Dehraduni basmati. Nothing was done in the past to popularize high-quality small- and medium-grained aromatic rice on domestic or international market.

9.7.3 DECLINE IN QUALITY

According to the farmer's perception, the basmati rice is losing aroma. different reasons for decline in aroma and grain quality include several factors, such as varietal impurity, use of chemical fertilizers, soil factors including micronutrient deficiency, cultivation practices, global warming, lack of incentives to farmers, and poor market support (Singh et al., 2000). However, one of the major factors for seed admixtures was the lack of systematic seed production program and maintenance breeding of the prevalent scented varieties in the state.

9.7.4 AROMA CRISIS

Indigenous aromatic rice is gradually losing their *bas* (aroma). This is the unanimous opinion of farmers. Some of factors for decline in aroma are:

- **Cultivation practices:** Although there is no experimental evidence, farmers perceived that there is better aroma in direct sown rice crop than in transplanted ones, particularly in Hansraj variety.
- **Rising temperature:** Both in scientific and farmers circles, it is accepted that comparatively lower temperature (20–27°C) at the time of flowering or grain filling engages the aroma content of seeds. However, this critical temperature may vary in different varieties. Gradual increase in temperature, deforestation, and Industrialization may have affected aroma content.

9.7.5 RAPID LOSS OF THE GERMPLASMS

Valuable germplasm of scented rice like basmati has been gradually lost in the course of time as there is no conservation strategy currently. Thus, in many of the areas, the number of high-quality rice germplasm which were in cultivation till a few years ago, are now extinct.

9.7.6 MARKETABILITY

Small and medium grain aromatic rice commands a very limited market base and fetches lower prices than the Basmati type. In a few cases, their prices could be equal to or even higher than the basmati type but the farmers and consumers are aware of their qualities. However, in general, long-grain aromatic rice is more attractive to the consumers. Again no concrete plan is attempted either by state or central government.to explore the export market for small and medium grain aromatic rice, which are superior to the basmati in all characteristics except grain length

9.7.7 *POLICIES*

- Some incentives and minimum price support to the farmers of native aromatic rice grower are suggested to grow native aromatic rice than shifting to other crops. A slight increase in basmati price as low as Rs. 3/kg shall change the basmati cultivation more lucrative venture even more profitable than sugarcane farming.
- Conservation rather than replacing these native varieties is the need of the hour. Government machineries ought to be more pragmatic in their outlook and approach. So far their major orientation has been to increase the productivity (quality?) and short-term profitability.
- Seed production and distribution programs for local rice cultivars.
- Restricting sale of land for non-agricultural use in prime scented rice growing areas must be the strictly enforced.

Exploring/developing the international market for organic basmati. This should be primarily the job of AIREA and APEDA.

9.8 EXPORT PERFORMANCE FROM INDIA

The growth in export of Basmati rice from India in terms of quantity and value has been steadily increased after the commencement of organic farming. It is quite evident that except for few years, the export of Basmati in terms of both quantity and value progressed well. The study revealed that there was a significant increase in the export of Basmati rice from India in terms of quantity as well as in value.

9.8.1 *ANALYSIS OF GROWTH*

The growth trends were estimated for the total Basmati exports for the period 1999–2019 by exponential function. It can be observed from Table 9.5 that the quantity and value of total Basmati exports registered a positive and significant growth in their exports with compound growth rate of 4.44% and 16.65%, respectively.

TABLE 9.5 Compound Growth Rate of Total Basmati Exports.

Particular	Basmati rice export	
	Quantity	Value
Intercept	4.62	2.08
Slope	0.018	0.076
	(0.004)	(0.003)
Growth	4.441*	16.655*
	(0.049)	(0.103)

*Significant at 5% level.

TABLE 9.6 Compound Growth Rate of Basmati Export from India.

Time duration	Export quantity	Value
1999–2009 (before commencement of Organic Farming)	11.45	18.76
2010–2019 (after commencement of Organic Farming)	17.68	35.78
Overall (1999–2019)	4.44	16.65

Significant at 5% level.

TABLE 9.7 Instability in Export of Basmati.

Year	Export quantity		Value	
	CV%	Instability index	CV%	Instability index
1999–2009 (before commencement of Organic Farming)	73.53	65.47	64.87	51.70
2010–2019 (after commencement of Organic Farming)	66.47	60.29	101.70	71.56
Overall (1999–2019)	95.89	86.70	110.78	98.40

Significant at 5% level.

During 1999–2019, the value of coefficient of variation was 95.89%. The nature of variation in export value was found to be quite reverse compared with variation in export quantity. The coefficient of variation in export value was 110.78% which was quite high as compared with CV% of export quantity, which indicates that the variation was found higher for export value then export quantity which might be due to change in international price, trade, and tariff.

9.8.2 COMPETITIVENESS OF INDIAN BASMATI

9.8.2.1 NOMINAL PROTECTION COEFFICIENT (NPC)

Competitiveness is the ability to produce and distribute product and/or services that can compete in the international markets, and which simultaneously increase the real income and living standards of its citizens. The significance of Basmati exports to the national exchequer was growing rapidly.

The competitiveness of the Indian Basmati is calculated on the basis of nominal protection coefficient (NPC) which is the ratio of domestic price (Pd) and international price (Pb). As already mentioned the ratio less than 1 indicates competitiveness. From Table 9.8, it is clear that Indian Basmati rice is competitive enough because the ratio turns out to be less than 1. In this case, the domestic price is taken as the average auction price at APMC.

Export price, domestic price, and NPC for Basmati are reported in Table 9.8 for the year 2013–2020. The NPC values in 2018–2019 and 2019–2020 were found to be 0.901 and 0.926, respectively indicating that the commodity could be exported.

TABLE 9.8 NPC for Indian Basmati.

Year	FOB price, Pb (Rs)	Domestic price, Pd (Rs./qtl)	NPC = Pd/Pb
2013–2014	7384	6000	0.812
2014–2015	7843	6200	0.790
2015–2016	8139	6100	0.749
2016–2017	8788	7050	0.802
2017–2018	8543	7600	0.889
2018–2019	8762	7900	0.901
2019–2020	8907	8250	0.926

9.8.2.2 DOMESTIC RESOURCE COST (DRC)

From DRC, resource cost ratio (RCR) was obtained which was taken as a measure of comparative advantage at a given point of time and these are presented in Table 9.9.

The marginal value of non-traded inputs of Basmati Farming system accounted to 47.72 as a result of the premium price of Saudi Arabia, the RCR ratio appeared to be lowest in Saudi Arabia market in the case of Basmati

(0.18) and organic basmati (0.25) and commanded a substantial comparative advantage over the remaining countries, regardless of the farming system.

The RCR ratio was found to be greater than 1 showing a comparative disadvantage in exporting the Basmati rice.

Generally, consumer preference and product differentiation lead to price differentials across markets. At any point of time, the relevant price of basmati rice in any country depends on the quality of the Basmati produced for export as well as that consumed in the market. Thus, it is clear that India has a larger comparative advantage in exporting Basmati rice to Saudi Arabia than to UAE and Iran markets largely because of the premium price received in the Saudi Arabia market.

TABLE 9.9 Resource Cost Ratio (RCR) Indices in Organic Basmati Production System.

	Particular	Value (Rs)	
		Basmati	**Organic basmati**
1	Seed	6.82	7.09
2	Labor	30.89	40.67
3	Fertilizer or green manure	10.01	16.46
A	Total	47.72	64.22
B	International reference price in Saudi Arabian market (US$/kg)	3.73	3.73
C	Exchange rate (1$ = Rs.)	68.07	68.07
D	Domestic resource cost = (A/B)	12.79	17.21
E	Resource cost ratio = (D/C)	0.18	0.25
F	International reference price in UAE market (US$/kg)	3.64	3.64
G	DRC = A/F	13.10	17.64
H	RCR = G/C	0.19	0.26
I	International reference price in Iran market (US$/kg)	3.18	3.18
J	DRC = A/I	15.00	20.19
K	RCR = J/C	0.22	0.29

9.8.2.3 EXPORT PERFORMANCE RATIO

The export performance ratio/revealed comparative advantage (RCA) and revealed symmetric comparative advantage (RSCA) for export in basmati rice and organic basmati were estimated to compare the export competitiveness of India and have been presented in Table 9.10.

The perusal of the table reveals that RCA in basmati rice was much higher, and unity and RSCA were positive and higher. This indicates that India was highly competitive in Basmati rice exports in the world during the period.

TABLE 9.10 RCA and RSCA of Basmati Rice.

Period	RCA			RSCA		
	Basmati	**Organic basmati**	**Total**	**Basmati**	**Organic basmati**	**Total**
TE 2013	11.54	5.58	10.74	0.73	0.54	0.71
TE 2016	14.16	10.76	12.84	0.75	0.71	0.75
TE 2019	18.72	21.73	19.05	0.80	0.81	0.80

Note: Data are for triennium ending (TE) average.

9.9 DEVELOPMENT OF UNIQUE SUPPLY CHAIN MODEL

In the context of mountain area, due to the mountain specification and inherent inbound and outbound logistics, the development of supply chain is always being a major challenge in any market linkages. The proposed model has been developed in such a way that layers in the supply chain are minimum without scaling down the vertical, horizontal, and backward linkages. Around 500 biovillagers will be interwoven directly into the supply chain, and there will also be good scope for expansion without increasing the layer. The major players in the chain will be farmers, producer groups, master trainers, supply chain officer, and retailers. Master trainers will act as connecting link among biovillages, UOCB or COF, and markets. This is a kind of model where information flows both up and down.

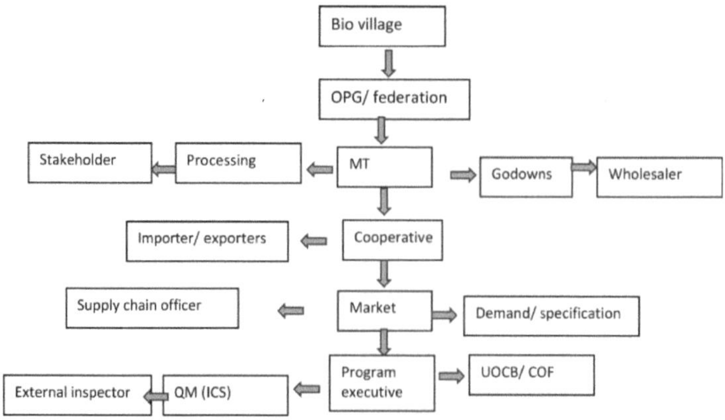

FIGURE 9.1 Unique supply chain model.

9.9.1 DIRECT MARKETING

A center for organic farming should develop a retail format, that is, retail outlet. It should be named related to organic such as "Organic harvest or *Harra Bhara*." The produce should be displayed in the counter with full address of where it has been cultivated. This means that it should contain all the information, including price, place of production, farmers name with address, and specific health information.

9.9.2 SWOT ANALYSIS BASMATI PROCESSING INDUSTRY

The underutilization of processing capacity has been meager. Thus, the SWOT analysis is undertaken to assess the strength, weakness, opportunities, and threats in the area. The results of the SWOT analysis of basmati processing industries are presented in Table 9.11.

TABLE 9.11 SWOT Analysis Basmati Processing Industry.

Strengths	Weakness
Superior technology	No brand promotion
Well-trained employee	Lack of well-developed distribution channel
Institution collaboration	Low brand awareness
Member of APEDA, AIREA FIEO	Low brand recognition
High-quality rice	
More nutritive rice	
Opportunities	Threats
Huge domestic market	Strengthening of distribution channel by the competitors
Absence of technologically superior rice mill	More promotional efforts by the competitors
Tie-up with organized retail store like HKB, VMM, and TKB	Establishment of more technologically superior rice mill with well-structured distribution network
Educated people	
High literacy ratio	
High per capita income	

9.9.3 STRATEGIES TO BE ADOPTED BY EXPORTERS

After identifying the constraints, exporters and the government must adopt these strategies so that they can trade organic basmati in an easy way. The price, quality, availability, logistics, and certification are the most focused areas, which need more attention.

9.9.3.1 STRATEGIES

The Uttarakhand government should carry out different measures in order to improve the trade image of Uttarakhand. According to APEDA, the demand for Indian organic basmati rice is increasing. Various export markets have great interest in Indian organic basmati rice.

In order to increase and improve the interest in the Indian organic (organic basmati rice) products, APEDA, UOCB, and other Indian institutions (e.g., Ministry of Agriculture and farmer welfare) should carry out the following measures.

- Uttarakhand lack in quality warehouse to store organic products, and infrastructure should be developed in order to minimize the postharvest loss of the product.
- Training programs for producers and certification agencies.
- Improving the quality of research and development.
- The National Center of Organic Farming set up by the Ministry of Agriculture and Farmer Welfare in Ghaziabad aimed at improving the different aspects of organic production should focus on whole chain, backward to forward linkages.
- Promoting certification programs.
- Developing national policies for organic production.
- Improving the quality of products, packaging, logistic, infrastructure, and technical support.
- Promoting Indian organic product at international fairs and promotional program in the media.
- Application for recognition under different countries' national organic regulation.
- Developing market links, such as consumer awareness program, marketing fairs, and Green restaurant.
- Development of unique supply chain model with emphasis on organic products.
- Introduction of direct marketing and brand promotion.
- Value addition and product development.
- Timely availability of organic inputs.
- Compensation for loss of profit during "0" and "1" year in conversion period.

- The government should provide financial support to those farmers who are willing to change from chemical to organic farming at least for first 2 years.
- The Biotourism concept should be implemented in relation to organic farming so that at an international level, awareness about the state as well as its organic products could increase.
- The government should introduce the ITC model for the procurement of organic products. So the sellers as well as buyers should be benefited equally by eliminating middle men involvement.

9.9.3.2 STRATEGIES FOR FARMER'S RESPONSE

- To organize small farmers to act as an organization that produces and markets their own products for the domestic and export markets.
- The APMC act would reduce dual taxation to the exporters by contract farming, and both buyers and producers will get benefits.
- Planning training programs for farmers and NGOs not only in organic agriculture method (production, harvest, postharvest techniques, and basic standards, etc.) but also in how to sell, promote, and diversify their markets and how to fulfill certification requirement (internal control and administration).
- To include small farmers (consortium) in the operative organizations of the whole chain (from production over processing to final sales outlet).
- Farmer should have access to financial support for organic production, administration, group certification, and marketing programs for their products.
- To offer continuous organic market information to farm organizations and direct contact to possible buyers.
- To promote organic products in the national and export markets where farmers can participate, for example, national and international fairs, commercial mission to specific export countries.

9.9.3.3 STRATEGIES FOR TRADER'S RESPONSE

- To organize commercial mission to specific countries in order to meet potential customer, improve and enhance relationship and to understand the market.
- To offer continuous organic market information and direct contact to possible buyers.

- Training program for traders in house to promote and sell organic products in the domestic and international market (e.g., improve client follow-up system, reliability, and to understand the customer).

9.9.3.4 STRATEGIES FOR CERTIFICATION

- National Certification Agency (Participatory Guarantee System (PGS)) provides compressive policy price (reduction in cost of certification), standardization, and simplified certification procedure with less documentation, it should play a major role in the domestic market.

9.9.3.5 STRATEGIES FOR GOVERNMENT END

- Improve organic national rules and regulation through PGS system.
- Achieve recognition on the international level.
- Promotion of organic farming through training programs, financial support, and subsidies.
- Improve the image of Indian organic products and trade (domestic and export promotion).
- Facilitate the implementation of a producer, traders, and processors national database.

Suggestions given in this chapter are based on the study conducted for Agri Export Development Unit. Suggestions to AEZ of Basmati rice are given in three areas, namely, general suggestions, suggestions to maintain supply chain, suggestion to deal with problems of Basmati rice exporters.

9.10 GENERAL SUGGESTIONS

1. The AEZ may train farmers to adopt the integrated crop management (ICM) to enable them to produce more rice from lesser and lesser input, thus making them competitive in the national and international markets.
2. Although, a large proportion of our rice cultivation is already under a sort of organic farming, the farmers do not apply much of the

external inputs. Also keeping in mind the burgeoning demand for organic food in European and other developed countries, the AEZ and agricultural universities may educate the farmers about the scientific methods of organic farming for basmati rice.

3. In the light of new trade opportunities setup by the market economy both at the national and international level, the government needs to consider the issues, such as geographical indication, enhanced productivity, contract farming, establishing referral laboratories, laboratories for residue analysis, and modernizing our processing units.
4. AEZ may give more focus on demonstration at farmers' field the IPM modules developed for basmati rice.
5. AEZ may organize trainings for workers in rice mills.
6. xtension services (training and education) may be provided to basmati growers and extension workers.
7. AEZ may take up coordination and provide technical expertise in the establishment of agriclinics.
8. AEZ may promote and facilitate production of only those varieties which are accepted in the international market in basmati.
9. AEZ may promote periodical replacement of existing varieties with new varieties having better disease and pest resistant, higher yielding, and better all-round basmati traits.
10. AEZ may monitor strict quality control at all stages, from seed to postharvest handling and export for the farmers.
11. AEZ may ensure regular exports to existing markets and development of new markets.

9.10.1 SUGGESTION TO MAINTAIN THE SUPPLY CHAIN

1. Exporting units and AEZ may sign agreements with selected farmers for participation under contract farming.
2. AEZ may facilitate setting up a market monitoring and development cell for constant study and data collection of existing markets and development and opening up of new markets.
3. AEZ may set up an independent quality monitoring facility equipped with latest techniques like DNA fingerprinting preferably at G.B. Pant University of Agriculture and Technology, Pantnagar, India.

4. Wide publicity about AEZ for basmati may be made in target districts through newspaper, handbills, tehsils, blocks, seed stores, etc., by the government and nodal agency.
5. AEZ may select interested farmers at village level and survey their lands regarding the suitability for organic basmati rice cultivation.
6. AEZ may do effort to take up production of only one variety of basmati in whole village to avoid admixtures and maintain the quality of basmati.
7. AEZ may identify areas for organic basmati cultivation.
8. AEZ may look that all benefits, subsidies, and incentives from state and central governments or from other agencies must be made available to all farmers on priority under single window basis with least paper work.
9. AEZ may arrange/provide cost of cultivation reducing and quality enhancing equipments (on custom hiring bases), such as Rotavators, Power tillers, Transplanters both manual and mechanicals, and Reapers/threshers on subsidized rates in time with least paper work.
10. Exporting units and AEZ may take steps to ensure buy-back of the farmers produce at agreed price by the designated agency under contract farming arrangement for initial few years.

9.10.2 SUGGESTION FOR ENHANCING EXPORT

1. AEZ may take up the matter to concerned authorities for setting up an office or appointing an officer in the state itself who has power to provide Phytosanitary Certificate.
2. AEZ may take up the matter to discussion with concerned department for setting up of an independent Hi-Tech Quality Testing Laboratory where tests like DNA testing can be done, also other quality tests can be done to settle dispute arising between farmers and exporters.
3. AEZ may take up the matter with the Mandi Parishad of Uttar Pradesh for the exemption of mandi fee, as in the case of basmati that is procured from Uttar Pradesh for the purpose of export.
4. For contract farming of basmati rice, efforts may be made to facilitate farmers and exporters. In this manner, exporters will be benefited as they will be getting the assured supply of the desired variety and quality; moreover, they will also get relief from the mandi fee and

farmers will get assured market and better price to their produce. In broader sense, this will give basmati rice exports from the state enhancing competitive advantages.

5. AEZ may take up the matter of abolition of quota system, as getting quota approved is a time-taking process; also, it is not feasible for exporters to get all orders in advance that they can get approval for that order.
6. AEZ may take up the matter of exemption of inter-state sale tax for 4%.
7. AEZ may discuss the problems of rice exporters with CONCOR (container corporation of India) to settle their problem regarding ICD.
8. AEZ may take up capacity building programs for entrepreneur exporters by organizing trainings on export credit guarantee scheme and documentation process for export.

9.11 COMMENTS FROM UOCB

Basmati is a long-grain aromatic rice grown for many centuries in geographical area, in the Himalayan foot hills of the Indian subcontinent, blessed with characteristics of extra-long slender grains that elongate at least twice their original size with a characteristic soft and fluffy texture upon cooking, delicious taste, superior aroma, and distinct flavors, Basmati rice is unique among other aromatic long-grain rice varieties.

Basmati paddy plays a major role in enhancing the income of farmers in Uttarakhand. In Uttarakhand, the major grown Basmati paddy verities are Pussa Basmati-1, Pussa Basmati-1121, Taraori Basmati, Type-3 (Dehraduni Basmati), and Basmati-370, mainly in four districts of Uttarakhand, namely, Dehradun, Haridwar, Udham Singh Nagar, and parts of Nainital.

Organic Basmati Model of Uttarakhand is a great success and farmers are getting as high as Rs. 35–40 per kg at gate price. Companies like Kohinoor foods are procuring organic basmati directly from farmers field thus removing most of the intermediaries from the Basmati rice sector.

During the year 2019–2020, the UOCB has facilitated the sale of 4570 quintals of Basmati paddy of worth Rs. 159 lakhs. Apart from UOCB, the sale of Basmati paddy is also being facilitated by other private agencies.

KEYWORDS

- **basmati rice**
- **export competitiveness**
- **backward and forward linkages**
- **international market**

REFERENCES

Agro Export Statistics. APEDA, 2014–2015.

Annual Report, 2015-16 ICAR-Indian Institute of Farming Systems Research, Modipuram, Meerut - 250 110, India

Alam; et al. Cost Benefit Analysis of Organically Produced Crops Grown in Dehradun and Udhamsingh Nagar; Center for Organic Farming Uttarakhand Organic Commodity Board Dehradun, 2007.

Annual Report UOCB, 2015–2016.

Arora. Marketing and Export of Horticultural Uttaranchal: Status, Potential and Strategies. *Indian J. Agric. Market.* **2005,** *19* (2), 194–198.

Balaji; Paltasingh. WTO and Indian Chilli Export: Issue of Growth, Instability and Competitiveness. *Indian J. Agric. Market.* **2014,** *28* (2), 35–49.

Brand. The Optimal Online Shop for Organic Products, 2005; vol *5* (31).

Bruneckiene, J.; Paltanaviciene D. Measurement of Export Competitiveness of the Baltic States by Composite Index. *Eng. Econ.* **2012,** *23* (1), 50–62.

Carolos. The International Market for Organic Products: Global Production and Consumption; 2001; vol *107*, pp 263–273.

Catry; et al. Market Share Strategy: The Concept and the Evidence. *J. Market.* **1975,** *39* (4), 59–60.

Chandrashekar. Changing Scenario of Organic Farming in India: An Overview. *Int. NGO J.* **2010,** *5* (1), 034–039.

Chauhan; Singh. Marketed Surplus of Paddy- A Regression Analysis. *Agric. Market.* **2002,** *XLV* (2), 25.

Clarkson; et al. Effects of India's Trade Policy on Rice Production and Exports. *Indian J. Econ. Bus.* **2011,** 1–23.

Gandhi. Merchandiser's Performance in Improving Supply Chain Competitiveness of Apparel Export Units, Ph.D. University of Delhi, Delhi, 2011.

Goyal; Agarwal. Export of India Fresh and Processed Fruits and Vegetable – A Study of Changing Pattern/Composition, Growth and Instability. *Indian J. Agric. Market.* **2013,** *27* (1), 76–84.

Gupta; Kumar. Growth Performance and Forecasts of Exports of Leather Industry in Punjab. *Foreign Trade Rev.* **2008,** *42* (3), 27–41.

Hiraga. Japanese Organic Market: Market Opportunities and Characteristics, 2002.

International Council for Research on International Economic Relations. Export Competitiveness of Indian Textile and Garment Industry, New Delhi; Verma, S.; New Delhi, 2002; pp 1–35.

Jung; Kang. The Actual State of Organic Products Marketing in Japan. *Korean J. Org. Agric.* **2003,** *11*, 43–60.

Kumar. WTO, Export Potential and Standards for Export of Cumin Seed with Special Reference to Gujarat State. *Indian J. Agric. Market.* **2014,** *28* (3), 130–141.

Kumar; et al. Performance, Competitiveness and Determinants of Tomato Export from India. *Agric. Econ. Res. Rev.* **2007,** *20* (CI), 551–562.

Kumar; et al. Growth and Export Performance of Indian Mangoes – An Economic Analysis. *Indian J. Agric. Market.* **2013,** *27* (1), 35–40.

Mahindru. In *Mannual of Basmati Rice*; Vedams ebooks (p) ltd.: Delhi, 1995.

Majumdar. Export Market of Indian Processed Food. *Indian J. Agric. Market.* **2016,** *30* (1), 49–57.

Meipporul; Bhanumathy. Export Performance of Coir in India. *Indian J. Agric. Mark.* **2010,** *24* (2), 188–192.

Organic and Biodynamic Farming, Government of India, Planning, 2016.

Pandey. The True Price of Rice. *Rice Today*, 2008, vol 7, p 18.

Panjabi. Supply Chain of Organic Basmati from Uttarakhand; United Nations Food and Agricultural Organization, 2008.

Patil; et al. Export Performance and Export Management of Indian Spices. *Indian J. Agric. Mark.* **2013,** *27* (3), 131–137.

Pittasboro. The Global Status, Prospects and Challenges of the Changing Organic Market, 2003, vol *33*.

Prasad. Current Science. Organic Farming vis-à-vis Modern Farming, 2005; vol 77, pp. 34–38.

Radman. Consumer Consumption and Perception of Organic Products in Croatia. *Br. Food J.* **2005,** *5* (17), 8–13.

Rice Knowledge Management Portal. Status Paper on Rice in Uttarakhand; Mani, S. C.; Hyderabad, 2012; pp 1–22.

Sadavatti. Exports of Basmati Rice and its Stability: Markov Chain Analysis. *Indian J. Agric. Mark.* **2006,** *20* (1), 11–18.

Salvador; et al. Market Opportunities and Challenges for Indian Organic products; Research Institute of Organic Agriculture (FIBL), 2003.

Sharma. Article on Basmati Rice: Production, Procurement, Processing and Marketing. *Res. J. Soc. Sci. Manag.* **2011,** *01* (08), 146–162.

Shreeshail; Satageri. Growth and Export Performance of Indian Tea. *Indian J. Agric. Mark.* **2013,** (3), 123–129.

Sikka; Narula. Enhancing Global Competitiveness of Indian Organic Produce: Opportunities, Challenges and Strategies. *Indian J. Agric. Mark.* **2006,** *20* (1), 19–24.

Sikka; et al. Problem and Prospects in Value Chain of Litchi Export from Uttaranchal. *Indian J. Agric. Mark.* **2005,** *19* (2), 207–217.

Singh; et al. Growth and Pattern of Export of Major Spices from India. *Indian J. Agric. Mark.* **2014,** *28* (2), 95–101.

Sir Ratan Tata Trust Publication. In *Impact Assessment Study of Center of Organic Farming I & II, Uttarkhand State*; Singh, J., Ed.; Mumbai, 2009; pp 1–8.

Slayton. Rice Crisis Forensic: How Asian Government Carelessly Set the World Rice Market on Fire. Working Paper no. 163, Center for Global Development: Washington, DC, 2009.

Sombhilla; et al. Medium and Long Term Prospects of Rice Supply and Demand in 21st Century in India. In *Development in Asian Rice Economy*; International Rice Research Institute: Phillipines, 2002; pp 59–94.

Subrahmanyam; Nagasree. Export of Organic Products from India: Present Status and Future Scope. *Indian J. Agric. Mark.* **2005,** *19* (3), 81–91.

Thulsiram; Alagumani. An Economic Analysis on Export Performance of Fresh Grapes from India. *Indian J. Agric. Mark.* **2015,** *29* (3), 33–39.

Thumar; et al. Growth and Instability in Production and Export Marketing of Garlic. *Indian J. Agric. Mark.* **2006,** *20* (2), 25–31.

Thumar; et al. Growth and Export Performance of Major Seed Spices of India. *Indian J. Agric. Mark.* **2012,** *26* (1), 21–35.

Umadevi; Eswaraprasad. Export Competitiveness of Indian Shrimp in the World Market. *Indian J. Agric. Mark.* **2006,** *20* (1), 90–99.

Vedamurthy; Pandey. Performance and Competitiveness of India's Onion Export. *Indian J. Agric. Mark.* **2010,** *24* (1), 86–95.

Yussefi; Willer. Organic Agriculture Worldwide, Statistics and Future Prospects, 2005.

CHAPTER 10

High-Value Plant Resources in Riparian Areas

POPPY RAJBONGSHI and TAPATI DAS

Department of Ecology and Environmental Science, Assam University, Silchar, India

ABSTRACT

In the present study, an attempt has been made to document riparian vegetation and their utilization patterns in terms of their ethnopharmacological and other uses. A survey was conducted in the Barak Valley in Assam, northeast India. The study revealed a total of 94 riparian plant species which possess a high value in terms of their antioxidant, anti-inflammatory, antimicrobial, anticancer, antidiarrheal, antifungal, and antidiabetic properties besides other uses like sources of essential and aromatic oil. The study also revealed the importance of riparian areas as a rich depository of plant resources which can be used for various purposes. The present study therefore suggests an in-depth phytochemical and pharmacological investigation of the riparian plant species to have scientific validation of the information collected from different published literatures. And, there should be further research regarding the potential of bioprospection of these plant resources.

10.1 INTRODUCTION

Riparian areas, the interface between land and surface water, are highly productive habitats which are colonized by taxonomically and functionally diverse communities of organisms including the riparian vegetation (Gregory et al., 1991; Naiman et al., 2005).

High-Value Plants: Novel Insights and Biotechnological Advances. Thangjam Anand Singh, Prakash Kumar Sarangi, and Ch. Brajakishore Singh (Eds.)
© 2025 Apple Academic Press, Inc. Co-published with CRC Press (Taylor & Francis)

Riparian ecosystem has received much attention in the recent years (Scott et al., 2009; Olokeogun et al., 2020) due to its role in providing many ecosystem services, such as preventing soil erosion, providing bank stability and habitat for different organism including the endemic species, besides regulating microclimatic condition and water quality of the nearby aquatic systems (Goldstein and Della Sala, 2020; Vari et al., 2021). Despite the importance of riparian ecosystem, they are known to be the most degraded and vulnerable ecosystems due to intensive human activities and other type of disturbances (Capon 2013; Goldstein and Della Sala, 2020). Studies on such ecosystem are still in infancy particularly in India, which might be due to a lack of awareness and also due to the complex nature of such ecosystem (Sunil et al., 2019).

The Barak Valley, Assam located in northeast India is dotted with numerous big and small water bodies' viz., ponds, rivers, lakes, and wetlands which are surrounded by diverse species of vegetation along the riparian region of such aquatic systems. However, there has been limited information on the utilization of these riparian plant resources or the services provided by them. Looking into the gap in information and also the importance of such studies in terms of ecosystem services through its various plant resources, the present study has been taken into consideration. In the present study, we attempted to document the importance of the recorded riparian plsant species in the study area for various purposes based on their potential uses/functions from existing published literatures.

10.2 MATERIALS AND METHODS

The present study was carried out in the floodplain of the Barak River basin in northeastern part of India. For carrying out the present study, a survey was done for 1 year (from December 2013–November 2014) along the riparian zone of different aquatic systems (10 nos.) located under different landscapes in the Cachar district of Barak Valley, Assam. We selected the lower angiosperm plant species and pteridophytes present in the riparian region of the selected aquatic systems for the present study. Plant species were identified following different standard keys of Kanjilal et al. (1934–1940), Kanjilal et al. (1939), Balakrishnan (1981–1983), and Hooker (1986). Besides, the herbarium of Botanical Survey of India, Shillong was consulted for identifying some of the plant species. Validation of nomenclature was done with the International Plant Names Index (IPNI) (www.ipni.org) and the plant list (www.theplantlist.org).

Documentation of the identified riparian plant species for their potential uses/functions was done following various published literatures.

10.3 RESULTS AND DISCUSSION

Overall, a total of 94 riparian plant species were recorded from the study area. Most of the plant species have anticancer, antioxidant, anti-inflammatory, antidiabetic, antimicrobial, wound healing, memory enhancing properties, whereas few species act as a source of aromatic and essential oil. The use pattern of the different species of the riparian vegetation showed the following trend: Antioxidant (59 species) > Anti-inflammatory (36 species) >Antimicrobial (33 species) > Antivenom (32 species) > Antibacteria (29 species) > Anticancer (23 species) > Antidiabetic (14 species) >Antifungal (9 species) > Wound healing (10 species) >= Anti-tumor (7 species) > Antiulcer> (6 species) > Antidiarrheal =Anti-stress = Antimalaria = Aromatic and essential oil (5 species each) > Antiaging = Antiviral (3 species) > Memory enhancing (2 species) (Table 10.1).

The compilation of information showed that the plant species found across the riparian areas of aquatic bodies possess various potentials for curing ailments of human being, besides other utilities. Through the present survey, we could observe that all of the recorded riparian plant species are of great importance and most of them possess antioxidant, anti-inflammatory, antimicrobial, anticancer, antidiarrheal, antifungal, antidiabetic, antihyperlipidemic, and neuropharmacological properties. These plant species might be utilized in the treatment of common fever, cough, diarrhea, diabetes, dysentery, headache, gynecological issues, and different skin diseases, as antidote for snake bite or scorpion sting, and others (Mir et al., 2018; Zihad et al., 2018; Pawar et al., 2018; Sagar et al., 2018; Nath et al., 2018; Bhattacharjee and Biswas, 2018; Das et al., 2019; Ferreira et al., 2019; Mohanan et al., 2020; Ayaz et al., 2020; Thengal et al., 2020, etc.) (Table 10.1).

Among the various riparian plant species, species, such as *Bacopa monnieri*, *Centella asiatica*, *Centipeda minima*, *Clerodendrum infortunatum*, *Cynodon dactylon*, *Enhydra fluctuans*, *Indigofera suffruticosa*, *Kyllinga nemoralis*, *Mikania micrantha*, *Peperomia pellucida*, *Sida acuta*, *and Urena lobata* are the most used species for various purposes (Table 10.1).

TABLE 10.1 List of Riparian Plant Species Present in the Study Area and Their Potential Uses/Functions.

Use/Function	Plants	Total number of plants	References
Anticancer	*Abelmoschus esculentus, Blechnum occidentale, Centella asiatica, Centipeda minima, Chromolaena odorata, Clerodendrum infortunatum, Colocasia esculenta, Cuscuta reflexa, Cyperus iria, Cyperus rotundus, Eclipta prostrata, Enhydra fluctuans, Heliotropium indicum, Indigofera suffruticosa, Ipomoea carnea, Lantana camara, Ludwigia octo-valvis, Melochia corchorifolia, Mikania micrantha, Oldenlandia diffusa, Oxalis corniculata, Peperomia pellucida, Urena lobata*	23	Simpson and Inglis, 2001; Gupta et al., 2004; Vieira et al., 2007; Nonato et al., 2009; Lai et al., 2010; Sannigrahi et al., 2010; Badwaik et al., 2011; Wei et al., 2011; Kundu et al., 2012; Dou et al., 2014; Fatima et al., 2014; Lin et al., 2014; Monte et al., 2014; Chunthorng-Orn et al., 2016; Solomon et al., 2016; Harini et al., 2017; Jiang et al., 2018; Sakonsinsiri et al., 2018; Wang et al., 2018; Ali et al., 2019; Kekuda et al., 2019; Simorangkir, et al., 2019; Zahara et al., 2019, Lee et al., 2020; Nelson et al., 2020; Al-Hakeim et al., 2021; Saeed et al., 2021; Teoh et al., 2021
Antidiabetic	*Abelmoschus esculentus, Chromolaena odorata, Cleome rutidosperma, Clerodendrum paniculatum, Colocasia esculenta, Cyanthillium cinereum, Cynodon dactylon, Ipomoea carnea, Kyllinga nemoralis, Mikania micrantha, Peperomia pellucida, Scoparia dulcis, Triticum aestivum, Urena lobata*	14	Daniel, 2005; Lin, 2005; Kumawat et al., 2010; Okoro et al., 2010; Ramachandran et al., 2010; Avvarai et al., 2011; Raju et al., 2011; Mishra et al., 2013; Purnomo et al., 2015; Ishak et al., 2016; Susilawati et al., 2017; Aligita et al., 2019; Warrier and Warrier, 2019; Zahara et al., 2019; Ajiboye et al., 2020; Nwaogwugwu et al., 2020; Saikia et al., 2020; Sheam et al., 2020; Varghese et al., 2021

TABLE 10.1 *(Continued)*

Use/Function	Plants	Total number of plants	References
Antioxidant	*Ageratum conyzoides, Alternanthera philoxeroides, Alternanthera pungens, Alternanthera sessilis, Bacopa monnieri, Blechnum occidentale, Centella asiatica, Centipeda minima, Chrysopogon zizanioides, Chromolaena odorata, Cleome rutidosperma, Clerodendrum infortunatum, Clerodendrum paniculatum,Commelina benghalensis,Commelina diffusa, Crinum asiaticum, Crotalaria pallida, Cuscuta reflexa, Cynodon dactylon, Cyperus iria, Cyperus rotundus, Dicranopteris linearis, Eclipta prostrata, Enhydra fluctuans, Evolvulus nummularius, Fimbristylis quinquangularis, Grangea maderaspatana, Heliotropium indicum, Hydrolea zeylanica, Indigofera suffruticosa, Ipomoea carnea, Kyllinga nemoralis, Leucas aspera,Lycopodiella cernua, Lygodium flexuosum, Melastoma malabathricum, Melochia corchorifolia, Mikania micrantha, Monochoria vaginalis, Ocimum basilicum, Oldenlandia corymbosa, Oxalis corniculata, Paspalum conjugatum, Peperomia pellucida, Persicaria barbata, Persicaria hydropiper, Polygonum plebeium, Pteris vittata,Rotala rotundifolia, Saccharum spontaneum, Scoparia dulcis, Senecio vulgaris, Senna tora, Sesamum indicum, Sida acuta, Spermacoce alata, Thysanolaena latifolia, Triticum aestivum, Urena lobata*	59	Yen and Chuang, 2000; Galati et al., 2001; Simpson and Inglis, 2001; Lee et al., 2005; Adeloye et al., 2007; Albayrak et al., 2008; Hasan et al., 2008; Luqman et al., 2009; Nonato et al., 2009; Pavithra et al., 2009; Lai et al., 2010; Badwaik et al., 2011; Cheikna et al., 2011; Hossain et al., 2011; Kaurinovic et al., 2011; Kilani-Jaziri et al., 2011; Wei et al., 2011; Zhang et al., 2011; Chandran et al., 2012; Chew et al., 2012; Das et al., 2012; Joffry et al., 2012; Pieme et al., 2012; Singh et al., 2012; Ali et al., 2013; Arriaga et al., 2013; Deori et al., 2013; Huang et al., 2013; Mishra et al., 2013; Mohan et al., 2013; Bhattacherjee et al., 2014; Fatima et al., 2014; Pukumpuang et al., 2014; Sindhu et al., 2014; Dev et al., 2015; Hasan et al., 2015; Singh et al., 2015; Chunthorng-Orn et al., 2016; Hoque et al., 2016; Kumari and Krishnan, 2016; Oh Sun-Min et al., 2016; Ramadas et al., 2016; Arciniegas et al., 2017; Harini et al., 2017; Jakhar et al., 2017; Sahoo et al., 2017; Campos et al., 2018; Islam et al., 2018; Jiang et al., 2018; Riris et al., 2018; Ruslan and Fidrianny, 2018; Lin et al., 2018; Porquis et al., 2018; Sakonsinsiri et al., 2018; Wang et al., 2018; Ali et al., 2019; Amritha and Jothi, 2019; Kekuda et al., 2019; Pena and Camara 2019; Nasrin et al., 2019; Nordin et al., 2019; Widodo et al., 2019; Zahara et al., 2019, Goswami et al., 2020; Zakaria et al., 2020; Saeed et al., 2021; Singh et al., 2021

TABLE 10.1 *(Continued)*

Use/Function	Plants	Total number of plants	References
Antibacterial	*Alternanthera philoxeroides,Alternanthera pungens, Bacopa monnieri, Blechnum occidentale, Centipeda minima, Chromolaena odorata, Cleome rutidosperma, Crotalaria pallida, Dicranopteris linearis, Ethulia conyzoides, Evolvulus nummularius, Indigofera suffruticosa, Ipomoea aquatic, Ipomoea carnea, Kyllinga nemoralis, Lantana camara; Leucas aspera; Lindernia ciliate, Lindernia crustacean, Ludwigia adscendens, Ludwigia octovalvis, Lygodium flexuosum; Mimosa pudica, Nephrolepis cordifolia, Persicaria hydropiper, Saccharum spontaneum, Senna alata, Solanum virginianum; Thysanolaena latifolia*	29	Adedayo et al., 2001; Chang et al., 2004; Ahmed et al., 2005; Liang et al., 2007; Nonato et al., 2009; Nisar et al., 2009; Pavithra et al., 2009; Khan et al., 2010; Lai et al., 2010; Rani et al., 2010; Chew et al., 2012; Nayak et al., 2013; Bhattacherjee et al., 2014; Fatima et al., 2014; Hussain et al., 2014; Praneetha et al., 2014; Lin et al., 2014; Sindhu et al., 2014; Le Thoa et al., 2015; Hoque et al., 2016; Oh Sun-Min et al., 2016; Ukil et al., 2016; Jakhar and Dahiya, 2017; Lin et al., 2017; Okoro et al., 2017; Padmavathy et al., 2017; Prashith kekuda et al., 2017; Raghavendra et al., 2017; Islam et al., 2018; Manvar et al., 2018; Pallerla et al., 2018; Das et al., 2019; Rasyid et al., 2020; Thomas et al., 2020
Antifungal	*Ageratum conyzoides, Cyperus iria, Indigofera suffruticosa, Ipomoea carnea, Lantana camara, Leucas aspera, Nephrolepis cordifolia, Oxalis corniculata, Senna alata*	9	Adedayo et al., 2009; Nisar et al., 2009; Rani et al., 2010; Badwaik et al., 2011; Javed et al., 2012; Fatima et al., 2014; Babu et al., 2016; Fayaz et al., 2017; Jiang et al., 2018; Saeed et al., 2021; Simpson and Inglis, 2001

TABLE 10.1 *(Continued)*

Use/Function	Plants	Total number of plants	References
Antimicrobial	*Ageratum conyzoides, Alternanthera philoxeroides, Alternanthera pungens, Bacopa monnieri, Centipeda minima,Clerodendrum infortunatum, Clerodendrum paniculatum,Cleome rutidosperma, Chromolaena odorata, Commelina diffusa, Crinum asiaticum, Cyperus iria, Cyperus rotundus, Eclipta prostrata, Enhydra fluctuans, Grangea maderaspatana, Indigofera suffruticosa,Ipomoea aquatica, Kyllinga nemoralis, Lantana camara, Leucas aspera, Melastoma malabathricum, Mikania micrantha, Oldenlandia corymbosa, Oxalis corniculata, Paspalum conjugatum,Peperomia pellucida, Senecio vulgaris, Senna alata, Senna tora, Solanum virginianum, Spermacoce alata, Urena lobata*	33	Simpson and Inglis, 2001; Bose et al., 2005; Mangathayaru et al., 2005; Leite et al., 2006; Soetardjo et al., 2007; Albayrak et al., 2008; Rahman and Rashid 2008; Sampathkumar et al., 2008; Ndip et al., 2009; Dhanasekaran et al., 2010; Pardhi et al., 2010; Badwaik et al., 2011; Kilani-Jaziri et al., 2011; Raju, et al., 2011; Saraf et al., 2011; Wei et al., 2011; Amin et al., 2012; Conserva and Ferreira, 2012; Singh et al., 2012; Deori et al., 2013; Hussain and Kumaresan, 2013; Li et al., 2013; Bhattacherjee et al., 2014; Mishra et al., 2016; Doughari and Okafor, 2007; Jakhar et al., 2017; Jiang et al., 2018; Wang et al., 2018; Garduque et al., 2019; Kekuda et al., 2019; Nasrin et al., 2019; Nordin et al., 2019; Goswami et al., 2020; Prabagar et al., 2020; Teoh et al., 2021; Saeed et al., 2021
Antiulcer	*Abelmoschus esculentus, Centella asiatica, Hydrolea zeylanica, Indigofera suffruticosa, Ipomoea aquatic, Peperomia pellucida*	6	Hamid et al., 2007; Abdulla et al., 2010; Muthukumaran and Pattabiraman 2010; Sahoo et al., 2017; Ortac et al., 2018
Antimalaria	*Chromolaena odorata, Cleome viscosa, Cyanthillium cinereum, Senna alata, Sida acuta*	5	Karou et al., 2003, Ukpai and Amaechi 2012; Kayembe et al., 2010; Elufioye and Onoja 2016
Antitumor	*Croton bonplandianum, Heliotropium indicum, Indigofera suffruticosa, Lantana camara, Mikania micrantha, Oldenlandia corymbosa, Urena lobata*	7	Vieira et al., 2007; Lee et al., 2011; Dou et al., 2014; Campos et al., 2018; Ghosh, et al., 2018; Mathappan et al., 2019; Al-Hakeim et al., 2021; Fayed, 2021
Antiaging	*Bacopa monnieri, Eclipta prostrata, Ludwigia octovalvis*	3	Lin et al., 2014; Simpson et al., 2015; Lee et al., 2018

TABLE 10.1 *(Continued)*

Use/Function	Plants	Total number of plants	References
Anti-inflammatory	*Acmella paniculate, Ageratum conyzoides, Bacopa monnieri, Blechnum occidentale, Centella asiatica, Centipeda minima, Chrysopogon aciculatus, Chrysopogon zizanioides, Cleome spinosa, Clerodendrum infortunatum, Clerodendrum paniculatum, Colocasia esculenta, Commelina benghalensis, Crotalaria pallida, Cuscuta reflexa, Cymbopogon citrates, Cynodon dactylon, Cyperus iria, Eclipta prostrata, Enhydra fluctuans, Indigofera suffruticosa, Heliotropium indicum; Ipomoea aquatica, Ipomoea carnea, Lycopodiella cernua, Lygodium flexuosum, Monochoria vaginalis, Oldenlandia corymbosa, Peperomia pellucida, Pogostemon auricularius, Pteris vittata, Scoparia dulcis, Sida acuta, Solanum americanum, Spermacoce ocymoides, Urena lobata*	36	Galati et al., 2001; Leite et al., 2003; Weng et al., 2003; Shah et al., 2007; Nonato et al., 2009; Zakaria et al., 2009; Das et al., 2010; Dhanasekaran et al., 2010; de Fatima et al., 2004; Figueirinha et al., 2010; Mutee et al., 2010; Hossain et al., 2011; Suresh et al., 2011; Tsai et al., 2011; Chandran et al., 2012; Das et al., 2012; Sundhararajan et al., 2012; Albarello et al., 2013; Deori et al., 2013; Huang et al., 2013; Boukhatem et al., 2014; Fatima et al., 2014; Boukhatem et al., 2014; Hossain et al., 2014; Williams et al., 2014; Chunthorng-Orn et al., 2016; Antonysamy et al., 2017; Arciniegas et al., 2017; Porquis et al., 2018; Rajagopal et al., 2018; Sakonsinsiri et al., 2018; Satyal et al., 2018; Zihad et al., 2018; Kekuda et al., 2019; Permawati et al., 2019; Saha and Paul 2019; Yeom et al., 2019; Abdul Rahim et al., 2021; Brimson et al., 2021; Grover et al., 2021; Saeed et al., 2021; Singh et al., 2021
Anti-stress	*Cynodon dactylon, Cyperus iria, Ipomea aquatic, Mikania micrantha, Sida acuta*	5	Muthukumaran and Pattabiraman, 2010; Chidrawar et al., 2011; Ittiyavirah and Sajid, 2013; Benjumea et al., 2016; Jiang et al., 2018
Antiviral	*Centipeda minima, Cuscuta reflexa, Phyllanthus niruri*	3	Mahmood et al., 1997; Awasthi, 1981; Tan et al., 2013
Antidiarrheal	*Clerodendrum infortunatum, Colocasia esculenta, Enhydra fluctuans, Fimbristylis quinquangularis, Urena lobata*	5	Yadav and Tangpu, 2007; Prajapati, et al., 2011; Ali et al., 2013; Hazarika et al., 2017; Mukta et al., 2020

TABLE 10.1 *(Continued)*

Use/Function	Plants	Total number of plants	References
Antivenom/ Antidote	*Alternanthera sessilis, Acmella paniculata, Ageratum conyzoides, Bacopa monnieri, Centella asiatica, Chromolaena odorata, Chrysopogon zizanioides, Commelina benghalensis, Crinum asiaticum, Croton bonplandianus, Crotalaria pallida, Cynodon dactylon, Cymbopogon citrates, Cyanthillium cinereum, Cyperus rotundus, Eclipta prostrata, Enhydra fluctuans, Eleusine indica, Heliotropium indicum, Leucas aspera, Melochiacorchorifolia, Mimosapudica, Ocimumbasilicum, Ocimum tenui-florum, Oldenlandia corymbosa, Oxalis corniculata, Oldenlandia diffusa, Peperomia pellucida, Scoparia dulcis, Senna alata, Senna tora, Xanthium strumarium*	32	Honda et al., 1996; Pithayanukul et al., 2004; Coe and Anderson, 2005; Mahishi et al., 2005; Mitra and Mukherjee, 2005; Samy et al., 2008; Makhija and Khamar, 2010; Meena, 2010; Santosh and Satya, 2010; Shukla et al., 2010; Sri and Reddi, 2011; Gupta and Peshin, 2012; Butt et al., 2015; Kala, 2015; Sulochana et al., 2015; Vasquez et al., 2015; Dharmadasa et al., 2016; Al-Asmari et al., 2017; Felix-Silva et al., 2017; Lavanya and Pattar, 2017; Reddy et al., 2019
Memory enhancing	*Bacopa monnieri, Centella asiatica*	3	Belwal et al., 2019; Brimson et al., 2021
Wound healing properties	*Ageratum conyzoides, Centella asiatica, Chromolaena odorata, Crinum asiaticum, Cynodon dactylon, Cyanthillium cinereum, Echinochloa colona, Heliotropium indicum, Ipomoea carnea, Oxalis corniculata*	10	Badwaik et al., 2011; Dash and Murthy, 2011; Arulprakash et al., 2012; Fatima et al., 2014; Sirinthipaporn and Jiraungkoorskul, 2017; Vijayaraghavan et al., 2017; Kumar, 2011; Fadillah and Santoso, 2019; Sumitra and Parul 2018; Ahmed et al., 2019; Zahara et al., 2019; Singh et al., 2021
Aromatic and essential oil	*Cymbopogon citrates, Chromolaena odorata, Lantana camara, Mikania micrantha, Ocimum tenuiflorum, Pogostemon auricularius, Sesamum indicum*	7	Nicollier and Thompson, 1981; Limberger et al., 2001; Warra, 2011; Machado et al., 2012; Boukhatem et al., 2014; Satyal et al., 2018; Sikka and Bartolome, 2018; Zahara et al., 2019; Dougnon and Ito, 2020; Saikia et al., 2020

10.4 CONCLUSIONS

The compilation of information related to the use of the riparian plant species from various literatures revealed that there is a huge scope for the utility of the riparian ecosystem through its diverse plant resources as they have the potential in improving human health. The study revealed the importance of riparian areas around different aquatic systems as a rich depository of plant resources which have various potential uses. The study suggests further investigation in this regard with an in-depth phytochemical and pharmacological analyses. The validation thus obtained would also give a direction toward conservation and management of riparian areas. This would also help in preserving rich and essential cultural heritage in a scientific and sustainable way. The study therefore suggests that there should be further research regarding the potential of bioprospection of the riparian plant species.

KEYWORDS

- **riparian vegetation**
- **utilization patterns**
- **remedial value**
- **Assam**
- **Northeast India**

REFERENCES

Abdul Rahim, R.; Jayusman, P. A.; Muhammad, N.; Mohamed, N.; Lim, V.; Ahmad, N. H.; Mohamed, I. N. Potential Antioxidant and Anti-inflammatory Effects of *Spilanthes acmella* and its Health Beneficial Effects: A Review. *Int.J. Environ. Res. Public Health* **2021,** *18* (7), 1–15.

Abdulla, M. A.; Al-Bayaty, F. H.; Younis, L. T.; Hassan, M. A. Anti-ulcer Activity of *Centella asiatica* Leaf Extract Against Ethanol-induced Gastric Mucosal Injury in Rats. *J. Med. Plants Res.* **2010,** *4* (13), 1253–1259.

Adedayo, O.; Anderson, W. A.; Moo-Young, M.; Kolawole, D. O. Antifungal Properties of Some Components of *Senna alata* Flower. *Pharm. Biol.* **1999,** *37* (5), 369–374.

Adedayo, O.; Anderson, W. A.; Moo-Young, M.; Snieckus, V.; Patil, P. A.; Kolawole, D. O. Phytochemistry and Antibacterial Activity of *Senna alata* Flower. *Pharm. Biol.* **2001,** *39* (6), 408–412.

Adeloye, O. A.; Akinpelu, A. D.; Ogundaini, O. A.; Obafemi, A. C. Studies on Antimicrobial, Antioxidant and Phytochemical Analysis of *Urena lobata* Leave Extract. *J. Phys. Nat. Sci.* **2007,** *1* (2), 1–9.

Ahmed, F.; Selim, M. S. T.; Shilpi, J. A. Antibacterial Activity of *Ludwigia adscendens*. *Fitoterapia* **2005,** *76* (5), 473–475.

Ahmed, A. S.; Taher, M.; Mandal, U. K.; Jaffri, J. M.; Susanti, D.; Mahmood, S.; Zakaria, Z. A. Pharmacological Properties of *Centella asiatica* Hydrogel in Accelerating Wound Healing in Rabbits. *BMC Complement. Altern. Med.* **2019,** *19* (1), 1–7.

Ajiboye, B. O.; Oloyede, H. O. B.; Salawu, M. O. Antidiabetic Activity of *Triticum aestivum* Seed–Based Diet on Alloxan-Induced Diabetic Rats. *J. Diet. Suppl.* **2020,** *17* (2), 133–149.

Al-Asmari, A.; Manthiri, R. A.; Abdo, N.; Al-Duaiji, F. A.; Khan, H. A. Saudi Medicinal Plants for the Treatment of Scorpion Sting Envenomation. *Saudi J. Biol. Sci.* **2017,** *24* (6), 1204–1211.

Albayrak, S.; Aksoy, A.; Hamzaoglu, E.; Ekici, L.; Budak, U. Antimicrobial and Antioxidant Activities of *Senecio* Species Growing in the Black Sea Region, Turkey. *Acta Botanica Gallica* **2008,** *155* (3), 447–456.

Al-Hakeim, H. K.; Al-Zabibah, R. S.; Alzihari, H. F.; Almensoori, A. K.; Al-Zubaidi, H. A.; Hassan, L. E. A. H. Anticancer and Antiangiogenic Activities of Alkaloids Isolated from *Lantana Camara* by Adsorption on the Magnetic Nanoparticles. *Karbala Int. J. Mod. Sci.* **2021,** *7* (1), 11.

Ali, M. R.; Billah, M. M.; Hassan, M. M.; Dewan, S. M. R.; Al-Emran, M. *Enhydra fluctuans* Lour: A Review. *Res. J. Pharm. Technol.* **2013,** *6* (9), 927–929.

Ali, I.; Suhail, M., Fazil, M.; Ahmad, B.; Sayeed, A.; Naqshbandi, M.; Azam, A. Anti-Cancer and Anti-oxidant Potencies of *Cuscuta reflexa* Roxb. Plant Extracts. *Am. J. Adv. Drugs Deliv.* **2019,** *7*, 107–114.

Aligita, W.; Muhsinin, S.; Susilawati, E.; Pratiwi, D. S.; Aprilliani, D.; Artarini, A., Adnyana, I. K. Antidiabetic Activity of Okra (*Abelmoschus esculentus* L.) Fruit Extract. *Rasāyan J. Chem.* **2019,** *12* (1), 157–167.

Amin, M. R.; Mondol, R.; Habib, M. R.; Hossain, M. T. Antimicrobial and Cytotoxic Activity of Three Bitter Plants *Enhydra fluctuans*, *Andrographis peniculata* and Clerodendrum viscosum. *Adv. Pharm. Bull.* **2012,** *2* (2), 207–2011.

Amritha, C. K.; and Jothi, E. T. Pharmaceutical Sciences. *Asian J. Res. Chem. Pharm. Sci.* **2019,** *7* (2) 439–445.

Antonysamy, J. M. A.; Ramakrishnan, P.; Perumal, S.; Shibila, T. Anti Inflammatory Activity of Selected Pteridophytes from Western Ghats. *Int. J. Complement. Alt. Med.* **2017,** *9* (4), 1–3.

Arciniegas, A.; Perez-Castorena, A. L.; Nieto-Camacho, A.; Kita, Y.; Vivar, A. R. D. Anti-hyperglycemic, Antioxidant, and Anti-inflammatory Activities of Extracts and Metabolites from *Sida acuta* and *Sida rhombifolia*. *Química Nova* **2017,** *40*, 176–181.

Arriaga, A. M.; Lemos, T. L.; Santiago, G. M.; Andrade-Neto, M.; Braga, M. A.; de Almeida, M. C.; Alves, P. B. Chemical Composition and Antioxidant Activity of *Indigofera suffruticosa*. *Chem. Natl. Comp.* **2013,** *49* (1), 150–151.

Arulprakash, K.; Murugan, R.; Ponrasu, T.; Iyappan, K.; Gayathri, V. S.; Suguna, L. Efficacy of *Ageratum conyzoides* on Tissue Repair and Collagen Formation in Rats. *Clin. Exp. Dermatol* **2012,** *37* (4), 418–424.

Avvarai, S. K.; Kattamanchi, G.; Doni, K.; Anugu, M. R.; Raju, C. Antidiabetic Activity of Ethanolic Extract of *Cynodon dactylon* Root Stalks in Streptozotocin Induced Diabetic Rats. *Int. J. Adv. Pharm. Res.* **2011,** *2* (8), 418–422.

Awasthi, L. P. The Purification and Nature of an Antiviral Protein from *Cuscuta reflexa* Plants. *Arch. Virol.* **1981,** *70* (3), 215–223.

Babu, A.; Mohamed, M. S. N.; Jaikumar, K.; Anand, D.; Saravanan, P. In-vitro Antifungal Activity of Leaf Extracts of *Leucas aspera* and *Leucas zeylanica. Int. J. Pharm. Sci. Res.* **2016,** *7* (2), 752.

Badwaik, H.; Singh, M. K.; Thakur, D.; Giri, T. K.; Tripathi, D. K. The Botany, Chemistry, Pharmacological and Therapeutic Application of *Oxalis corniculata* Linn-a Review. *Int. J. Phytomed.* **2011,** *3* (1), 01.

Balakrishnan, N. P. Flora of Jowai and Vicinity Meghalaya: A Contribution Towards a Detailed Knowledge of the Flora of the Northeastern Region of India, 1st Ed.; Botanical Survey of India: Howrah, 1981–1983.

Belwal, T.; Andola, H. C.; Atanassova, M. S.; Joshi, B.; Suyal, R.; Thakur, S.; Rawal, R. S. Gotu Kola (*Centella asiatica*). In *Nonvitamin and Nonmineral Nutritional Supplements*; Academic Press, Elsevier: United State, 2019; pp 265–275.

Benjumea, D. M.; Gomez-Betancur, I. C.; Vásquez, J.; Alzate, F.; García-Silva, A.; Fontenla, J. A. Neuropharmacological Effects of the Ethanolic Extract of *Sida acuta. Revista Brasileira de Farmacognosia* **2016,** *26*, 209–215.

Bhattacherjee, A.; Ghosh, T.; Sil, R.; Datta, A. Isolation and Characterisation of Methanol-Soluble Fraction of *Alternanthera philoxeroides* (Mart.)–Evaluation of Their Antioxidant, α-glucosidase Inhibitory and Antimicrobial Activity in In Vitro Systems. *Natl. Prod. Res.* **2014,** *28* (23), 2199–2202.

Bose, A.; Gupta, J. K.; Ghosh, T.; Dash, G. K. Antimicrobial Activity of Certain Extracts of *Cleome rutidosperma. Indian J. Natl. Prod.* **2005,** *21* (3), 39–41.

Bose, A.; Mondal, S.; Gupta, J. K.; Ghosh, T.; Debbhuti, D.; Si, S. Antioxidant and Free Radical Scavenging Activities of *Cleome rutidosperma. Orient. Pharm. Exp. Med.* **2008,** *8* (2), 135–145.

Boukhatem, M. N.; Ferhat, M. A.; Kameli, A.; Saidi, F.; Kebir, H. T. Lemon Grass (*Cymbopogon citratus*) Essential Oil as a Potent Anti-inflammatory and Antifungal Drugs. *Libyan J. Med.* **2014,** *9* (1), 1–10.

Brimson, J. M.; Brimson, S.; Prasanth, M. I.; Thitilertdecha, P.; Malar, D. S.; Tencomnao, T. The Effectiveness of *Bacopa monnieri* (Linn.) Wettst. As a Nootropic, Neuroprotective, or Antidepressant Supplement: Analysis of the Available Clinical Data. *Sci. Rep.* **2021,** *11* (1), 1–11.

Butt, M. A.; Ahmad, M.; Fatima, A.; Sultana, S.; Zafar, M.; Yaseen, G.; Kayani, S. Ethnomedicinal Uses of Plants for the Treatment of Snake and Scorpion Bite in Northern Pakistan. *J. Ethnopharmacol.* **2015,** *168*, 164–181.

Campos, J. K.; Araujo, T. F. D. S.; Brito, T. G. D. S.; da Silva, A. P.; Cunha, R. X. D.; Martins, M. B.; Lima, V. L. D. M. Indigofera Suffruticosa mill. (Anil): Plant Profile, Phytochemistry, and Pharmacology Review. *Adv. Pharmacol. Sci.* **2018,** 1–6.

Capon, S. J.; Chambers, L. E.; Mac Nally, R.; Naiman, R. J.; Davies, P.; Marshall, N.; Williams, S. E. Riparian Ecosystems in the 21st Century: Hotspots for Climate Change Adaptation? *Ecosystems* **2013,** *16* (3), 359–381.

Chandran, R.; Thangaraj, P.; Shanmugam, S.; Thankarajan, S.; Karuppusamy, A. Antioxidant and Anti-inflammatory Potential of *Monochoria vaginalis* (burm. f.) c. presl.: A Wild Edible Plant. *J. Food Biochem.* **2012,** *36* (4), 421–431.

Chea, A.; Hout, S., Long, C.; Marcourt, L.; Faure, R.; Azas, N.; Elias, R. Antimalarial Activity of Sesquiterpene Lactones from *Vernonia cinerea. Chem. Pharm. Bull.* **2006,** *54* (10), 1437–1439.

Cheikna, Z.; Savadogo, A.; Somda, M. K.; Koudou, J.; Traore, A. S. In Vitro Evaluation of the Antimicrobial and Antioxidant Properties of Extracts from Whole Plant of Alternanthera Pungens HB & K. and Leaves of Combretum sericeum G. Don. *Int. J. Phytomed.* **2011,** *3* (2), 182.

Chew, A. L.; Jessica, J. J. A.; Sasidharan, S. Antioxidant and Antibacterial Activity of Different Parts of *Leucas aspera. Asian Pac. J. Trop. Biomed.* **2012,** *2* (3), 176–180.

Chidrawar, V. R.; Chitme, H. R.; Patel, K. N.; Patel, N. J.; Racharla, V. R.; Dhoraji, N. C.; Vadalia, K. R. Effects of *Cynodon dactylon* on Stress-Induced Infertility in Male Rats. *J Young Pharm.* **2011,** *3* (1), 26–35.

Chunthorng-Orn, J.; Dechayont, B.; Phuaklee, P.; Prajuabjinda, O.; Juckmeta, T.; Itharat, A. Cytotoxic, Anti-inflammatory and Antioxidant Activities of *Heliotropium indicum* Extracts. *J. Med. Assoc. Thailand Chotmaihet Thangphaet* **2016,** *99*, 102–109.

Coe, F. G.; and Anderson, G. J. Snakebite Ethnopharmacopoeia of Eastern Nicaragua. *J. Ethnopharmacol.* **2005,** *96* (1-2), 303–323.

Conserva, L. M., Jesu Costa Ferreira, J. *Borreria* and *Spermacoce* Species (Rubiaceae): A Review of their Ethnomedicinal Properties, Chemical Constituents, and Biological Activities. *Pharmacogn. Rev.* **2012,** *6* (11), 46–45.

Daniel, M. In *Medicinal Plants: Chemistry and Properties*; Science Publishers (Taylor & Francis): United Kingdom, 2006.

Das, S. R. C.; Ahmed, A. B.; Shil, D.; Chanda, I. Isolation and In Vitro Antioxidant Activity of Flavonoid from Lindernia Crustacea (L) F. Muell. *Asian J. Pharm. Clin. Res.* **2019,** *12* (6), 289–293.

Das, B.; Talekar, Y. P.; Apte, K. G.; Chauhan, R. A Preliminary Study of Antiinflammatory Activity and Antioxidant Property of *Lygodium flexuosum*, A Climbing Fern. *Int. J. Pharm. Pharm. Sci.* **2012,** *4* (4), 358–361.

Dash, G. K.; Murthy, P. N. Studies on Wound Healing Activity of *Heliotropium indicum* Linn. Leaves on Rats. *Int. Sch. Res. Notices* **2011,** 1–8.

Dash, G. K.; Murthy, P. N. Wound Healing Effects of *Ageratum conyzoides* Linn. *Int. J. Pharma. Bio. Sci.* **2011,** *2* (2), 369–383.

Datta, S.; Seal, T. Anti-Diabetic, Anti-Inflammatory and Antioxidant Properties of Four Underutilized Ethnomedicinal Plants: An in Vitro Approach, 2020; pp 1–14.

de Fatima Arrigoni-Blank, M.; Dmitrieva, E. G.; Franzotti, E. M.; Antoniolli, A. R.; Andrade, M. R.; Marchioro, M. Anti-inflammatory and Analgesic Activity of *Peperomia pellucida* (L.) HBK (Piperaceae). *J. Ethnopharmacol.* **2004,** *91* (3), 215–218.

Deori, C.; Roy, D. K.; Talukdar, S. R.; Pagag, K.; Sarma, N. Diversity of the Genus *Clerodendrum* Linnaeus (Lamiaceae) in Northeast India with Special Reference to Barnadi Wildlife Sanctuary, Assam. *Pleione* **2013,** *7* (2), 473–488.

Dev, U. K.; Hossain, M. T.; Islam, M. Z. Phytochemical Investigation, Antioxidant Activity and Anthelmintic Activity of *Mikania micrantha* Leaves. *World J. Pharm. Res.* **2015,** *4* (5), 121–133.

Dhanasekaran, S.; Palayan, M.; Kumar, S. S. Evaluation of Anti-microbial and Anti-inflammatory Activity of Methanol Leaf Extract of *Ipomoea aquatica* Forsk. *Res. J. Pharm. Biol. Chem. Sci.* **2010,** *1* (2), 258–264.

Dharmadasa, R. M.; Akalanka, G. C.; Muthukumarana, P. R. M.; Wijesekara, R. G. S. Ethnopharmacological Survey on medicinal Plants Used in Snakebite Treatments in Western and Sabaragamuwa Provinces in Sri Lanka. *J. Ethnopharmacol.* **2016,** *179*, 110–127.

Dou, X.; Zhang, Y.; Sun, N.; Wu, Y.; Li, L. The Anti-tumor Activity of *Mikania micrantha* Aaqueous Extract In Vitro and In Vivo. *Cytotechnology* **2014,** *66* (1), 107–117.

Doughari, J. H.; Okafor, B. Antimicrobial Activity of *Senna alata* Linn. *East Central Afr. J. Pharm. Sci.* **2007,** *10* (1), 17–21.

Dougnon, G.; Ito, M. Sedative Effects of the Essential Oil from the Leaves of *Lantana camara* Occurring in the Republic of Benin Via Inhalation in Mice. *J. Natl. Med.* **2020,** *74* (1), 159–169.

Elufioye, T. O.; Onoja, J. O. In Vivo Anti-malarial Activity of Cleome Viscosa L. Whole Plant. *Res. J. Phytochem.* **2016,** *10* (1), 30–38.

Fadillah, M.; Santoso, P. The Sirangak (*Cyanthillium cinereum*; Asteraceae) Oil Accelerates Sliced-Wound Healing by Enhancing the Hematological Endurance in Male Albino Mice. *J. Phys. Conf. Ser.* **2019,***1317* (1), 1–7.

Fatima, N.; Rahman, M. M.; Khan, M. A.; Fu, J. A Review on *Ipomoea carnea*: Pharmacology, Toxicology and Phytochemistry. *J. Complement. Integr. Med.* **2014,** *11* (2), 55–62.

Fayaz, M.; Bhat, M. H.; Fayaz, M.; Kumar, A.; Jain, A. K. Antifungal Activity of *Lantana camara* L. Leaf Extracts in Different Solvents Against Some Pathogenic Fungal Strains. *Pharmacologia* **2017,** *8* (3), 105–112.

Fayed, M. A. *Heliotropium*; A Genus Rich in Pyrrolizidine Alkaloids: A Systematic Review Following its Phytochemistry and Pharmacology. *Phytomed. Plus* **2021,** 1–9.

Felix-Silva, J.; Silva-Junior, A. A.; Zucolotto, S. M.; Fernandes-Pedrosa, M. D. F. Medicinal Plants for the Treatment of Local Tissue Damage Induced by Snake Venoms: An Overview from Traditional use to Pharmacological Evidence. *Evid. Based Complement. Alter. Med.* **2017,** 1–52.

Galati, E. M.; Miceli, N.; Taviano, M. F.; Sanogo, R.; Raneri, E. Anti-inflammatory and Antioxidant Activity of *Ageratum conyzoides*. *Pharm. Biol.* **2001,** *39* (5), 336–339.

Garduque, D. A. P.; Mateo, K. R. G.; Oyinloye, S. M. A. In *Antimicrobial Efficacy of Carabao Grass (Paspalum conjugatum) Leaves on Staphylococcus aureus*, Abstract Proceedings International Scholars Conference, 2019; vol *7* (1), pp 384–397.

Ghosh, T.; Biswas, M. K.; Roy, P.; Guin, C. A Review on Traditional and Pharmacological Uses of *Croton bonplandianum* with Special Reference to Phytochemical Aspect. *Eur. J. Med. Plants* **2018,** 1–10.

Goldstein, M. I.; DellaSala, D. A. In *Encyclopedia of the World's Biomes*; Oxford: Amsterdam; Elsevier: Cambridge, 2020.

Goswami, S.; Das, R.; Ghosh, P.; Chakraborty, T.; Barman, A., Ray, S. Comparative Antioxidant and Antimicrobial Potentials of Leaf Successive Extract Fractions of Poison Bulb, *Crinum asiaticum* L. *Ind. Crops Prod.* **2020,** *154*, 1–9.

Gregory, S. V.; Swanson, F. J.; McKee, W. A.; Cummins, K. W. An Ecosystem Perspective of Riparian Zones. *BioScience* **1991,** *41* (8), 540–551.

Gupta, Y. K.; Peshin, S. S. Do Herbal Medicines have Potential for Managing Snake Bite Envenomation? *Toxicol. Int.* **2012,** *19* (2), 89.

Gupta, S.; Zhang, D.; Yi, J.; Shao, J. Anticancer Activities of *Oldenlandia diffusa. J. Herb. Pharmacother*. **2004,** *4* (1), 21–33.

Hamid, R. A.; Zakaria, N.; Zuraini, A. Anti-ulcer Activity of Aqueous Ethanol Extract of *Peperomia pellucida* in Sprague Dawley Rats. *Planta Med.* **2007,** *73*,455–459.

Harini, V.; Vijayalakshmi, M.; Sivaraj, C.; Arumugam, P. Antioxidant and Anticancer Activities of Methanol Extract of Melochia Corchorifolia L. *Int. J. Sci. Res*. **2017,** *6* (1), 1310–1316.

Hasan, S. M.; Hossain, M. M.; Faruque, A.; Mazumder, M. E. H.; Rana, M. S.; Akter, R. Alam, M. A. Comparison of Antioxidant Potential of Different Fractions of Commelina Benghalensis Linn. *Bang. J. Life. Sci.* **2008,** *20* (2), 9–16.

Hasan, A. N.; Roy, P.; Bristy, N. J.; Paul, S. K.; Wahed, T. B.; Alam, M. N. Evaluation of In Vitro Antioxidant and Brine Shrimp Lethality Bioassay of Different Extracts of *Polygonum plebeium* R. Br. *Int. J.* **2015,** *3* (12), 97–107.

Hazarika, A.; Saha, D. Preliminary Phytochemical Screening and Evaluation of Anti-diarrhoeal Activity of Ethanolic Extract of Leaves of *Clerodendrum infortunatum. Int. J. Curr. Pharm. Res*. **2017,** *9* (4), 143–146.

Honda, G.; Yeşilada, E.; Tabata, M.; Sezik, E.; Fujita, T.; Takeda, Y.; Tanaka, T. Traditional Medicine in Turkey VI. Folk Medicine in West Anatolia: Afyon, Kütahya, Denizli, Mugla, Aydin Provinces. *J. Ethnopharmacol*. **1996,** *53* (2), 75–87.

Hooker, J. D. In *The Flora of British India*; L. Reeve and Co.: London, 1986.

Hoque, N.; Sohrab, M. H.; Debnath, T.; Rana, M. S. Antioxidant, Antibacterial and Cytotoxic Activities of Various Extracts of *Thysanolaena maxima* (Roxb) Kuntze Available in Chittagong Hill Tracts of Bangladesh. *Int. J. Pharm. Pharm. Sci*. **2016,** *8* (7), 168–172.

Hossain, M. S.; Alam, M. B.; Chowdhury, N. S.; Asadujjaman, M.; Zahan, R.; Islam, M. M.; Islam, A. Antioxidant, Analgesic and Anti-inflammatory Activities of the Herb *Eclipta prostrata. J. Pharmacol. Toxicol*. **2011,** *6* (5), 468–480.

Hossain, F.; Saha, S.; Islan, M. M.; Nasrin, S., Adhikari, S. Analgesic and Anti-infammatory Activity of Commelina Benghalensis Linn. *Turk. J. Pharm. Sci*. **2014,** *11* (1), 25–32.

Huang, S. S.; Chiu, C. S.; Lin, T. H.; Lee, M. M.; Lee, C. Y.; Chang, S. J.; Deng, J. S. Antioxidant and Anti-inflammatory Activities of Aqueous Extract of *Centipeda minima. J. Ethnopharmacol*. **2013,** *147* (2), 395–405.

Hussain, M.; Khan, M. R. U.; Raza, S. M.; Aziz, A.; Bakhsh, H.; Majeed, A.; Mumtaz, F. Assessment of Antibacterial Potential of *Saccharum spontaneum* Linn. (Family: Poaceae), Against Different Pathogenic Microbes-an In Vitro Study, 2014; vol *3* (3), pp 36–41.

Hussain, A. Z.; Kumaresan, S. Phytochemical and Antimicrobial Evaluation of *Oldenlandia corymbosa. Asian J. Plant Sci. Res*. **2013,** *3* (4), 155–158.

Ishak, A. H.; Shafie, N. H.; Esa, N. M.; Bahari, H. Nutritional, Phytochemical and Pharmacological Properties of Mikania micrantha Kunth. *Pertanika J. Scholarly Res. Rev.* **2016,** *2* (3).

Islam, M. Z.; Hossain, M. T.; Hossen, F.; Mukharjee, S. K.; Sultana, N.; Paul, S. C. Evaluation of Antioxidant and Antibacterial Activities of *Crotalaria pallida* Stem Extract. *Clin. Phytosci*. **2018,** *4* (1), 1–7.

Ittiyavirah, S. P.; Sajid, K. P. Anti Stress Activity of Mikania Micrantha Kunth Roots in Wistar Albino rats. *J. Sci. Innov. Res*. **2013,** *2* (6), 999–1005.

Jakhar, S.; Dahiya, P. Antimicrobial, Antioxidant and Phytochemical Potential of Alternanthera Pungens HB and K. *J. Pharm. Sci. Res*. **2017,** *9* (8), 1305–1311.

Javed, S.; Bashir, U. Antifungal Activity of Different Extracts of *Ageratum conyzoides* for the Management of Fusarium Solani. *Afr. J. Biotechnol.* **2012,** *11* (49), 11022–11029.

Jiang, Y.; Ownley, B. H.; Chen, F. Terpenoids from Weedy Ricefield Flatsedge (*Cyperus iria* L.) are Developmentally Regulated and Stress-Induced, and have Antifungal Properties. *Molecules* **2018,** *23* (12), 1–13.

Joffry, S. M.; Yob, N. J.; Rofiee, M. S.; Affandi, M. M. R.; Suhaili, Z.; Othman, F.; Zakaria, Z. A. *Melastoma malabathricum* (L.) Smith Ethnomedicinal Uses, Chemical Constituents, and Pharmacological Properties: A Review. *Evid. Based Complement. Altern. Med.* **2012,** 1–12.

Kala, C. P. Herbal Treatment for Snakebites in Uttarakhand State of India. NISCAIR-CSIR, India, 2015; pp 56–61.

Kanjilal, U. N.; Kanjilal, P. C.; Das, A.; De, R. N. In *Flora of Assam*; Govt. Press: Shillong, India, 1939.

Kanjilal, U. N.; Kanjilal, P. C., Das, A., De, R. N., Bor, N. L. In *Flora of Assam*; Govt. Press: Shillong, 1934–1940.

Karou, D.; Dicko, M. H.; Sanon, S.; Simpore, J.; Traore, A. S. Antimalarial Activity of *Sida acuta* Burm. f. (Malvaceae) and *Pterocarpus erinaceus* Poir. (Fabaceae). *J. Ethnopharmacol.* **2003,** *89* (2-3), 291–294.

Kaurinovic, B.; Popovic, M.; Vlaisavljevic, S.; Trivic, S. Antioxidant Capacity of Ocimum basilicum L. and Origanum vulgare L. extracts. *Molecules* **2011,** *16* (9), 7401–7414.

Kayembe, J. S.; Taba, K. M.; Ntumba, K.; Tshiongo, M. T. C.; Kazadi, T. K. In Vitro Anti-Malarial Activity of 20 Quinones Isolated from Four Plants used by Traditional Healers in the Democratic Republic of Congo. *J. Med. Plants Res.* **2010,** *4* (11), 991–994.

Kekuda, T. P.; Shree, V. D.; Noorain, G. S.; Sahana, B. K., Raghavendra, H. L. Ethnobotanical Uses, Phytochemistry and Pharmacological Activities of *Clerodendrum infortunatum* L. (Lamiaceae): A Review. *J. Drug Deliv. Ther.* **2019,** *9* (2), 547–559.

Khan, A. V.; Ahmed, Q. U.; Shukla, I.; Khan, A. A. Antibacterial Efficacy of *Bacopa monnieri* Leaf Extracts Against Pathogenic Bacteria. *Asian Biomed.* **2010,** *4* (4), 651–655.

Kilani-Jaziri, S.; Bhouri, W.; Skandrani, I.; Limem, I., Chekir-Ghedira, L.; Ghedira, K. Phytochemical, Antimicrobial, Antioxidant and Antigenotoxic Potentials of *Cyperus rotundus* Extracts. *South Afr. J. Bot.* **2011,** *77* (3), 767–776.

Kumar, K. K. S. Evaluation of Wound Healing Activity of Leaves of *Crinum asiaticum. Int. J. Phytother.* **2011,** *1* (1), 16–20.

Kumari, M. E. V. N.; Krishnan, V. Antimicrobial Activity of *Alternanthera Sessilis* (L) R. BR. Ex. DC and *Alternanthera Philoxeroides* (Mart). Griseb. *World J. Res. Rev.* **2016,** *3* (3), 78–81.

Kumawat, N. S.; Chaudhari, S. P.; Wani, N. S.; Deshmukh, T. A.; Patil, V. R. Antidiabetic Activity of Ethanol Extract of *Colocasia esculenta* Leaves in Alloxan Induced Diabetic Rats. *Int. J. PharmTech Res.* **2010,** *2* (2), 1246–1249.

Kundu, N.; Campbell, P.; Hampton, B.; Lin, C. Y.; Ma, X.; Ambulos, N.; Fulton, A. M. Antimetastatic Activity Isolated from *Colocasia esculenta* (taro). *Anti-cancer Drugs* **2012,** *23* (2), 200–211.

Lai, H.; Lim, Y. Evaluation of Antioxidant Activities of the Methanolic Extracts of Selected Ferns in Malaysia. *Int. J. Environ. Sci. Dev.* **2011,** *2* (6), 442–447.

Lavanya, R.; Pattar, P. V. Traditional Knowledge on Folk Medicine by Rural Women in Chikkanayakanahalli Taluk, Tumkur District, Karnataka. *Int. J. Herb. Med.* **2017,** *5*, 05–10.

Le Thoa, N. T.; Nam, P. C.; Nhat, D. M. Antibacterial Activities of the Extracts of *Mimosa pudica* L. An In-Vitro Study. *Int. J. Adv. Sci. Eng. Sci. Technol.* **2015,** *5* (5), 358–361.

Lee, H. Z.; Bau, D. T.; Kuo, C. L.; Tsai, R. Y.; Chen, Y. C.; Chang, Y. H. Clarification of the Phenotypic Characteristics and aAnti-tumor Activity of *Hedyotis diffusa*. *Am. J. Chinese Med.* **2011,** *39* (01), 201–213.

Lee, M. M. L.; Chan, B. D.; Wong, W. Y.; Qu, Z.; Chan, M. S.; Leung, T. W., Tai, W. C. S. Anti-cancer Activity of *Centipeda minima* Extract in Triple Negative Breast Cancer via Inhibition of AKT, NF-κB, and STAT3 Signaling Pathways. *Front. Oncol.* **2020,** *10*, 1–15.

Lee, S. L.; Song, B. R.; Shin, H. S.; Lee, Y. J.; Park, S. N. Antioxidant, Antiaging and Antimicrobial Effects of Ethanolic Extract and Ethyl Acetate Fraction from *Eclipta prostrata*. *J. Soc. Cosmet. Sci. Korea* **2018,** *44* (3), 349–362.

Lee, S. J.; Umano, K.; Shibamoto, T.; Lee, K. G. Identification of Volatile Components in Basil (*Ocimum basilicum* L.) and Thyme Leaves (*Thymus vulgaris* L.) and Their Antioxidant Properties. *Food Chem.* **2005,** *91* (1), 131–137.

Leite, S. P.; Silva, L. L. S.; Catanho, M. T. J. A.; Lima, E. O.; Lima, V. L. M. Anti-inflammatory Activity of Indigofera Suffruticosa Extract. *Rebrasa* **2003,** *7* (1), 47–52.

Leite, S. P.; Vieira, J. R. C.; de Medeiros, P. L.; Leite, R. M. P.; de Menezes Lima, V. L.; Xavier, H. S.; de Oliveira Lima, E. Antimicrobial Activity of *Indigofera suffruticosa*. *Evid. Based Complement. Alter. Med.* **2006,** *3* (2), 261–265.

Li, Y.; Li, J.; Li, Y.; Wang, X. X.; Cao, A. C. Antimicrobial Constituents of the Leaves of *Mikania micrantha* HB K. *Plos One* **2013,** *8* (10), 1–10.

Liang, H. X.; Bao, F. K.; Dong, X. P.; Zhu, H. J.; Lu, X. J.; Shi, M.; Lu, Q.; Cheng, Y. X. Two New Antibacterial Sesquiterpenoids from *Centipeda minima*. *Chem. Biodiver.* **2007,** *4* (12), 2810–2816.

Limberger, R. P.; Aboy, A. L.; Bassani, V. L.; Moreno, P. R.; Ritter, M. R.; Henriques, A. T. Essential Oils from Four Mikania Species (Asteraceae). *J. Essent. Oil Res.* **2001,** *13* (4), 225–228.

Lin, K. W. Ethnobotanical Study of Medicinal Plants used by the Jah Hut Peoples in Malaysia. *Indian J. Med. Sci.* **2005,** *59* (4), 156–161.

Lin, W. S.; Chen, J. Y.; Wang, J. C.; Chen, L. Y.; Lin, C. H.; Hsieh, T. R.; Wang, P. Y. The Anti-aging Effects of *Ludwigia octovalvis* on *Drosophila melanogaster* and SAMP8 mice. *Age* **2014,** *36* (2), 689–703.

Lin, L. M.; Xiong, S. H.; Zhao, L. J.; Tang, J.; Zhang, Z. M.; Li, Y. M.; Xia, B. H. Extraction, Characterization, Antioxidant, and Immunostimulatory Activities of Polysaccharides from Hedyotis Corymbosa. *Evid. Based Complement. Alter. Med.* **2018,** 1–10.

Luqman, S.; Kumar, R.; Kaushik, S.; Srivastava, S.; Darokar, M. P.; Khanuja, S. P. Antioxidant Potential of the Root of *Vetiveria zizanioides* (L.) Nash; *CSIR*, 2009**;** pp 122–125

Machado, R. R.; Valente Júnior, W.; Lesche, B.; Coimbra, E. S.; Souza, N. B. D.; Abramo, C.; Kaplan, M. A. C. Essential Oil from Leaves of Lantana Camara: A Potential Source of Medicine Against Leishmaniasis. *Revista Brasileira de Farmacognosia* **2012,** *22*, 1011–1017.

Mahishi, P.; Srinivasa, B. H.; Shivanna, M. B. Medicinal Plant Wealth of Local Communities in Some Villages in Shimoga District of Karnataka, India. *J. Ethnopharmacol.* **2005,** *98* (3), 307–312.

Mahmood, N.; Piacente, S.; Burke, A.; Khan, A. L.; Pizza, C. Constituents of Cuscuto reflexa are anti-HIV Agents. *Antivir. Chem. Chemother.* **1997,** *8* (1), 70–74.

Makhija, I. K.; Khamar, D. Anti-Snake Venom Properties of Medicinal Plants. *Der Pharmacia Lettre*. **2010,** *2* (5), 399–411.

Mangathayaru, K.; Lakshmikant, J.; Sundar, N. S.; Swapna, R. G. X. F.; Grace, X. F.; Vasantha, J. Antimicrobial Activity of Leucas aspera Flowers. *Fitoterapia* **2005,** *76* (7-8), 752–754.

Manvar, M. N. Antibacterial Activity of Leaves and Flowers of Ipomoea Aquatica Forsk. (Convolvulaceae). *Asian J. Pharm. Res*. **2018,** *8* (2), 94–98.

Mathappan, R.; Selvarajan, K. K.; Sujeet, S.; Tribedi, S. Evaluation of Antitumor Activity of *Urena lobata* Against Ehrlich Ascites Carcinoma Treated Mice. *Orient. Pharm. Exp. Med.* **2019,** *19* (1), 21–26.

Meena, K. L.; Yadav, B. L. Some Ethnomedicinal Plants of Southern Rajasthan, 2010; pp 169–172.

Mishra, M. R.; Mishra, A.; Pradhan, D. K.; Panda, A. K.; Behera, R. K.; Jha, S. Antidiabetic and Antioxidant Activity of Scoparia dulcis linn. *Indian J. Pharm. Sci*. **2013,** *75* (5), 610.

Mishra, V. K.; Singh, G.; Passari, A. K.; Yadav, M. K.; Gupta, V. K.; Singh, B. P. Distribution and Antimicrobial Potential of Endophytic Fungi Associated with Ethnomedicinal Plant Melastoma malabathricum L. *J. Environ. Biol*. **2016,** *37* (2), 229–237.

Mitra, S.; Mukherjee, S. K. Ethnobotanical Usages of Grasses by the Tribals of West Dinajpur District, West Bengal, 2005.

Mohan, Y.; Jesuthankaraj, G. N.; Ramasamy Thangavelu, N. Antidiabetic and Antioxidant Properties of *Triticum aestivum* in Streptozotocin-Induced Diabetic Rats. *Adv. Pharmacol. Sci*. **2013,** 1–9.

Monte, L. G.; Santi-Gadelha, T.; Reis, L. B.; Braganhol, E.; Prietsch, R. F.; Dellagostin, O. A.; Pinto, L. S. Lectin of *Abelmoschus esculentus* (okra) Promotes Selective Antitumor Effects in Human Breast Cancer Cells. *Biotechnol. Lett*. **2014,** *36* (3), 461–469.

Mukta, U. H.; Roy, R.; Daula, A. S. U.; Ferdous, M.; Chowdhury, A.; Mia, S.; Basher, M. A. Phytochemical Analysis, Antioxidant and Antidiarrhoeal Activities of Methanol Extract of Fimbristylis miliacea (L.) Vahl. *J. Pharmacogn. Phytoth.* **2020,** *12* (1), 10–18.

Mutee, A. F.; Salhimi, S. M.; Yam, M. F.; Lim, C. P.; Abdullah, G. Z.; Ameer, O. Z.; Asmawi, M. Z In Vivo Anti-inflammatory and In Vitro Antioxidant Activities of *Peperomia pellucida*. *Int. J. Pharmacol*. **2010,** *6* (5), 686–690.

Muthukumaran, P.; and Pattabiraman, K. Anti-Ulcer Effects of Ipomoea aquatica forsk Leaves Against Gastric Ulcers in Rats. *Res. J. Pharmacogn. Phytochem*. **2010,** *2* (6), 471–474.

Naiman, R. J.; Bechtold, J. S.; Drake, D. C.; Latterell, J. J.; O'keefe, T. C.; Balian, E. V. Origins, Patterns, and Importance of Heterogeneity in Riparian Systems. In *Ecosystem Function in Heterogeneous Landscapes*; Springer: New York, NY, 2005; pp 279–309.

Nasrin, M.; Afroz, F.; Sharmin, S.; Rana, M. S.; Sohrab, M. H. Cytotoxic, Antimicrobial and Antioxidant Properties of Commelina diffusa Burm. F. *Pharmacol. Pharm*. **2019,** *10* (02), 82.

Ndip, R. N.; Ajonglefac, A. N.; Wirna, T.; Luma, H. N.; Wirmum, C.; Efange, S. M. N. In-vitro Antimicrobial Activity of Ageratum conyzoides (Linn) on Clinical Isolates of Helicobacter pylori. *Afr. J. Pharm. Pharmacol*. **2009,** *3* (11), 585–592.

Nelson Kumar, V.; Sahoo, N. K.; Sahu, M.; Hara Sudhan, H.; Pullaiah, C. P., Muralikrishna, K. S. In Vitro Anticancer Activity of Eclipta alba Whole Plant Extract on Colon Cancer Cell HCT-116. *BMC Complement. Med. Therap*. **2020,** *20* (1), 1–8.

Nicollier, G.; Thompson, A. C. Essential Oil and Terpenoids of Mikania Micrantha. *Phytochemistry* **1981,** *20* (11), 2587–2588.

Nisar, M.; Tariq, S. A.; Marwat, I. K.; Shah, M. R.; Khan, I. A. Antibacterial, Antifungal, Insecticidal, Cytotoxicity and Phytotoxicity Studies on Indigofera. *J. Enzyme Inhib. Med. Chem.* **2009,** *24* (1), 224–229.

Nonato, F. R.; Barros, T. A. A.; Lucchese, A. M.; Oliveira, C. E. C.; dos Santos, R. R.; Soares, M. B. P.; Villarreal, C. F. Anti-inflammatory and Anti-nociceptive Activities of *Blechnum occidentale* L. Extract. *J. Ethnopharmacol.* **2009,** *125* (1), 102–107.

Nordin, N. A.; Mohd Zin, N. B.; Fazilah, N. F.; Wasoh, H.; Ariff, A. B.; Halim, M. Phytochemicals, Antioxidant and Antimicrobial Properties of Senna Alata and Senna Tora Leaf Extracts Against Bacterial Strains Causing Skin Infections. *Bacterial Empire* **2019,** *2* (1), 19–25.

Nwaogwugwu, J.; Uhegbu, F. O.; Okereke, S. C.; Nosiri, C. I.; Amaka, E. N.; Nwamaka, I. J. (Hematological Changes and Antidiabetic Activities of *Colocasia esculenta* (L.) Schatt Stem Tuber Aqueous Extract in Alloxan Induced Diabetic Rats. *J. Pharm. Res. Int.* **2020,** 1–9.

Oh, S. M.; Oh, K. Y.; Ahn, S. C. A Study of Antioxidant and Antibacterial Activates of the Extraction of Perscaria hydropiper L. *Culi. Sci. Hosp. Res.* **2016,** *22* (6), 14–23.

Okoro, I. O.; Umar, I. A.; Atawodi, S. E.; Anigo, K. M. Antidiabetic Effect of *Cleome rutidosperma* DC and Senecio biafrae (Oliv. & Hiern) Extracts in Streptozotocin-Induced Diabetic Rats. *Int. J. Pharm. Sci. Res.* **2014,** *5* (6), 2480–2497.

Oladeji, O. S.; Adelowo, F. E.; Oluyori, A. P.; Bankole, D. T. Ethnobotanical Description and Biological Activities of Senna alata. *Evid. Based Complement. Altern. Med.* **2020,** 1–12.

Olokeogun, O. S.; Ayanlade, A.; Popoola, O. O. Assessment of Riparian Zone Dynamics and its Flood-Related Implications in Eleyele Area of Ibadan, Nigeria. *Environ. Syst. Res.* **2020,** *9* (1), 1–11.

Ortac, D.; Cemek, M.; Karaca, T.; Buyukokuroglu, M. E.; Ozdemir, Z. O., Kocaman, A. T.; Gones, S. In Vivo Anti-ulcerogenic Effect of Okra (*Abelmoschus esculentus*) on Ethanol-Induced Acute Gastric Mucosal Lesions. *Pharm. Biol.* **2018,** *56* (1), 165–175.

Padmavathy, A.; Rasny, M. R. M.; Reyadh, R.; Khan, J. Evaluation of Antibacterial Activity of Different Extracts of *Ipomoea aquatica* Leaves Against *Escherichia coli* and *Salmonella typhi*. *J. Manag. Sci.* **2017,** *15*, 51–68.

Pallerla, P.; Vanapatla, S.; Reddy Yellu, N.; Kumar Bobbala, R. Effect of *Lindernia ciliata* (Colsm.) Pennell. Against Ethanol Induced Oxidative Damage in HEPG2 Cells. *Int. J. Pharm. Chem. Res.* **2018,** *10* (4), 117–120.

Pardhi, P.; Jain, A. P.; Ganeshpurkar, A.; Rai, G. Anti-microbial, Anti-oxidant and Anthelmintic Activity of Crude Extract of Solanum xanthocarpum. *Pharmacogn. J.* **2010,** *2* (11), 400–404.

Pavithra, P. S.; Sreevidya, N.; Verma, R. S. Antibacterial and Antioxidant Activity of Methanol Extract of *Evolvulus nummularius*. *Indian J. Pharmacol.* **2009,** *41* (5), 233–236.

Pena, A. K. F. D.; Camara, J. S. Antioxidant Activity of Talahib (*Saccharum spontaneum*) Leaves Ethanolic Extract Through DPPH Assay. *Philipp. J. Natl. Soc. Sci.* **2019,** *1* (1), 1–2.

Permawati, M.; Anwar, E.; Arsianti, A.; Bahtiar, A. Anti-inflammatory Activity of Nanoemulgel Formulated from *Ageratum conyzoides* (L.) L. and *Oldenlandia corymbosa* L. Extracts in Rats. *J. Natl. Remed.* **2019,** *19* (3), 124–134.

Pieme, C. A.; Ngogang, J.; Costache, M. In Vitro Antiproliferative and Anti-oxidant Activities of Methanol Extracts of *Urena lobata* and *Viscum album* Against Breast Cancer Cell Lines. *Toxicol. Environ. Chem.* **2012,** *94* (5), 987–999.

Pithayanukul, P.; Laovachirasuwan, S.; Bavovada, R.; Pakmanee, N.; Suttisri, R. Anti-venom Potential of Butanolic Extract of *Eclipta prostrata* Against Malayan Pit Viper Venom. *J. Ethnopharmacol.* **2004,** *90* (2-3), 347–352.

Porquis, H. C.; Ang, A. M. G.; Doblas, G. Z.; Amoroso, V. B.; Jacalan, D. R. Y.; Batbatan, C. G.; Cruz, R. Y. D. Anti-inflammatory, Antioxidant and Cytotoxicity Studies on Lycopodiella cernua (L.) J. Sm. in Bukidnon, Philippines. *Asian J. Biol. Life Sci.* **2018,** *7* (2), 47.

Prabagar, S.; Nanthakumar, J.; Thuraisingam, S.; Prabagar, J. Investigation of Antimicrobial Activity of Aqueous Extracts of Eclipta Prostrata. *Int. J. Biol. Sci. Technol.* **2020,** *11* (1), 1–6.

Prajapati, R.; Kalariya, M.; Umbarkar, R.; Parmar, S.; Sheth, N. *Colocasia esculenta*: A Potent Indigenous Plant. *Int. J. Nutr. Pharmacol. Neurol. Dis.* **2011,** *1* (2), 90–96.

Praneetha, P.; Rani, V. S.; Reddy, Y. N.; Kumar, B. R. Hepatoprotective Studies on Methanolic Extract of Whole Plant of *Lindernia ciliata*. *Bangladesh J. Pharmacol.* **2014,** *9* (4), 567–74.

Pukumpuang, W.; Chansakaow, S.; Tragoolpua, Y. Antioxidant Activity, Phenolic Compound Content and Phytochemical Constituents of Eclipta Prostrata (Linn.) Linn. *Chiang Mai. J. Sci.* **2014,** *41* (3), 568–576.

Purnomo, Y.; Soeatmadji, D. W.; Sumitro, S. B.; Widodo, M. A. Anti-diabetic Potential of *Urena lobata* Leaf Extract Through Inhibition of Dipeptidyl Peptidase IV Activity. *Asian Pac. J. Trop. Biomed.* **2015,** *5* (8), 645–649.

Rahman, M. S.; Rashid, M. A. Antimicrobial Activity and Cytotoxicity of Eclipta Prostrata. *Orient. Pharm. Exp. Med.* **2008,** *8* (1), 47–52.

Rajagopal, P. L.; Linsha, K. T.; Kumar, P. S.; Parthasarathy, I. A.; Sreejith, K. R.; Aneeshia, S. Anti-inflammatory Activity of the Leaves of *Urena lobata* Linn. *World Wide J. Multidiscip. Res. Dev.* **2018,** *4* (11), 59–61.

Raju, S.; Kavimani, S.; Sreeramulu, R. K. *Kyllinga nemoralis* (Hutch & Dalz) (Cyperaceae): Ethnobotany, Phytochemistry and Pharmacology. *Pharmacogn. J.* **2011,** *3* (24), 7–10.

Raghavendra, H. L.; Ragesh, M. R.; Avinash, H. C.; Ankith, G. N.; Karthik, N. K. Antimicrobial, Insecticidal, and Antiradical Activity of *Solanum Virginianum* L. (Solanaceae). *Asian J. Pharm. Clin. Res.* **2017,** *10* (11), 163–167.

Ramachandran, S.; Sandeep, V. S.; Srinivas, N. K.; Dhanaraju, M. D. Anti-diabetic Activity of Abelmoschus Esculentus Linn on Alloxan-Induced Diabetic Rats. *Res. Rev. BioSci.* **2010,** *4* (3), 121–123.

Ramadas, D.; Ravishankar, M.; Shwetha, S.; Chikkanna, D. Phytochemical Studies and Antioxidant Activity of Bacopa monnieri Plant Leaves Proteins. *Indo Am. J. Pharm. Res.* **2016,** *6* (1), 4302–4307.

Rani, D.; Khare, P. B.; Dantu, P. K. In Vitro Antibacterial and Antifungal Properties of Aqueous and Non-aqueous Frond Extracts of *Psilotum nudum*, *Nephrolepis biserrata* and *Nephrolepis cordifolia*. *Indian J. Pharm. Sci.* **2010,** *72* (6), 818.

Rasyid, S. A.; Surya, R. A.; Natalia, W. O. R. The Antibacterial Activity of Tembelekan Leaf (*Lantana camara* L.) and Kopasanda Leaf (*Chromolaena odorata* L.) Extracts Against Staphylococcus aureus. *Infect. Dis. Rep.* **2020,** *12* (1), 65–67.

Reddy, A. M.; Babu, M. V. S.; Rao, R. R. Ethnobotanical Study of Traditional Herbal Plants used by Local People of Seshachalam Biosphere Reserve in Eastern Ghats, India. *Herba Polonica* **2019,** *65* (1), 40–54.

Riris, I. D.; Simorangkir, M.; Silalahi, A. Antioxidant, Toxicity and Antibacterial of Ompu-ompu (*Crinum asiaticum*-L) Ethanol Extract. *Rasayan J. Chem.* **2018,** *11* (03), 1229–1235.

Ruslan, K.; Happyniar, S.; Fidrianny, I. Antioxidant Potential of Two Varieties of Sesamum Indicum L. Collected from Indonesia. *J. Taibah Univ. Med. Sci.* **2018,** *13* (3), 211–218.

Saeed, M.; Sharif, A.; UlHassan, S.; Akhtar, B.; Muhammad, F.; Malik, M. *Cyperus Iria* Aqueous-Ethanol Extract Ameliorated Hyperglycemia, Oxidative Stress and Regulated Inflammatory Cytokines in Streptozotocin Induced Diabetic Rats, 2021; pp 1–22.

Saha, S.; Paul, S. A Review on Phytochemical Constituents and Pharmacological Properties of *Enhydra fluctuans* Lour. *J. Pharmacogn. Phytochem.* **2019,** *8* (2), 887–893.

Sahoo, A. K.; Kanhar, S. Antioxidant and Antiulcer Potential of *Hydrolea zeylanica* (L.) Vahl Against Gastric Ulcers in Rats. *Int. J. Complement. Alter. Med.* **2017,** *10* (1), 1–8.

Saikia, S.; Tamuli, K. J.; Narzary, B.; Banik, D.; Bordoloi, M. Chemical Characterization, Antimicrobial Activity, and Cytotoxic Activity of *Mikania micrantha* Kunth Flower Essential Oil from North East India. *Chem. Papers* **2020,** 1–14.

Sakonsinsiri, C.; Kaewlert, W.; Armartmuntree, N.; Thanan, R.; Pakdeechote, P. Anti-Cancer Activity of Asiatic Acid Against Human Cholangiocarcinoma Cells Through Inhibition of Proliferation and Induction of Apoptosis. *Cell. Mol. Biol.* **2018,** *64* (10), 28–33.

Sampathkumar, P.; Dheeba, B.; Vidhyasagar, V.; Arulprakash, T.; Vinothkannan, R. Potential Antimicrobial Activity of Various Extracts of Bacopa monnieri (Linn.). *Int. J. Pharmacol.* **2008,** *4* (3), 230–232.

Samy, R. P.; Thwin, M. M.; Gopalakrishnakone, P.; Ignacimuthu, S. Ethnobotanical Survey of Folk Plants for the Treatment of Snakebites in Southern Part of Tamilnadu, India. *J. Ethnopharmacol.* **2008,** *115* (2), 302–312.

Sannigrahi, S.; Mazumder, U. K.; Mondal, A.; Pal, D.; Mishra, S. L.; Roy, S. Flavonoids of Enhydra Fluctuans Exhibit Anticancer Activity Against Ehrlich's Ascites Carcinoma in Mice. *Natl. Prod. Commun.* **2010,** *5* (8), 1239–1242.

Santosh, K.; Satya, N. Herbal Remedies of Wetlands Macrophytes in India. *Int. J. Pharma Bio Sci.* **2010,** *1* (2), 6–29.

Saraf, A.; Quereshi, S.; Sharma, K.; Khan, N. A. Antimicrobial Activity of Lantana Camara L. *J. Exp. Sci.* **2011,** *2* (10), 50–54.

Satyal, P.; Chuong, N. T. H.; Pham, V. T.; Hung, N. H.; Hien, V. T.; Setzer, W. N. Chemical Composition of the Essential Oils of *Pogostemon auricularius*, A Vietnamese Medicinal Plant. *Natl. Prod. Commun.* **2018,** *13* (5), 617–620.

Scott, M. L.; Nagler, P. L.; Glenn, E. P.; Valdes-Casillas, C.; Erker, J. A.; Reynolds, E. W.; Jones, C. L. Assessing the Extent and Diversity of Riparian Ecosystems in Sonora, Mexico. *Biodivers. Conserv.* **2009,** *18* (2), 247–269.

Sharief, M. U.; Kumar, S.; Diwakar, P. G.; Sharma, T. V. R. S. Traditional Phytotherapy Among Karens of Middle Andaman; CSIR, 2005; pp 429–436.

Sheam, M.; Haque, Z.; Nain, Z. Towards the Antimicrobial, Therapeutic and Invasive Properties of Mikania Micrantha Knuth: A Brief Overview. *J. Adv. Biotechnol. Exp. Ther.* **2020,** *3* (2), 92–101.

Shukla, A. N.; Srivastava, S.; Rawat, A. K. S. An Ethnobotanical Study of Medicinal Plants of Rewa District, Madhya Pradesh, CSIR, 2010; pp 191–202.

Sikka, S. C.; Bartolome, A. R. Perfumery, Essential Oils, and Household Chemicals Affecting Reproductive and Sexual Health, In *Bioenvironmental Issues Affecting Men's Reproductive and Sexual Health*, Academic Press, 2018; pp 557–569.

Simorangkir, D.; Masfria, M.; Harahap, U.; Satria, D. Activity Anticancer n-hexane Fraction of Cyperus Rotundus l. Rhizome to Breast Cancer MCF-7 Cell Line. *Open Access Maced. J. Med. Sci.* **2019,** *7* (22), 3904–3906.

Simpson, D. A.; Inglis, C. A. Cyperaceae of Economic, Ethnobotanical and Horticultural Importance: A Checklist. *Kew Bull.* **2001,** 257–360.

Simpson, T.; Pase, M.; Stough, C. *Bacopa monnieri* as an Antioxidant Therapy to Reduce Oxidative Stress in the Aging Brain. *Evid. Based Complement. Alternat. Med.* **2015,** 1–15.

Sindhu, T.; Rajamanikandan, S.; Srinivasan, P. In Vitro Antioxidant and Antibacterial Activities of Methanol Extract of Kyllinga Nemoralis. *Indian J. Pharm. Sci.* **2014,** *76* (2), 170–174.

Singh, M.; Malhotra, K.; Singh, C. Phytochemical Evaluation and Antioxidant Activity of Different Samples of Pteris vittata in Doon Valley, Uttarakhand Region. *Int. J. Pure App. Biosci.* **2015,** *3*, 296–304.

Singh, D.; Mathela, C. S.; Pande, V.; Panwar, A. Antioxidant and Antimicrobial Activity of Grangea Maderaspatana (L.) Poir. *Extract. JDDT* **2013,** *1,* 46–52.

Singh, V.; Singh, A.; Singh, I. P.; Kumar, B. D. Phytomedicinal Properties of *Cynodon dactylon* (L.) pers.(durva) in its Traditional Preparation and Extracts. *Phytomed. Plus* **2021,** *1* (1), 1–11.

Sirinthipaporn, A.; Jiraungkoorskul, W. Wound Healing Property Review of Siam Weed, *Chromolaena odorata. Pharmacogn. Rev.* **2017,** *11* (21), 1–35.

Smida, I.; Sweidan, A.; Souissi, Y.; Rouaud, I.; Sauvager, A.; Torre, F.; Tomasi, S. Anti-acne, Antioxidant and Cytotoxic Properties of Ludwigia Peploides Leaf Extract. *Int. J. Pharmacogn. Phytochem. Res.* **2018,** *10* (7), 271–278.

Soetardjo, S.; Jong, P. C.; Noor, A. M.; Lachimanan, Y. L.; Sreenivasan, S. Chemical Composition and Biological Activity of the *Centipeda minima* (Asteraceae). *Malaysian J. Nutr.* **2007,** *13* (1), 81–87.

Solomon, S.; Muruganantham, N.; Senthamilselvi, M. M. Anticancer Activity of Abelmoschus esculentus (flower) Against Human Liver Cancer. *Int. J. Pharmacol. Biol. Sci.* **2016,** *6*, 154–157.

Sri, B. S.; Reddi, T. S. Traditional Phyto-antidotes used for Snakebite by Bagata Tribe of Eastern Ghats of Visakhapatnam District, Andhra Pradesh. *Int. Multidiscip. Res. J.* **2011,** *1* (6), 1–9.

Sulochana, A. K.; Raveendran, D.; Krishnamma, A. P.; Oommen, O. V. Ethnomedicinal Plants used for Snake Envenomation by Folk Traditional Practitioners from Kallar Forest Region of South Western Ghats, Kerala, India. *J. Intercult. Ethnopharmacol.* **2015,** *4* (1), 47–51.

Sumitra, S.; Parul, S. Medicinal Potential of Weed Echinochloa colona (l.) Link: A Review. *Int. J. Pharmacogn. Phytochem. Res.* **2018,** *10* (3), 98–102.

Sundhararajan, R.; Ravichandiran, V. Anti-inflammatory Activity of Spermacoce Ocymoides Burm. F. *J. Pharm. Res.* **2012,** *5* (1), 719–721.

Sunil, C.; Somashekar, R. K.; Nagaraja, B. C. Influence of *Terminalia arjuna* on the Riparian Landscapes of the River Cauvery of South India. *Landscape Res.* **2019,** 1–15.

Susilawati, Y.; Nugraha, R.; Krishnan, J.; Muhtadi, A.; Sutardjo, S.; Supratman, U. A New Antidiabetic Compound 8, 9-dimethoxy Ellagic Acid from Sasaladaan (Peperomia pellucida L. Kunth). *Res. J. Pharm. Biol. Chem. Sci.* **2017,** *8*, 269–274.

Tan, W. C.; Jaganath, I. B.; Manikam, R.; Sekaran, S. D. Evaluation of Antiviral Activities of Four Local Malaysian Phyllanthus Species Against Herpes Simplex Viruses and Possible Antiviral Target. *Int. J. Med. Sci.* **2013,** *10* (13), 1817.

Teoh, L.; Gnanasegaran, N.; Adnan, A. F. M.; Taha, R. M. The Comparative Antimicrobial and Anticancer of Chemical Extract from In Vitro and In Vivo Peperomia Pellucida Plantlet. *J. Appl. Biol. Biotechnol.* **2021,** *9* (2), 115–123.

Thomas, T.; Kurup, G. M.; Sadasivan, C. Antibacterial Activity of *Dicranopteris linearis* Under In Vitro Conditions. *STARS: Int. J. Sci.* **2020,** *1* (2), 191–195.

Tsai, J. C.; Peng, W. H.; Chiu, T. H.; Lai, S. C.; Lee, C. Y. Anti-inflammatory Effects of Scoparia dulcis L. and Betulinic Acid. *Am. J. Chinese Med.* **2011,** *39* (05), 943–956.

Ukil, S.; Laskar, S.; Roy, R. N. Physicochemical Characterization and Antibacterial Activity of the Leaf Oil of *Crotalaria pallida* Aiton. *J. Taibah Univ. Sci.* **2016,** *10* (4), 490–496.

Ukpai, O. M.; Amaechi, E. C. Evaluation of In Vivo Antimalarial Activity of the Ethanolic Leaf Extracts of Chromolaena Odorata and Cymbopogon Citratus in Mice. *Nigerian J. Biotechnol.* **2012,** *24* (2012) 27–34.

Varghese, S.; Kannappan, P.; Kanakasabapathi, D.; Madathil, S.; Perumalsamy, M. Antidiabetic and Antilipidemic Effect of Clerodendrum Paniculatum Flower Ethanolic Extract. An In Vivo Investigation in Albino Wistar Rats. *Biocatal. Agric. Biotechnol.* **2021,** *36,* 102095.

Vari, A.; Podschun, S. A.; Eros, T.; Hein, T.; Pataki, B.; Ioja, I. C., Baldi, A. Freshwater Systems and Ecosystem Services: Challenges and Chances for Cross-Fertilization of Disciplines. *Ambio* **2021,** *51* (1), 1–17.

Vasquez, J.; Alarcon, J. C.; Jimenez, S. L.; Jaramillo, G. I.; Gomez-Betancur, I. C.; Rey-Suarez, J. P.; Romero, J. O. Main Plants used in Traditional Medicine for the Treatment of Snake Bites n the Regions of the Department of Antioquia, Colombia. *J. Ethnopharmacol.* **2015,** *170,* 158–166.

Vieira, J. R. C.; De Souza, I. A.; Do Nascimento, S. C.; Leite, S. P. *Indigofera suffruticosa*: An Alternative Anticancer Therapy. *Evid. Based Complement. Altern. Med.* **2007,** *4* (3), 355–359.

Vijayaraghavan, K.; Rajkumar, J.; Bukhari, S. N. A.; Al-Sayed, B.; Seyed, M. A. *Chromolaena odorata*: A Neglected Weed with a Wide Spectrum of Pharmacological Activities. *Mol. Med. Rep.* **2017,** *15* (3), 1007–1016.

Wang, J. H.; Luan, F.; He, X. D.; Wang, Y.; Li, M. X. Traditional Uses and Pharmacological Properties of *Clerodendrum* Phytochemicals. *J. Tradit. Complement. Med.* **2018,** *8* (1), 24–38.

Warra, A. A. Sesame (sesamum indicum l.) Seed Oil Methods of Extraction and its Prospects in Cosmetic Industry: A Review. *Bayero J. Pure Appl. Sci.* **2011,** *4* (2), 164–168.

Warrier, K. C.; Warrier, R. R. Sacred Groves: Repositories of Medicinal Plants. In *Prospects in Conservation of Medicinal Plants*, **2019**; pp 72–102.

Wei, L. S.; Wee, W.; Siong, J. Y. F.; Syamsumir, D. F. Characterization of Anticancer, Antimicrobial, Antioxidant Properties and Chemical Compositions of *Peperomia pellucida* Leaf Extract. *Acta Med. Iran.* **2011,** 670–674.

Widodo, H.; Sismindari, S.; Asmara, W.; Rohman, A. Antioxidant Activity, Total Phenolic and Flavonoid Contents of Selected Medicinal Plants used for Liver Diseases and its Classification with Chemometrics. *J. Appl. Pharm. Sci.* **2019,** *9* (6), 99–105.

Williams, R.; Münch, G.; Gyengesi, E.; Bennett, L. *Bacopa monnieri* (L.) Exerts Anti-inflammatory Effects on Cells of the Innate Immune System In Vitro. *Food Funct.* **2014,** *5* (3), 517–520.

Yadav, A. K.; Tangpu, V. Antidiarrheal Activity of *Lithocarpus dealbata* and *Urena lobata.* Extracts: Therapeutic Implications. *Pharm. Biol.* **2007,** *45* (3), 223–229.

Yen, G. C.; Chuang, D. Y. Antioxidant Properties of Water Extracts from Cassia tora L. in Relation to the Degree of Roasting. *J. Agric. Food Chem.* **2000,** *48* (7), 2760–2765.

Yeom, Y. E.; Kim, M. A.; Kim, J.; Lee, C. M. Anti-inflammatory Effects of the Extract of *Solanum nigrum* L. on an Acute Ear Edema Mouse Model. *Mater. Technol.* **2019,** *34* (14), 851–857.

Zahara, M. In *Description of Chromolaena Odorata LRM King and H. Robinson as Medicinal Plant: A Review*, IOP Conference Series: Materials Science and Engineering; IOP Publishing, 2019; vol *506* (1), p 012022.

Zakaria, Z. A.; Kamisan, F. H.; Kek, T. L.; Salleh, M. Z. Hepatoprotective and Antioxidant Activities of *Dicranopteris linearis* Leaf Extract Against Paracetamol-induced Liver Intoxication in Rats. *Pharm. Biol.* **2020,** *58* (1), 478–489.

Zakaria, Z. A.; Sulaiman, M. R.; Morsid, N. A.; Aris, A.; Zainal, H.; Mohd Pojan, N. H.; Hanan Kumar, G. Antinociceptive, Anti-inflammatory and Antipyretic Effects of *Solanum nigrum* Aqueous Extract in Animal Models. *Meth. Find. Exp. Clin. Pharmacol.* **2009,** *31* (2), 81–88.

Zhang, L. J.; Yeh, S. F.; Yu, Y. T.; Kuo, L. M. Y.; Kuo, Y. H. Antioxidative Flavonol Glucuronides and Anti-hbsag Flavonol from *Rotala rotundifolia*. *J. Tradit. Complement. Med.* **2011,** *1* (1), 57–63.

CHAPTER 11

Processing and Potential of Broom Grass (*Thysanolaena latifolia*)

ACHINTYA KUMAR SINHA[1], MIHIR KUMAR REANG[2], and SOURABH DEB[2]

[1]*Gorkha Basti, Kunjaban, Agartala, Tripura, India*

[2]*Department of Forestry and Biodiversity, Tripura University, Agartala, India*

ABSTRACT

Broom grass grows in dense clumps known as "tussocks" producing numerous upright or arching stems that are unbranched and have joints at regular intervals. Under natural conditions, it regenerates through seeds. Its inflorescence is a huge and drooping panicle of about 60–90 cm. Culms of 75 cm are generally considered ideal for broom making. Due to the high demand in the past few years, some parts of Tripura started cultivating broom grass as an economic source. The quality of broom grass depends on several factors. The best time for harvesting panicles is after anthesis but before pollination. Panicles can be harvested by plucking or cutting at the base of the tussock. Plucked ones make better brooms compared to those made of cuttings. Harvesting by cutting at the base of the tussock, on the other hand, is good for the regeneration of the stock. The marketing chain of broom grass originates from collectors and producers and normally moves through a number of intermediaries consisting of middlemen, traders, retailers, and then consumers. Large-scale artificial cultivation of broom grass should be taken up in degraded forest land for stabilization of steep hill slopes as well as for augmenting the supply of its pinnacles for broom production.

High-Value Plants: Novel Insights and Biotechnological Advances. Thangjam Anand Singh, Prakash Kumar Sarangi, and Ch. Brajakishore Singh (Eds.)
© 2025 Apple Academic Press, Inc. Co-published with CRC Press (Taylor & Francis)

11.1 INTRODUCTION

Broom grass (*Thysanolaena latifolia* Roxb. ex Hornem.) is a reed-like tufted perennial grass with leaves resembling those of bamboo. It belongs to the family *Poaceae*. It is known as *Arjun Phul* (flower) in Tripura. The species grows in southeast Asia and India, where it occurs in abundance in the northeastern states and in the states of Orissa, Chhattisgarh, and Andhra Pradesh. It grows naturally in open areas with sparse tree cover, particularly in logged-over sites and hill slopes, damp lowland areas, steep banks along ravines, and on sandy banks of rivers. It is an excellent soil binder. It grows in tussocks, producing numerous upright or arching, unbranched stems with joints at regular intervals. Its inflorescence, a drooping panicle 60–90 cm long, resembles a foxtail. The species acquires its economic importance principally for its bushy panicles, which are used in making soft brooms. The leaves and the tips of broom grass are also used as cattle feed during times of fodder scarcity. It has been cultivated commercially on a large scale in Meghalaya since 1995. There has been some initiative for planting broom grass in Tripura since 2012.

Thysanolaena latifolia is a huge tufted grass. The inflorescence is huge, the axis and branches at first rounded, ultimately capillary, not sharply angled; spikelets are numerous, often in pairs on a common peduncle, each pedicel distinct; lower glume clasping, ovate-acute; lower lemma lanceolate-acuminate, subhyaline, with one or two long setose hairs near the margin; upper lemma lanceolate-acuminate, three-nerved, green between the nerves; hyaline thence to the margin, with stiff setose hairs along the hyaline portion on both sides; palea narrow, two-nerved, hyaline scale; stamens two (3); stigmata two, plumose; reddish brown, the rachilla continues as a flattened process with an expanded tip, beyond and behind the upper lemma. The aspect of the spikelet changes with the onset of anthesis when the upper lemma emerges and its setose hairs gradually adopt a stance at right angles to the lemma's surface (Bisht and Ahlawat, 1998). An analysis of growth parameters, namely, leaf number, leaf area index, basal area, canopy cover, and tussock height, revealed that broom grass is a fast-growing perennial grass that attains maximum growth in 4 years (Tiwari et al., 2012).

There are two distinct types or grades of harvested broom grass, which depend on the mode of harvest. If the panicle stick is pulled out gently by hand, leaving out the sheaths on the lower part, it is of plucking grade. On the other hand, if it is harvested by cutting the panicle stick at the base, it is nonplucking, often known as the cutting grade. Datta et al. (2008) reported the presence of broom grass in the tilla land of Tripura. It is found that broom grass has two types depending on the geographical region in Tripura, and the local collector called one "Hard" and the other "Soft" with their flowering types. The "Hard" type is particularly found in the Sakhan hill and Jampui hill areas. The collector named it as hard because the broom plants from this area, as soon as they start flowering, mature and becomes difficult to pull out the panicles. So we need to use the cutting method during harvesting. The "Soft" type is the normal type of broom grass that grows in all parts of Tripura. The soft types of broom grass take time to mature their seeds, which makes it easy for the collector to pull out the panicles during harvesting.

11.2 METHODOLOGY

Relevant literature on broom grass in general and with special reference to the state of Tripura was scanned for an overview of the sector. Some of the basic statistics were collected and compiled from official records, particularly from the State Forest Department, the NTFP Center of Excellence of Tripura, the JICA project, the Tripura Bamboo Mission, TRPC (Tripura Rehabilitation Plantation Corporation Limited), and randomly selected Forest Field Offices. This species was selected based on secondary information, inputs from consultation with stakeholders in the NTFP sector, such as cultivators, harvesters, processors, traders, consumers, and facilitators, including government agencies and other institutions and organizations. The selection of species and products was also done in consideration of their market share and their importance in the context of culture and tradition. Primary data were gathered from stakeholders through general discussion, a structured questionnaire, and PRA. Photographs of products and processes were taken for better appreciation of the content and for clarity. The final manuscript was prepared through a rigorous process of data processing, analysis, and interpretation based on inputs from primary and secondary sources.

11.3 RESULT AND DISCUSSION

11.3.1 CULTIVATION OF BROOM GRASS

Due to the high demand in the past few years, some people in Tripura started the cultivation of broom grass as a part of their economic source. The cultivators mostly choose the hilly slope area nearby the stream or river for cultivation of broom grass and to get the best quality broom panicles. The selected area for cultivation is cleared by cutting down the woody trees and other plant species growing within the area, and this area is left for some weeks to dry the felled woody trees and other plant species before proceeding with burning. This process is usually done in the months of February and March in Tripura. When the felled woody trees and other plant species get dried properly in the cleared area, they start burning them thereafter, usually in the months of March and April. The burned soil contains potash, which increases the nutrient content of the soil. The cultivator collects the rhizome/ roots of wild and naturally regenerated broom grass plants from the nearby forest area in huge quantities. Once the collection is done, the rhizomes are taken to the selected area for cultivation. The cultivators start the cultivation of rhizomes in the months of April or May before the rainy season begins by creating a mound of soil for each rhizome with spacing about 5–6 feet apart so that they have plenty of space to grow afterward. Weeding is done twice or thrice, depending on the weed population, before harvesting, once when the rhizomes give rise to a new plant (after a few months of plantation), depending on the growth of the plants or the rainy season, and before the plant gets mature. It is done so that the unwanted plant does not disturb the growth of the broom grass plant or the nutrients supplied to the soil.

11.3.2 HARVESTING OF BROOM GRASS

The initiative for planting broom grass is a recent phenomenon in Tripura and is yet to catch up in a big way, unlike in Meghalaya, where most harvesting is done from natural stock. The time of harvest is very important, as that largely determines the quality. The flowering normally commences in the month of December and continues until February. To get the best quality of panicles for soft brooms, harvesting should be done within 1 week of flowering, that is, after anthesis and before pollination (Banumathi, 2013). Jackobsen (2006) reported the best quality broom production in 3-years-old fallow in the uplands of northcentral Vietnam. The panicles should be soft and green at the time of harvest for the best quality. Delay in harvesting after maturity of panicles leads to discoloration from the preferred green to a reddish tint, and these fetch very low prices in the market and also affect the quality of the broom. The harvesting season for broom grass in Tripura starts from the third week of December to the first week of March. The best season for its harvest in Tripura is between the third week of December and the third week of January. This, however, depends to an extent on whether it rains in December–January or not. Thus, to get the best quality of panicles for broom grass, one should also depend on physical inspection of the panicles

(Fig. 11.1a and b). Harvesting is done mainly in two different ways, one by gently pulling out the panicle by hand from the tussock. This product is the plucking grade of broom grass. The other way is to harvest the panicle by cutting the stem at the base with a sickle. The later product is the nonplucking grade. The stem of the nonplucking grade is covered by rough sheaths, unlike the plucking grade, which has a smooth and bare stem and is much lighter in weight. The process of pulling out the panicles is done when the flower or panicles are suitable for removal before they are fully mature and tough. Pulling out the panicles with stems adversely affects regeneration. In Zimbabwe, traditionally, broom grass has been dug (because the roots help to hold the brooms together), but there is increasing concern that this practice is not sustainable and that it is preferable to cut the grass and leave the roots intact (Mutimukuru et al., 2002). The recommended procedure is to first cut the tussock bearing the panicle at about 10 cm above the ground with a sickle, hold the base firmly with one hand, and gently pull out the panicle with the naked stem by the other hand. This, however, is hardly practiced in Tripura despite sensitization and campaigns by different organizations obviously because nobody, not even the JFM communities, has any sense of owning the broom grass in forestland, and they are only interested in the immediate benefit. The method of cutting the stem is mostly used in the hilly areas of Jampui hill and Sakhan hill, as the broom grass panicles from these areas become hard as soon as they start bearing panicles and become difficult to pull out. It is also applied in harvesting during the late harvesting season when the flower or panicles get mature.

After harvesting, the plants' residues are burned so that their roots give rise to new plants for the next season. In once-cultivated areas, the process of rhizome cultivation is not required for the next season, and other processes like burning and weeding are repeated again for harvesting in the next season (Table 11.1).

TABLE 11.1 Monthly Activities of Broom Production in Tripura.

Activities	Jan	Feb	Mar	Apr	May	Jun	Jul	Aug	Sep	Oct	Nov	Dec
Slashing		■	■									
Burning			■	■								
Cultivation				■	■							
Weeding							■	■	■	■		
Harvesting	■	■										■

The color in the box denote the season.

11.3.3 PROCESSING OF BROOM

The area for drying broom panicles is selected according to the quantity of broom grass. It may be a small or large area, like a harvested paddy field. Broom grass is loosely laid out and spread in thin layers on the top of a raised or laid bamboo *machan* or rack for drying under the sun (Fig. 11.1c–f). Sundrying should continue for 3–4 days for pull-out panicles and 4–5 days for cut-stem panicles, depending on the sunlight. Some of them soak the bottom part of the pulled-out panicles in boiled (hot) water for a few seconds before proceeding with the drying process so that it gets dried faster than usual. It takes around 2–3 days to dry. The full and halved bamboos on the racks are usually spaced out 12 in or 30 cm apart for good aeration. The broom grass on the platform is turned over once every 2 h for uniform drying. A deposit of dew on the grass can spoil the color and quality. The grass, therefore, should be taken out of the godown for sunning by 9 a.m. and returned to the godown by 4 p.m. In the godown, the broom grass should be stored leaning on supports made of bamboo or otherwise, with the soft panicle upward. In the absence of a godown, the grass on the top of the *machan* should be covered with thick polythene sheets between 4 p.m. and 9 a.m. the next morning. Fruits and seeds, if any, on the inflorescence should be removed during drying by shaking and tapping the grass gently against any hard object like a stone or a wood log. Once the drying process is complete, the panicles are bundled into an average size of 7 cm in diameter, containing about 210 panicles, and stored with the panicles upward for sale, supply to local traders, or delivery to processors for broom making. However, 3.2 kg of fresh soft-type or pull-out panicles covert into 1 kg when it gets dry, and it takes 2.5 kg of fresh hard-type or cut panicles, including stems, to get 1 kg of dry broom grass. The average weight of fresh hard and soft broom grass to get 1 kg dry is around 2.8 kg.

The local collectors harvest the broom panicles and sell the fresh products to the local buyer. Some of the locals even made the final product and sold it in the markets. Dried broom grass tied in a variety of forms as bundles is used for sweeping. Ordinary brooms are made of the nonplucking grade of grass that is sold in smaller local markets. Nonplucking grade panicles require 2.5 kg to get 1 kg of dried broom grass, and this 1 kg of dried broom grass panicles can produce around four final product brooms. These are made by bundling around 25–30 nonplucking-grade sticks of broom grass and tying the stems in the lower half with plastic tape and iron wire. Thin bamboo sticks are also inserted from the stem end to tighten the bundle further, and the broom is ready for use. These brooms are relatively cheaper, but they do not last long because the sheaths attached to the stems of non-plucking grade of grass slip off, loosening the bundle after 3–4 months of use. The plucking-grade broom grass with thin and smooth stems and long green panicles is used for making sophisticated household brooms for export outside the state and also for sale in Agartala, other urban markets, and seasonal fairs in the capital town and district headquarters. The grass is bundled as in the case of the ordinary brooms described above, but no bamboo sticks are needed for packing as the stems in this case are solid and without sheaths. The stems are bound by closely intertwining colorful plastic tapes in attractive designs. To top it all off, a brightly colored small plastic cap is fitted into the lower end of the stems for a strong grip. In this conventional type of broom, three separate grades are recognized: King, Queen, and Jack. The King grade, being the top one, has more intricate weaving with thinner colored plastic tapes compared to the Queen grade. The Jack does not show the binding material, and the lower part of the stems, being the handle, is entirely covered with a colored plastic pipe. Some Self-Help Groups (SHGs) have started the manufacture of more sophisticated brooms with bamboo handles of different lengths since 2009, with training from artisans from Shillong, organized by the Forest Department and Tripura Rehabilitation Corporation Limited. For this, the stems are cut short. They are made with bamboo poles as handles of different lengths for different purposes, like a very long handle for cleaning the ceiling, a medium length for cleaning framed paintings and artifacts on the walls, and a very short handle for cleaning the bed or table tops. The bamboo poles used are slim and around 6–7 cm in girth. The species of bamboo used for handles of brooms in Tripura are usually Muli (*Melocanna baccifera*) and Kanak kaich (*Thyrsostachys oliveri*). Apart from the tools required for conventional brooms, an appropriate type of glue is necessary for strong bonding between a bundle of stems with the bamboo handle and also a hand saw and secateurs. Over and above that, a small burner is essential for flame treatment to create spiraling dark-brown shining stains on bamboo handles for another level of customization.

The productivity of an average artisan from JFMCs and SHGs linked to JFMCs, mostly tribal, is very low and is around 50% of the productivity of an artisan engaged by regular processors in their factory; in the factory, they turn out around 125 brooms of the King grade or about 150 of the Queen grade per person per day. The artisans with JFMCs and SHGs are likely to improve their productivity if they continue to make brooms every season with occasional refresher training organized by the Forest Department through JFMCs.

FIGURE 11.1 Harvesting and processing of broom grass: (a) harvesting, (b) bundling, (c) drying, (d) storing, (e) processing, and (f) final product.

11.4 MARKETING OF BROOM

For centuries, brooms have been used for cleaning houses, ovens, fireplaces, yards, and streets, as ritual tools, as well as for some special functions. It is one of the most important tools one needs to have. For a long period, brooms were domestically produced and hand-made of tree branches, bushes, etc. The broom is an important tool in keeping the living area clean. For economic reasons, keeping cleanness in houses, areas around houses, farmyards, street, etc., brooms are a daily necessity. Together with the use of technical instruments for cleaning streets and yards, brooms are still in use, so the requirement and demand for brooms in the market are always high throughout the year (Zairemmawii, 2016). Brooms made of the nonplucking grade are almost entirely sold through retailers in the local market by the local traders (processors). SHGs linked to JFMC's brooms sell them to local traders and also directly to customers in local markets. On an average, 12 truckloads of conventional brooms are exported to Siliguri, Mumbai, and other destinations, and that accounts for 144 MT of dried broom grass. According to a rough estimate based on interaction with traders, 160,000 brooms are sold locally for use within the state, and that would account for 48 MT of dried broom grass. Assuming an average collection of 450 MT of broom grass per year (based on the records for 2011–2012 and 2012–2013), the conversion of raw grass into finished products within the state is 42%. An average of 15–16 truckloads of dried broom grass are exported outside the state, mainly to Mumbai, Kolkata, Bihar, Jharkhand, Ranchi, UP, etc. (Sharma, 2010).

The middlemen or traders who are going to buy all the broom grass from local harvesters or collectors in a particular area need to sign an agreement with the village committee of that area, like JFMC (Joint Forest Management Committees) and BMC (Biodiversity Management Committee). The only person who signed the agreement is permitted to supply or export the broom grass to other places. The middlemen or traders play an important role in marketing the broom grass for export. They sign an agreement with the village's committee, collect all freshly harvested broom panicles from the local collectors, and do the processing work like drying and supplying the dried broom grass to other places. They export the dried broom grass after charging all prices for the taxes paid to village committees, drying, and transportation fees (Table 11.2).

TABLE 11.2 Investments by the Middlemen/Traders to Get 1 kg of Dried Broom Grass.

Charges	Amount (Rupees)
Fresh	$2.8 \times 25 = 70$
BMC	5% per kg = 1.25
JFMC	2% per kg = 0.5
GP	3
TP	2
Transportation	1
Drying (Labor)	5
Total	82.75

After calculating their total investment, the middlemen or traders set the final price of dried broom grass at around Rs. 90–96 per kg. Broom grass was leased out by the Forest Department to traders and processors against a lump sum lease rent from 1960 until 1984–1985. The Tripura Apex Marketing Cooperative Society, now the State Cooperative Marketing Federation of Tripura Limited (MARKFED), was entrusted with the responsibility for harvesting broom grass from forestland on a lease basis since 1985–1986 to ensure a fair price to collectors according to government policy. The LAMPS (Large-Sized Multipurpose Society) and the PACS (Primary Agricultural Credit Society), the grassroots-level marketing and credit societies, procure broom grass from harvesters for supply to MARKFED. The LAMPS and PACS are institutional members of MARKFED. The broom grass business was, however, hardly ever a financially gainful activity for MARKFED, and in the best of times, like in 2005–2006, the cost of procurement and sales had just matched out (Sharma, 2010). That broom grass was not a good business for MARKFED was further demonstrated by the fact that it had not accepted the lease in 1999–2000 and that the lease rent had to be lowered more than once; from Rs. 2.25 lakhs in 1989–1990, it had to be scaled down to Rs. 0.50 lakh in 2006–2007.

A few SHG and JFM Committees have commenced the collection of broom grass, the making of brooms, and their sale in the local markets since 2008–2009. The State Forest Department too encouraged such involvement of JFM Committees in broom making. The annual collection of broom grass "used to be 250–300 MT," and at that stage, on a proposal from the NTFP Center of Excellence of the Tripura-JICA project (NCE), the Forest Department allowed NCE to organize the collection, processing, and marketing of broom grass by "JFM members" of the Pencharthal range during 2010–2011 as a pilot project (NCE Leaflet: From Grass to Glory—A Success Story).

Under agroforestry in north Tripura, realizing the strength of the site in Depacherra village under Kanchanpur Forest Division, the PRA approach explained the benefits of cultivating broom grass along with perennial arhar. Jhumias were motivated to cultivate this cash crop. In the upper-storey component, 2500 plants of broom grass were planted under intercropping with perennial arhar, chilli, brinjal, and banana in an area of 1 ha. In the second year, chilli and brinjal plants were not there because of the substantial shade of the main crops. The number of arhar plants was also reduced to 625/ha (25% due to wind damage) after the first harvest and also due to competition with broom grass under huge but centrifugal expansion in the second year (Kaushik, 2014).

Under the TRPC project on broom grass (2019), they bought the raw, fresh broom grass at Rs. 25 per kg and the dried one at Rs. 55 per kg from the local harvester. They trained the beneficiaries to manufacture the final products and hired them to do so with a payment of Rs. 8 per piece of broom. After the final products were manufactured, they supplied them to the local markets like Agartala, Dharmanagar, etc., and to markets outside the state like Kerala and Karnataka at a price of Rs. 35 per piece.

These sophisticated products with decorated bamboo handles are sold mostly at fairs and exhibitions within Tripura, but a clientele has been growing, indicating their future potential within the state. There is a good market for this sophisticated version of soft brooms with handles in the metros and big cities within the country and abroad, provided the state is ready to manufacture them on a large scale in industry mode, ensuring high-quality standards. The cost of production of King-grade brooms and conventional household brooms is presented in Table 11.3.

TABLE 11.3 Cost of Production and Revenue from the Sale of One Truckload (40,000) of Brooms from Tripura in Mumbai.

Costs	Per broom (Rs.)	For 40,000 brooms (Rs.)
Grass 400 grams @ 38[a]/kg	15.20	608,000
Cost of plastic binding materials	4.0	160,000
Cost of iron wire and pins	0.50	20,000
Plastic cap	2.90	116,000
Labor cost	1.60	64,000
Transportation	3.38	135,200
Total cost	27.58	1,103,200

[a]Average cost calculated by taking into account weight loss in storage and damages.

Broom grade: King grade (400 gm of broom grass per broom).

The local harvesters, on an average, can only collect around 10 kg of broom grass in a day, which means they earn only Rs. 250 per day (Rs. 25 × 10 = 250). This is less than their daily earnings, so most of the locals/villagers do not prefer harvesting broom grass. According to the data mentioned earlier, in marketing, the middlemen/traders and the one who makes the final product make a good profit in this business.

The unemployed local villagers knew that trading and marketing broom grass had a good income but restricted it due to the following reasons:

i. The locals or villagers can't store the harvested broom grass in huge quantities because they require daily earnings for their livelihood, so they sell the fresh collection to the middlemen on a daily basis.
ii. They lack knowledge of marketing channels. Even selling in the rural market has no stability, as most of the households in villages have their own.
iii. Lack of finance to buy, collect the freshly harvested broom grass from the local harvesters, and pay taxes to the village committees like JFMC and BMC.
iv. Required a forest trade license for trading, which is not easy for the villagers.

11.5 CONCLUSION AND RECOMMENDATIONS

The state government may cause an independent review of the existing processes in the business of harvesting broom grass from forestland, its procurement, processing, value addition, and marketing for the purpose of developing an appropriate protocol for fair trade. Community members of JFMCs may be entitled to free access to NTFP, subject to regulation by the concerned JFMCs, to secure sustainable yield with due regard to conservation. Similarly, right holders under the Scheduled Tribes and Other Traditional Forest Dwellers (Recognition of Forest Rights) Act, 2006 (FRA) are also entitled to free access to NTFP for the land allocated in their favor under FRA. In place of forest royalty, NTFP Conservation and Development Fees (NCDF) may be charged at an appropriate rate payable to the concerned Beat Officer-cum-Member Secretary of the JFMC. This deposit should be kept in the revolving fund (Village Development Fund) of the JFMC for appropriate uses, including postharvest cultural operations in broom grass areas, grants of microcredit to SHGs constituted with members of JFM households for broom making, etc., as may be resolved in the Annual or Extraordinary General Body (GB) meeting of the JFMC, subject to guidelines according to resolutions in the Executive Committee (EC) or GB meetings of the Forest Development Agency (FDA) at the forest division level. The BO-cum-Member Secretary should be authorized to issue transit passes to traders (processors) for broom grass or value-added products made thereof against receipts. The revolving fund account should be operated under the joint signature of the Member Secretary and the President of the JFMC. Range officers shall examine the

VDF accounts during the routine annual inspection of the beat office and submit copies of the inspection report to the DFO for necessary action. The part of the inspection report dealing with VDF shall be discussed in the GB of the EC of the concerned JFMC for necessary resolution and action thereon. The Forest Department may call for tender centrally from registered traders (processors) of broom grass and/or broom making with the requirement of earnest money (equivalent to 2.5% of the estimated market value of harvestable broom grass) deposit for all the other ranges, excluding those allocated to MARKFED and except those in protected areas, with the target to finalize allotment by the month of August. The traders (processors) in tender shall quote the minimum rate of procurement price payable to collectors for plucking and nonplucking grades of broom grass. The earnest money deposit will be automatically converted to a security deposit after the issue of the work order. The Forest Department shall organize wide publicity of the procurement price of plucking and nonplucking grades of broom grass for each range, and it will be the duty of the concerned beat officer to report cases of payments at lower than the approved rates and other irregular and/or illegal or damaging actions by the traders or others in the matter of harvesting, processing, and trade of broom grass and value-added products thereon. The Forest Department shall organize training and motivation of SHGs constituted of members of JFMC households in broom making with financial and infrastructure support from concerned departments of the state government. Range officers shall help and guide the SHGs linked to JFMCs under their jurisdiction on the procurement of plastic tapes, plastic caps, etc. for the manufacture of brooms at reasonable rates and also in the matter of marketing. The concerned DFO shall entrust a senior officer, preferably at the level of additional DFO, to monitor progress and intervene when necessary, and the said officer shall also organize continued motivation of traders (processors) for interaction with concerned SHGs for quality control, followed by procurement of finished products at a fair price from them for mutual benefit in the long run. The cultivation of broom grass shows lots of scope for the people living in rural areas as a source of income, so the government needs to encourage the farmers to cultivate broom grass as cash crops. There is no fixed price for fresh, dry, and final products of broom grass planned by the government. Therefore, it would be beneficial to collectors and producers if the government could set a fixed price for brooms.

KEYWORDS

- ***Thysanolaena latifolia***
- **Tripura**
- **broom grass**
- **sustainable harvest**
- **cultivation**
- **economics**

REFERENCES

Banumathi, G. *Study on Productivity of Broom Grass in Naturally Regenerated Area Taken up in Subalshing and Pecharthal Range*; Forest Research Division Report, Agartala, 2013.

Bisht, N. S.; Ahlawat, S. P. *Broom Grass* (Report); State Forest Research Institute, Itanagar, Arunachal Pradesh, India; Information Bulletin no.6., 1998.

Datta, B. K.; Saha, R.; Roy, M.; Majumder, K. Grasses of West Tripura District, Tripura, India. *Pleione* **2008,** *2* (1), 98–105.

Jakobsen, J. The Role of NTFPs in a Shifting Cultivation System in Transition: A Village Case Study from the Uplands of North Central Vietnam. *Danish J. Geog.* **2006,** *106* (2), 103–114.

Kaushik, P. K.; Saha, A. Broom Grass under Agroforestry – A Success story of Participatory Research and Extension. *Manjari* 2014; vol *1* (1).

Mutimukuru, T.; Nyirenda, R.; Matose, F. *Learning Amongst Ourselves: Towards Adaptiveness by Stakeholders in Forest Management Through Social Learning in Mafungautsi*, Manuscript, CIFOR, Harare, 2002.

Sharma, D. K. *Non-Timber Forest Products of Tripura*. NTFP Centre of Excellence, Tripura JICA Project, Forest Department, Tripura, India, 2009; vol I.

Tiwari, B. K.; Shukla, R. P.; Lynser, M. B.; Tynsong, H. Growth Pattern, Production, and Marketing of *Thysanolaena maxima (Roxb.) Kuntze*: An Important Non-timber Forest Product of Meghalaya, India. *For. Trees Liveli.* **2012,** *21* (3) 176–187.

Zairemmawii. *A Study of Marketing of Broom in Aizawl*. M.Phil. Desertation, MZU, 2016.

CHAPTER 12

Lichens in the Himalayan Region: Present and Future Perspectives

AWADHESH KUMAR,[1] NURPEN MEITEI THANGJAM,[1] TBC LALDINGLIANI,[1] and CH BRAJAKISHOR SINGH[2]

[1]*Department of Horticulture, Aromatic, and Medicinal Plants, School of Earth Sciences and Natural Resource Management, Mizoram University, Aizawl, Mizoram, India*

[2]*Institute of Bioresources and Sustainable Development, Takyelpat, Imphal, Manipur, India*

ABSTRACT

Lichens are a symbiotic relationship between fungi (mycobionts) and their algal partners (photobionts). They are distributed throughout the world. Their sensitiveness to environmental pollution and climate change makes them more responsible for environmental biomonitoring. The use of lichens all over the world is a long-standing practice in human civilization; they are used for different purposes like food, fodder, spices, dyes, cosmetics, and perfumery. Parmotrema tinctorum is used as a spice and flavoring agent; Lobaria pulmonaria, Parmelia saxatilis, and Parmotrema crinitum are used for dyes by many ethnic groups; whereas Pseudevernia furfuracea and Evernia prunastri are highly demanded in the perfumery industries. Other lichens, such as Roccella tinctoria, are used for litmus paper; Xanthoria parietina, X. elegans, and Cetraria islandica provide protection against UV radiation. Cetraria islandica, Cladonia rangiferina, Evernia prunastri, Everniastrum cirrhatum, Hypogymnia physodes, Parmotrema chinense, Peltigera canina, and Usnea longissima were reported in traditional medicine and biological purposes, including Charila and Ushna as ayurvedic and unani products from

High-Value Plants: Novel Insights and Biotechnological Advances. Thangjam Anand Singh, Prakash Kumar Sarangi, and Ch. Brajakishore Singh (Eds.)
© 2025 Apple Academic Press, Inc. Co-published with CRC Press (Taylor & Francis)

lichens. The above facts are enough to highlight the impact of the Himalayan region, where thousands of lichens are growing luxuriantly. Therefore, it is high time to explore more lichen species for commercial uses and livelihood generation in the Himalayan region. Parallely, awareness programs are also required to develop conservation strategies for lichens.

12.1 INTRODUCTION

The term "Lichen" was given by Theopharastus (the father of botany) in 371–286 BC for superficial growth on the bark of the Olea europaea (olive) tree. The word lichens is of Greek origin and is pronounced as "Lie Kens" (Hawksworth and Hill, 1984). Lichens are considered a constant and self-sustaining relationship between fungi as mycobionts and their algal partners as photobionts. As the mycobiont is solely involved in the symbiotic relationship, it typically controls the relationship (Ranković and Kosanić, 2015). Lichens are also known as one of the most positive symbiotic relations between a fungus and green and/or blue-green alga, which can survive everywhere on this planet (Rai et al., 2015). The field of botany that deals with the study of lichens is called lichenology, and the scientist or researcher who works in the field of lichenology is called lichenologist.

Lichens don't have the proper natural system of classification; they are classified based on the nature and kinds of fruiting bodies of the fungal partner. Thus, they are integrated into the system of fungi. Austrian botanist Zahlbruckner classified lichens into two main groups based on the structure of fruiting bodies of fungal partners as ascolichens (belonging to the fungal member of lichen ascomycotina) and basidiolichens (belonging to the fungal member of lichen basidiomycotina) (Alexopoulos and Mims, 1979). Later, they were classified again into three different main groups: basidiolichen (basidiomycetes), deuterolichen (deuteromycetes), and ascolichen (ascomycetes) (Büdel and Henssen, 1983; Web2).

Lichens are perennial plants that are very slow in growth, especially the mycobiont. The supply of nutrients and food produced through photosynthesis by the photobiont at the growing sites is utilized by the fungal partner (mycobiont). The growth forms of lichens are centrifungal, apical, and marginal (Shukla et al., 2014). Mycobiont dominated 90% of the thallus body of lichens that makes the color, shape, and structure, along with photobiont with 10%. Subsequently, the outer part of the thallus is fungal and covers the algal, hence they are categorized in the mycota (fungi) kingdom, and those lichens are known as lichenized fungi. 95% of the 20,000 species reported as lichen were from the ascomycetes group of fungi, and the rest, 3 and 2%, were from the basidiomycetes and deuteriomycetes groups, respectively (Nash and Nash, 1996).

12.2 HABIT AND HABITAT OF LICHENS

Lichens can grow everywhere; they are highly diverse in distribution and capable of living from hot deserts to high-altitude Antarctica. They can tolerate extremes of climate (cold, heat, and drought). Lichens grow on any type of substratum, like rocks, soil, and tree bark, and on any nonliving objects too. On the basis of their occurrence, lichens are categorized by different names: the lichens that are growing on tree trunk and bark (corticolous), twig-inhabiting (ramicolous), on decaying or dead wood (legnicolous), on rocks (saxicolous), on moss (muscicolous), on soil (terricolous), and on evergreen leaves (foliicolous). The lichen that grows commonly on other plants is called as epiphytic lichen. The growth and abundance of the lichens also depend on moisture, elevation, being free from pollution, and a favorable substratum.

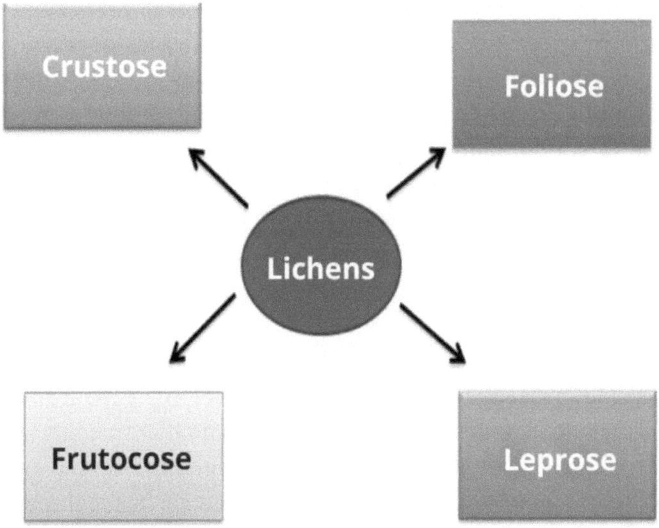

FIGURE 12.1 Types of lichens.

Considering the morphology of lichen thallus, it is found that lichens are growing in different growth forms, such as crustose (used to grow attached to the substratum), followed by foliose (used to grow like leafy and slackly attached to the substratum), and lastly fruticose (used to grow like bushes that are hanging or upright growing on the substratum) (Mishra and Upreti, 2016). In addition to this, several other growth forms of lichens exist, like leprose (powdery or granular), placodioid (thallus is closely attached to the substratum at center and lobate with the lack of rhizines), dimorphic (the thallus has quite similar characteristics with foliose and fruticose lichens), and squamulose (the form of the thallus is in minute lobes with dorsiventral differentiation) (Web 3) (Fig. 12.1). These leprose and crustose lichens, including a few placodioid and squamulose lichens, are generally called as microlichens, because they are smaller in size and require microscopic studies to identify. On the other hand, foliose, dimorphic, and fruticose lichens are named macrolichens. These lichens have a larger thallus, which is quite adequate to identify by using a hand lens or a stereozoom microscope (Nayaka, 2014).

12.3 DIVERSITY AND DISTRIBUTION OF LICHENS

Lichens are widely distributed in all geographical regions of the world. The total number of lichen species and lichenized fungi is estimated about 17,000–20,000 species around the world (Galloway, 1992). However, Aptroot reported that the biodiversity of fungi is very high, but the projected number of species is varied, and it was estimated that there are 30,000 recognized species with known ascomycetes (Aptroot, 1997). Later, the latest update of lichen species reported that the number of accepted species is 19,387 in 995 genera and 115 families. The Parmeliaceae family has the highest number of species with 2765 spp. and 77 genera, followed by the Graphidaceae with 2161 spp. and 79 genera (Lücking et al., 2017).

India was counted as the fifth nation in the world for its biodiversity, and overall, 2303 lichen species were projected under 305 genera and 74 families. Foliose lichens of family Parmeliaceae dominate the diversity with 345 spp. widely distributed in subtropical, temperate, and alpine regions; crustose of family Graphidaceae with 278 spp. distributed in tropical and subtropical regions of India; followed by the families Thelotremataceae, Pyrenulaceae, Caliciaceae, Lecanoraceae, Physciaceae, Trypetheliaceae, Teloschistaceae, and Collemataceae with their descending order as 130, 116, 103, 99, 97,

77, 75, and 67 spp., respectively. Likewise, genus Graphis dominates high diversity with 111 spp., followed by genera Pyrenula and Lecanora with 84 spp., and mostly these lichens are widely distributed in tropical and subtropical regions of India (Singh and Sinha, 2010; Rashmi and Rajkumar, 2014). Moreover, Sinha et al. reported the addition of 411 species to the 2010 annotated checklist; therefore, the total number of lichens recorded from India became 2714 species, with the family Graphidaceae having 106 spp. and the genus Graphis with 34 spp., followed by the orders arthoniales and opegrapha with 52 and 17 spp., respectively (Sinha et al., 2018).

In the context of the eastern Himalayas, Singh et al. reported 1047 species in 592 crustose, 321 foliose, 125 fruticose, and 9 squamulose growth forms, which are widely distributed in tropical, subtropical, temperate, subalpine, and alpine regions, comprising about 41.22% of the total lichen diversity of India (Singh et al., 2018). However, the eastern Himalayan region, including Sikkim and the seven other northeastern states of India, is still unexplored, and most of the researchers related to lichens are attempting to explore the new accounts from the country.

The literature of some recent publications showed that research about lichen exploration began only in the last 16 years (Singh and Pinokiyo, 2008; Pinokiyo et al., 2008; Rout et al., 2010; Daimari et al., 2014; Devi et al., 2015; Upreti et al., 2015; Logesh et al., 2015; Singh and Singh, 2016; Thangjam et al., 2019; Thangjam et al., 2020).

Coccocarpia palmicola (Spreng.) Arv. & D.J. Galloway

Dirinaria aegialita (Afzel. ex Ach.) B.J. Moore

FIGURE 12.2 *(Continued)*

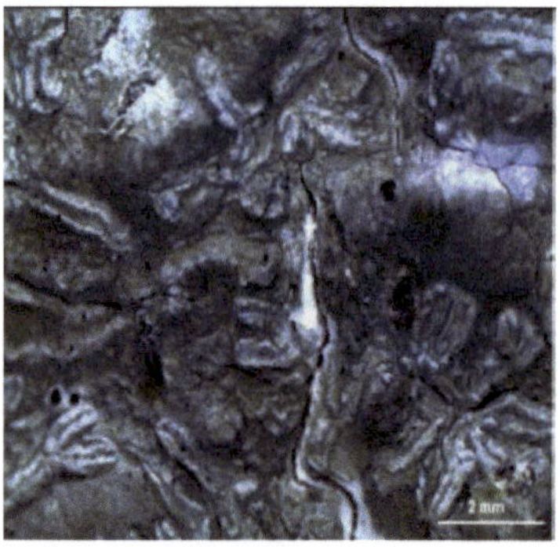

Diorygma hieroglyphicum (Pers.) Staiger & Kalb

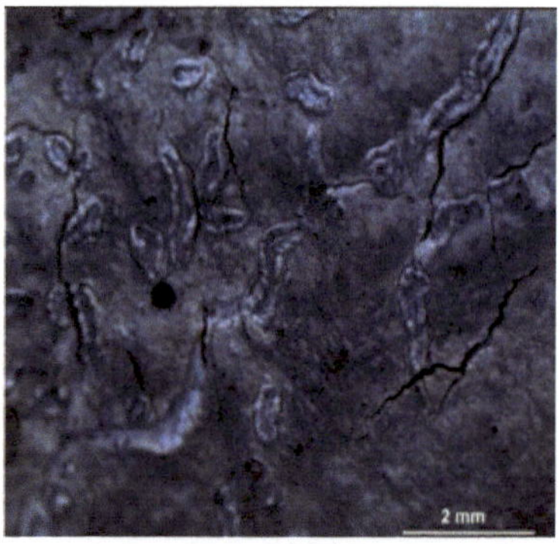

Diorygma soozanum (Zahlbr.) M. Nakan.& Kashiw.

Everniastrum cirrhatum (Fr.) Hale ex Sipman.

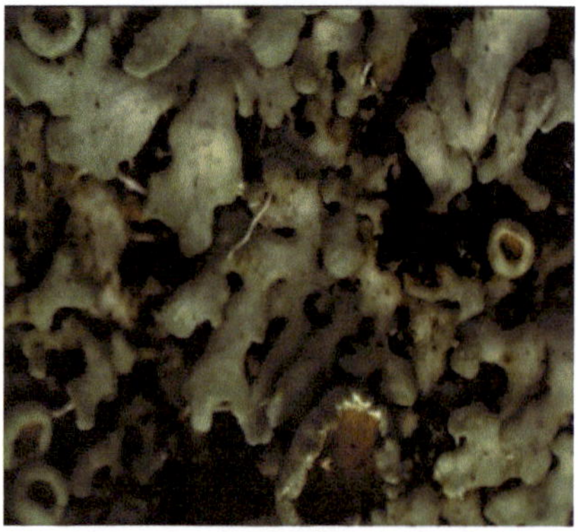

Heterodermia diademata (Taylor) D.D. Awasthi.

Heterodermia speciosa (Wulfen) Trevis.

Hypotrachyna awasthi Hale & Patw.

FIGURE 12.2 *(Continued)*

Parmotrema austrosinense (Zahlbr.) Hale

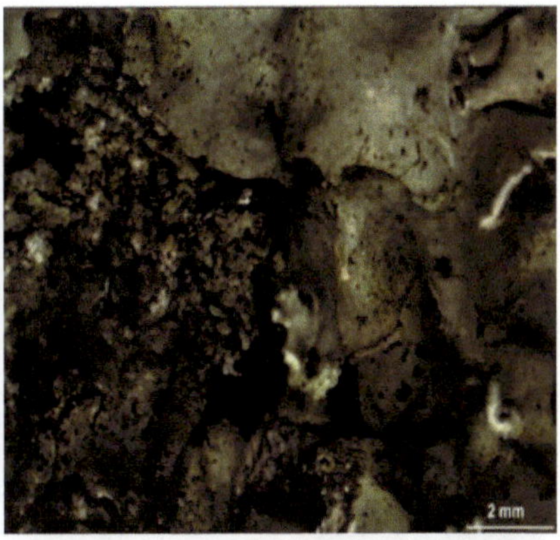

Parmotrema praesorediosum (Nyl.) Hale

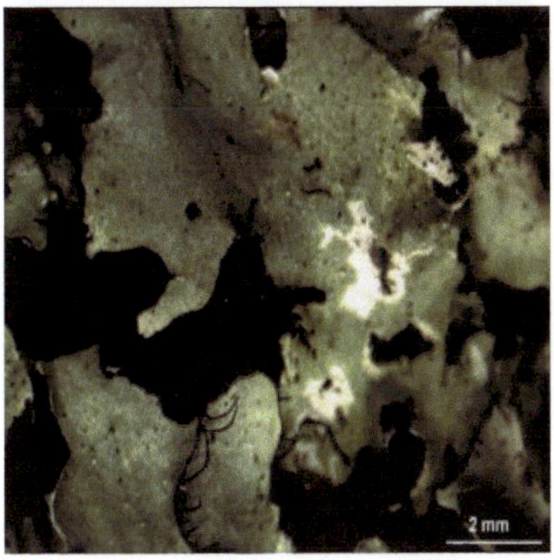

Parmotrema reticulatum (Taylor) M. Choisy.

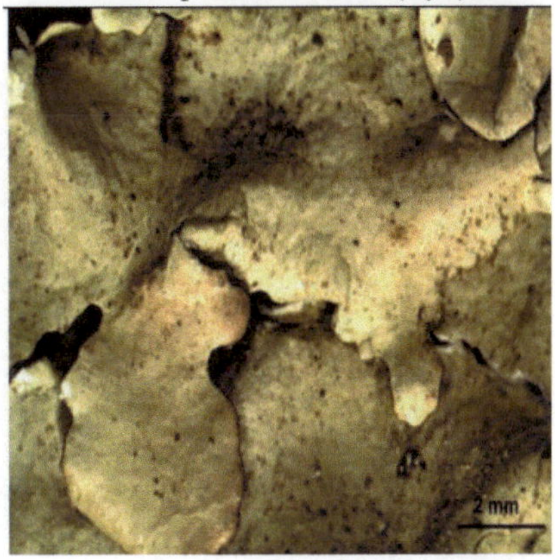

Parmotrema saccatilobum (Taylor) Hale.

Parmotrema tinctorum (Despr. ex Nyl.) Hale.

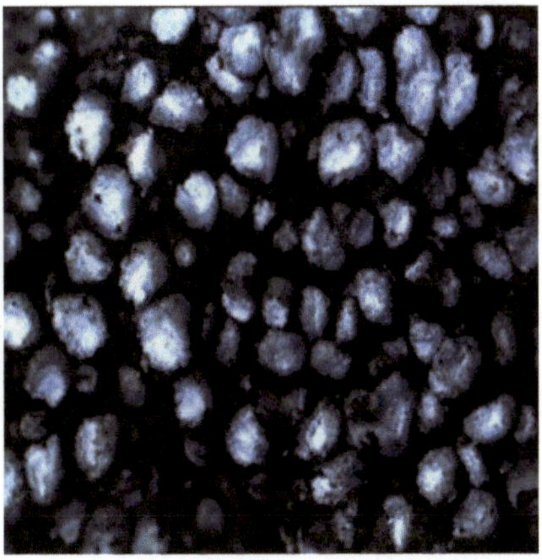

Pertusaria leucosorodes Nyl.

FIGURE 12.2 *(Continued)*

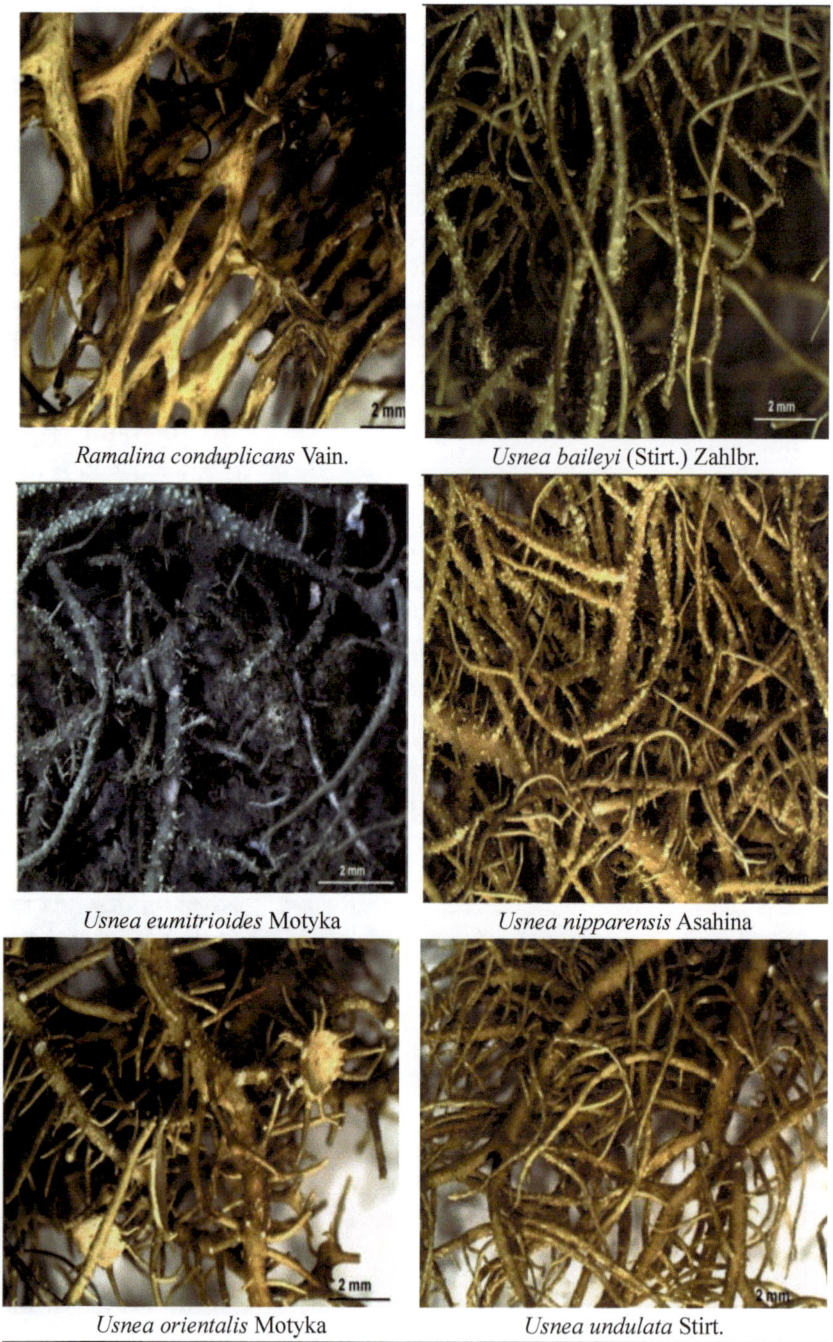

FIGURE 12.2 Some of the common lichens found in the eastern Himalayan region.

12.4 ECOLOGICAL IMPORTANCE OF LICHENS

Lichens are known for their ability to adapt to diverse climatic conditions; they can withstand high temperatures to the cold Arctic and Antarctic regions and are widely distributed in all geographical regions around the world. Hence, it is true that most of the diversity of the lichens is dependent on the climatic conditions. As per recent research, the monitoring and measurement of ecosystems by using natural systems as an indicator for checking them has now become a trend. In the same line, lichens are also used as an indicator for monitoring ecological conditions and helping to provide a warning at the earliest possible time. Lichens are highly sensitive to the environment and pollution because they don't possess a vascular system, which is used to synthesize their food by capturing their requirements like water and nutrients from the surrounding environment. Therefore, lichens are important tools to check out the details of environmental changes such as climate, pollution, and other biological processes (Kuldeep and Prodyut, 2015). The studies of forest ecosystems using lichen communities as an indicator also provide information that is concerned with biodiversity, air quality, and the sustainability of timber production (McCune, 1998).

The biodiversity of lichens and their conservation have become important goals for researchers in their exploration, documentation, conservation, and sustainable utilization. It's also an important factor for the socioeconomic development of society. Lichens are one of the most reasonable ecological indicators because of their known sensitiveness to different types of environmental stressors like pollution and climate change. Due to this, lichens are also used as terrestrial biomonitors of environmental contamination in industrial areas and urban and rural habitats. There are also several other methods being proposed for lichens as monitors of heavy metal pollution and climate change in different parts of the world (Upreti, 1995; Upreti and Nayaka, 2008).

12.5 ECONOMICAL IMPORTANCE OF LICHENS

Lichens have been used in households since time immemorial for their vast properties. It was believed that lichens are widely used in different traditional systems of medicine (TSM) such as ayurveda, Chinese medicine (TCM), homeopathy, and unani for the treatment of various ailments (Saklani and Upreti, 1992). They play a very important role in the ancient and traditional system of medicinal practices. Lichens are also used for different purposes, namely, food, fodder, spices, dyes, cosmetics, and perfumery. Kumar and Upreti indicated the early use of lichens in various cultural or ritual functions for the first time in India. Shipal in the Atharveda (1500 B.C.) is the first record of the use of lichen as medicine. The very popular vernacular name "Charila" is widely used in ayurveda, an ancient system of medicine in India. The Sanskrit synonyms for lichens are Shailaya and Shila Pushp (Shila, rock; Pushp, flower). Many ayurvedic and unani products prepared from different species of lichens are marketed in India under the trade names Charila and Ushna. Lichens are also used as holy lichens and burned in holy sacrificial fires known as "Hawan or Homa," often mixed with other aromatic herbs (Kumar and Upreti, 2001).

Lichen was used as a traditional medicine in various parts of the world, in the temperate and arctic regions in particular; it was believed that the Europeans started using lichens in the 15th and 16th centuries. A total of 52 lichen taxa were listed as traditional medicines practiced by different cultures, and the genus Usnea is among the most widely used in the world (Crawford, 2019). The major problems people face while eating lichens are that the secondary lichen compounds are mostly acidic and give acrid flavors, and they are also complex carbohydrates that are very difficult to digest and irritating to the digestive tract. Some lichens, such as usnic and vulpunic acids, are also found to be poisonous to humans if ingested in large amounts. People have traditionally used lichens in various preparations for consumption by removing the secondary compounds and hydrolyzing the polysaccharides. Usually, boiling and steaming techniques are used before the preparation of any dish, and they are mostly practiced in Europe, India, and North America. The lichen polysaccharides can be hydrolyzed by boiling into digestible form (Ivanova and Ivanov, 2009).

Saklani and Upreti initiated the studies of ethnomedicinal properties of lichen species by collecting important information about their traditional uses from the various parts of India (tribal and nontribal communities) (Saklani and Upreti, 1992). Further, Upreti et al. reported 15 lichen species commonly used in India by different tribes in day-to-day life for their medicinal as well as aesthetic values. The studies also highlighted that the western Himalayas are the main location for the collection of lichens and their distribution, either in raw form or processed, to other parts of the country (Upreti et al., 2005). Further, Upreti and Chatterjee had provided information on more than 50 lichen taxa that have ethnomedicinal values and are still in use in different forms globally (Upreti and Chatterjee, 2007). Identically, Nayaka et al. reported 137 lichens possessing medicinal properties from India, of which 36 species were traditionally used, 55 species for antimicrobial activity, 57 for antioxidant property, and 37 for anticancer activity. Their findings also showed that Cetraria islandica, Cladonia rangiferina, Evernia prunastri, Everniastrum cirrhatum, Hypogymnia physodes, Parmotrema chinense, Peltigera canina, and Usnea longissima were the common lichen species explored more for their uses in traditional medicine as well as some other biological purposes (Nayaka et al., 2010). But, due to overcollection, these species are now at an alarming stage.

The grounded power of the whole lichen parts was used as a spice ingredient in food. Many species were also used as flavoring agents in the preparation of various food items. Further, the parmelioid lichens, such as Parmotrema tinctorum, were known to be used by the ethnic people of India and Nepal as spice and flavoring agents for the preparation of meat and vegetables, and the product of lichen spice is named Al-Sheba in Arabic in Arab countries (Abo-Khatwa et al., 1996; Upreti et al., 2005; Anupama et al., 2017).

In addition to this, it was also reported that the perfume produced from some of the lichens has been popular since the last 800 years in a town called Kannauj in Uttar Pradesh; now, too, perfume preparation is carried out under the name "Otto" (the Hina Attar) in the present generation (Upreti et al., 2005). Parmotrema nilgherrense, Everniastrum nepalense, Ramalina subcomplanata, and Usnea lucea are more popular in the regions of Ramanagar and Tanakpur of the Uttar Pradesh hills for commercial use.

12.6 SECONDARY METABOLITES OF LICHENS

Lichens are known for their unique secondary metabolites and can synthesize a variety of metabolites (Richardson, 1988; Lawrey, 1989; Elix and Stocker-Wörgötter, 1996). Lichens' compounds are classified into three groups, namely, aliphatic lichen, aromatic lichen substances, and carbohydrates, which comprise aliphatic, cycloaliphatic, aromatic, and terpenic compounds (Huneck, 1999). Looking at their biosynthetic pathways, lichens are further classified into three pathways to produce their compounds, named polymalonate (polyketide), shikimic acid, and mevalonic acid pathways, and most of the metabolites are derived from the polyketide synthetic pathway, whereas others are derived from the shikimic acid and mevalonic acid pathways (Boustie and Grube, 2005). These secondary metabolites of lichens are produced by the mycobiont (fungi), which are present in the form of amorphous or crystallized material on the outer surface of the hyphae. Primary metabolites are common to other plants, like amino acids, polyols, carotenoids, polysaccharides, and vitamins. So far, 800 secondary metabolites of lichens have been updated in recent scientific publications; lichen bioactive secondary metabolites have a major role in the pharmaceutical field for their active phytochemical constituents and their applications (Ranković and Kosanić, 2019).

Basically, various kinds of lichen acids are derived from different biosynthetic pathways, as mentioned above; examples are the shikimic acid biosynthetic pathways, which are the source of pulvinic acid derivatives (yellow pigments), the mevalonic acid pathways, which are a source of terpenes, and the polymalonate pathway, which is considered the biggest pathway because the majority of the lichen compounds are produced by this pathway. The compounds derived from the polymalonate pathway include depside, depsidones, usnic acid, anthraquinones, xanthine, and aliphatic acids. The mevalonic acid pathway produces steroids and triterpenes like zeorin. Usnic acid is one of the known medicinally important lichen acids (Ivanova and Ivanov, 2009).

As mentioned earlier, lichens have a very unique kind of secondary metabolites with wide ranges of biological activity, but the knowledge of the isolation of lichens' secondary metabolites was developed in the last one and a half decades with the improvement of culture techniques because it is easy for the extraction of metabolites (Oksanen, 2006). Müller had compiled a review of the pharmacologically relevant secondary metabolites of lichens, and it was found that the presence of secondary metabolites like aliphatic acids, pulvinic acid derivatives, depsides and depsidones, dibenzofurans, anthraquinones, naphthoquinones, and epidithiopiperazinediones has a high effectiveness as antibiotic, antimycobacterial, antiviral, antiinflammatory, analgesic, antipyretic, antiproliferative, and cytotoxic effects (Müller, 2001). But still, the potential of lichens is not yet fully explored, and it does not reach the level of pharmaceutical companies.

The secondary metabolites of lichens are not only important for their existence and growth, but they are also used to establish good relationships with the environment. They have the ability to protect themselves from different factors such as herbivores, pathogens, UV radiation, etc. Since many lichens are known for their edible properties, most of their metabolites are also explored for biological activities (Molnár and Farkas, 2010). Usnic acid is the most common acid that has a wider range of properties; moreover, there are some other metabolites that have a highly potent medicinal value, such as atranorin, diffractaic acid, lecanoric acid, lobaric acid, stictic acid, salazinic acid, fumarprotocetraric acid, physodic acid, and orsellinates. These secondary metabolites of lichens have the properties of antioxidants, antimicrobials, and various pharmacological activities (White et al., 2014).

12.7 PHARMACOLOGICAL IMPORTANCE OF LICHENS

Lichens have been highly valued since time immemorial for their various medicinal activities, including the role of medicinal practitioners. Lichens are listed in various pharmacopias, particularly for their antimicrobial and antioxidant properties. The study of lichens and their antibiotic properties was initiated by Burkholder et al. (1944) and Vartia (1950), who tested nine crystalline lichenic acids against 11 species of bacteria and proved that lichens have antimicrobial properties against gram-positive bacteria. Further, Vartia studied secondary metabolites such as usnic acids, the lichesterinic acid group, orcinol-type depsides, and depsidones to check out their ability to inhibit the growth of bacteria (Vartia, 1973). Later on, many prominent scientists started working on different lichens to investigate the potency of lichens against bacteria. After successive results over bacterial pathogens, lichens were investigated for their antifungal properties. Burzlaff has evaluated the growth inhibition of Penicillium and Rhizopus through ethanolic and aqueous extracst of Parmelia molliuscula (Burzlaff, 1950).

Due to their unique nature and habitat, lichens are always a point of attraction for taxonomists as well as pharmacologists for identification and evaluation of different microbes. Compared to other plant species, lichens are very small and tiny, but this group is a hub of secondary metabolites. Though lichens are formed by fungi and algae, the chemical constituents so far extracted are unique in nature. Some other known chemical constituents of lichens, such as atranorin, fumarprotocetraric acid, gyrophoric acid, lecanoric acid, physodic acid, protocetraric acid, stictic acid, and usnic acid, possess high therapeutic and pharmacological properties. Usnic acid is one of the most commonly used compounds in the treatment of different ailments, like dermatitis and infected burns, as a traditional medicine (Rowe et al., 1999). Moreover, lichens as a whole and their identified metabolites also have numerous biological activities such as antimicrobial, antiprotozoal, antiviral, antiproliferative, anti-inflammatory, analgesic, antipyretic, antitermite, antioxidant, cytotoxic, enzyme inhibitory, insecticidal, wound healing, and antitumor (Yılmaz et al., 2004; Kosanić et al., 2013; Rajan et al., 2016).

12.8 IMPORTANCE OF LICHENS IN HIMALAYAN REGION

Considering the above facts, it is very important to know the diversity and role of lichens in the Himalayan region. Since then, 1047 lichen species have been identified from the eastern Himalayan region, along with 159 species from Mizoram, northeast India (Logesh et al., 2015). In addition to that, Thangjam et al. have increased the number of lichen biota in the northeast by reporting 22 lichen species from the Mizoram state of northeast India, along with Pyrenula dissimulans (Pyrenulaceae) as the new record for Indian lichen biota (Thangjam et al., 2020). Having this much lichen biota in the Himalayan region is a source of pride for the people residing in this hilly terrain, but it is challenging for their conservation. Let us look at some of the important economic uses of lichens that make them more appealing for cultivation in this region.

12.8.1 DYES FOR WOOL AND FABRIC

Many commercial lichens are now identified for their grey or pale green color when they are dry. When these lichens are moistened and absorb water, their algal component gives them a deeper hue; though the fungus component is often colorless, it also gives them some color, which overall gives them a vivid color. During the process of dye formation, ammonia is used in water along with lichen cut pieces; after several weeks, these lichens provide a variety of colors such as brown, green, orange, purple, red, and yellow. Mostly, these dyes are used for coloring clothes and baskets; the best known lichen species used for dye making include Lobaria pulmonaria, Parmelia saxatilis, Parmotrema crinitum, and Umbilicaria (Ball, 2008). These lichen species are very common and found in the Himalayan region.

12.8.2 LICHENS AS FOOD FOR HUMANS

In India, Lecanora esculenta from the Parmelia species, also called rathapu or "rock flower" in Telugu, was used as a curry and taken by the inhabitants of the Middle East deserts, as stated in the ANBG and ANH (2012). And there was also a report for Umbilicaria (rock tripe), Bryoria, used by native people of North America. Sometimes they used Cladina sp. and Cetraria islandica for survival (Web 1). These species are all known to the Himalayan region. Parmotrema sp. is very common in the Indian spice system and is usually used in "garam masala" while cooking meat.

12.8.3 LICHENS USED AS PERFUME

Two lichen species, Pseudevernia furfuracea and Evernia prunastri, are in high demand for making perfume in the global market. These lichens have a very pleasing aroma, that's why researchers are now using them to make different kinds of body deodorants. Pseudevernia furfuracea has already been reported by Goni et al. in Jammu and Kashmir in the Himalayan region (Goni et al., 2014).

12.8.4 LICHENS USED AS SUNSCREEN

Some of the lichen species, such as Xanthoria parietina, X. elegans, and Cetraria islandica, have been evaluated for their very specific properties, like protection from UV radiation. These lichens contain a yellow pigment called parietin, which has the property to absorb ultraviolet radiation and act as a sunscreen to protect the algal cells inside the lichen. Now, the scientists are working on these lichen species to synthesize sunscreens for humans (Solhaug et al., 2003; Nybakken et al., 2004). Since this study is being conducted to extract the bioactive compound, that is, parietin, for human sunscreen formation, it will need more raw materials in the future. Therefore, the Himalayan region could be a major source for these lichens.

12.8.5 LICHENS USED FOR LITMUS PAPER

Lichen species Roccella tinctoria is used for the manufacturing of litmus paper. Since it is reported from the Malabar coastal region of India, it is required to explore more Himalayan regions to identify this for the development of an economical setup by exporting the said genera.

12.8.6 LICHENS AS A TOOL FOR POLLUTION INDICATORS, CONTRIBUTION TO BIOLOGICAL DIVERSITY, AND NUTRIENT AND WATER CYCLING

Lichens are very sensitive to some specific pollutants present in the atmosphere. Lichens and their algal and cynobacterial partners are especially vulnerable to air pollutants like sulfur dioxide, ammonia, fluorine, nitric, and sulfuric acids. Once the level of these gases reaches a threshold level, it interrupts biological processes like photosynthesis and respiration in lichens, which indicate that the air present in the atmosphere is highly polluted with the mentioned gases. Lichens are more sensitive to nutrients, especially nitrogen, which is an indication of the fast growth of smaller weeds over larger and more ecologically valuable plant species. Therefore, the contribution of lichen to biological diversity is very important (United States Department of Agriculture (USDA), 2016). The luxuriant growth of lichen species indicates a healthy and pollution-free atmosphere and a favorable climate for the regeneration of different forest communities. The contribution of lichens to fixing atmospheric nitrogen into usable forms is not ignorable. Lichens play a very significant role in mineral and hydrological cycles, most importantly nitrogen fixation. Moreover, lichens are ecologically very important for food, shelter, and nesting material for wildlife such as deer, elk, moose, caribou, mountain goats, bighorn sheep, mice, squirrels, pikas, bats, chipmunks, etc. (USDA, 2016).

Overall, it showed that lichens have very peculiar uses in the whole Himalayan region. The richness of biodiversity makes it a prime location for tourists. Since the availability of different lichen biota in the Himalayan region showed the presence of a pollution-free atmosphere, it is very well understood that lichens are a major economic source in the Himalayan region.

12.9 CONCLUSION

These colorful and bushy forms of lichens are always a point of attraction for common people, nature lovers, and researchers. Very astonishingly, they used to observe and feel more joy after looking at the various lichens growing on trees, rocks, and soils. The present study also showed the importance of lichens in different fields of human life, including their impact on biodiversity. The vast distribution and number of lichen biota reported from the Himalayan region is a source of pride for the people who are living in this eco-friendly environment, which has tremendous scope in the future. The richness of the lichen distribution in this region shows its healthy and pollution-free atmosphere. The abundance of lichens has its own place in the Himalayan region, but its commercialization should also be explored by the local residents. Therefore, utilization of lichens for their commercial value to generate livelihood needs to be started as soon as possible. The creation of herbal gardens to attract lovers of forest safaris is an additional source of income for the residents of the Himalayan region. Moreover, conservation strategies as well as their pharmacological activities for the isolation of novel bioactive compounds will lead to herbal drugs.

ACKNOWLEDGMENTS

Authors are thankful to the Department of Horticulture, Aromatic and Medicinal Plants, Mizoram University for providing the necessary requirements to complete the study. Authors are also thankful for the financial support provided by the IERP project (Sanction order No. GBPI/IERP/18-19/02; dt.29/03/2019).

KEYWORDS

- **Himalayan region**
- **lichens**
- **livelihood generation**
- **Parmotrema**
- **traditional medicine1**

REFERENCES

Abo-Khatwa, A. N.; Al-Robai, A. A.; Al-Jawhari, D. A. Lichen Acids as Uncouplers of Oxidative Phosphorylation of Mouse-Liver Mitochondria. *Nat. Toxins*. **1996,** *4* (2), 96–102.

Alexopoulos, C. J.; Mims, C. W. Introducción a la micología (No. QK603. A4318 3A ED.); Eudeba, 1979.

Anupama, T. V.; Sheela, K. B.; Saji, G. Phytochemical Screening and Proximate Composition of Lichen *Parmotrema tinctorum* (Nyl.) Hale (Parmeliaceae) from Wayanad, Kerala. *Int. J. Chem. Stud.* **2017,** *5* (5), 1003–1007.

Aptroot, A. Species Diversity in Tropical Rainforest Ascomycetes: Lichenized Versus Non-lichenized; Foliicolous Versus Corticolous. *Abstracta Botanica* **1997,** 37–44.

Ball, P. In *Bright Earth: the Invention of Colour*; Vintage Books, London, 2008.

Boustie, J.; Grube, M. Lichens—A Promising Source of Bioactive Secondary Metabolites. *Plant Genet. Resour.* **2005,** *3* (2), 273–287.

Büdel, B.; Henssen, A. *Chroococcidiopsis* (Cyanophyceae), A Phycobiont in the Lichen Family Lichinaceae. *Phycologia* **1983,** *22* (4), 367–375.

Burkholder, P. R.; Evans, A. W.; McVeigh, I.; Thornton, H. K. Antibiotic Activity of Lichens. *Proc. Nat. Acad. Sci. U S. Am.* **1944,** *30* (9), 250.

Burzlaff, D. F. The Effect of Extracts from the Lichen, *Parmelia molliuseula,* Upon Seed Germination and Upon Growth Rate of Fungi. *J. Colorado-Wyoming Acad. Sci.* **1950,** *4 (2).*

Crawford, S. D. Lichens used in Traditional Medicine. In *Lichen Secondary Metabolites, Bioactive Properties and Pharmaceutical Potential*; Springer: Champ, 2019; pp 31–97.

Daimari, R.; Hazarika, N.; Hoque , R. R.; Nayaka, S.; Upreti, D. K. New Records of Epiphytic Lichens from Three Districts of Assam, India. *Indian For.* **2014,** *140* (8), 807–811.

Devi, A. B.; Mohabe, S.; Reddy, M. A.; Nayaka, S. Efficacy of a Potential Lichen *Parmotrema andinum* (Müll. Arg.) Hale Against Pathogenic Microorganisms. *J. New Biol. Rep.* **2015,** *4* (2), 149–156.

Elix, J. A.; Stocker-Wörgötter, E. Biochemistry and Secondary Metabolites. *Lichen Biol.* **1996,** *1*, 154–180.

Galloway, D. J. Lichens of Laguna San Rafael, ParqueNacional'Laguna San Rafael', Southern Chile: Indicators of Environmental Change. *Global Ecol. Biogeogr. Lett.* **1992,** *2,* 37–45.

Goni, R.; Mishra, G. K.; Upreti, D. K.; Sharma, N. Occurance of Lichen species *Pseudevernia furfuracea* (l.) Zorf from Jammu & Kashmir, India. *Phytotaxonomy* **2014,** *14,* 114–116.

Hawksworth, D. L.; Hill, D. J. In *The Lichen-Forming Fungi*; Blackie Academic and Professional: New York, 1984.

Huneck, S. The Significance of Lichens and their Metabolites. *Naturwissenschaften* **1999,** *86* (12), 559–570.

Ivanova, D.; Ivanov, D. Ethnobotanical use of Lichens: Lichens for Food Review. *Scr. Sci. Med.* **2009,** *41* (1), 11–16.

Kosanić, M.; Ranković, B.; Stanojković, T. Investigation of Selected Serbian Lichens for Antioxidant, Antimicrobial and Anticancer Properties. *J. Anim. Plant Sci.* **2013,** *23* (6), 1628–1633.

Kuldeep, S.; Prodyut, B. Lichen as a Bio-Indicator Tool for Assessment of Climate and Air Pollution Vulnerability. *Int. J. Environ. Sci.* **2015,** *4* (12), 107–117.

Kumar, K.; Upreti, D. K. *Parmelia* spp. (Lichen) - In Ancient Medicinal Plant Lore of India. *Econ. Bot.* **2001,** *55* (3), 458–459.

Lawrey, J. D. Lichen Secondary Compounds: Evidence for a Correspondence between Antiherbivore and Antimicrobial Function. *Bryologist* **1989,** 326–328.

Logesh, A. R.; Chinlampianga, M.; Shukla, A. C.; Upreti, D. K. Studies on Lichens of Mizoram, Northeast India. *Proc. Natl. Acad. Sci. India Sect. B Biol. Sci.* **2015,** *87* (2), 1–13.

Lücking, R.; Hodkinson, B. P.; Leavitt, S. D. The 2016 Classification of Lichenized Fungi in the Ascomycota and Basidiomycota–Approaching One Thousand Genera. *Bryologist* **2017,** *119* (4), 361–417.

McCune, B.; Rogers, P.; Ruchty, A.; Ryan, B. Lichen Communities for Forest Health Monitoring in Colorado, *USA*; A Report to the USDA Forest Service, 1998; pp 1–28.

Mishra, G. K.; Upreti, D. K. Diversity and Distribution of Macro-Lichen in Kumaun Himalaya, Uttarakhand. *Int. J. Adv. Res.* **2016,** *4* (2), 912–925.

Molnár, K.; Farkas, E.; Current Results on Biological Activities of Lichen Secondary Metabolites: A Review. *Zeitschrift für Naturforschung C.* **2010,** *65 (*3-4), 157–173.

Müller, K. Pharmaceutically Relevant Metabolites from Lichens. *Appl. Microbiol. Biotechnol.* **2001,** *56* (1-2), 9–16.

Nash, T. H.; Nash, T. H. In *Lichen Biology*; Cambridge University Press: London, 1996.

Nayaka, S. Methods and Techniques in Collection, Preservation and Identification of Lichens. In *Plant Toxonomy and Biosystematics: Classical and Modern Methods*, 2014; pp. 101–105.

Nayaka, S.; Upreti, D. K.; Khare, R. Medicinal Lichens of India. Drugs from plants; Avishkar Publishers, Distributors: Jaipur, 2010; pp 1–54.

Nybakken, L.; Solhaug, K. A.; Bilger, W.; Gauslaa, Y. The Lichens *Xanthoria elegans* and *Cetraria islandica* Maintain a High Protection Against UV-B Radiation in Arctic Habitats. *Oecologia* **2004,** *140*, 211–216.

Oksanen, I. Ecological and Biotechnological Aspects of Lichens. *Appl. Microbiol. Biotechnol.* **2006,** *73* (4), 723–734.

Pinokiyo, A.; Singh, K. P.; Singh, J. S. Diversity and Distribution of Lichens in Relation to Altitude within a Protected Biodiversity Hot Spot, North-east India. *Lichenologist* **2008,** *40* (1), 47–62.

Rai, H.; Khare, R.; Baniya, C. B.; Upreti, D. K.; Gupta, R. K. Elevational Gradients of Terricolous Lichen Species Richness in the Western Himalaya. *Biodivers. Conserv.* **2015,** *24* (5), 1155–1174.

Rajan, V. P.; Gunasekaran, S.; Ramanathan, S.; Murugaiyah, V.; Samsudin, M.; Din, L. Biological Activities of Four *Parmotrema* Species of Malaysian Origin and their Chemical Constituents. *J. Appl. Pharm. Sci.* **2016,** *6* (8), 36–43.

Ranković, B.; Kosanić, M. Lichens as a Potential Source of Bioactive Secondary Metabolites. In *Lichen Secondary Metabolites*; Ranković, B., Ed.; Springer: Switzerland, 2015; pp 1–26.

Ranković, B.; Kosanić, M. Lichens as a Potential Source of Bioactive Secondary Metabolites. In *Lichen Secondary Metabolites*; Ranković, B., Ed.; Springer: Cham, 2019; pp 1–29.

Rashmi, S.; Rajkumar, H. G. Preliminary Phytochemical Screening of Different Solvent Extracts of Lichens from Kodagu District, Karnataka. *J. Pharmacogn. Phytochem.* **2014,** *3* (4), 209–212.

Richardson, D. H. S. Medicinal and Other Economic Aspects of Lichens. In *CRC Handbook of Lichenology*, 1988; vol *3*, pp 93–108.

Rout, J.; Das, P.; Upreti, D. K. Epiphytic Lichen Diversity in a Reserve Forest in Southern Assam, Northeast India. *Trop. Ecol.* **2010,** *51* (2), 281.

Rowe, J. G.; Gimenez, M. G.; Rodriguez, M. S. Some Lichen Products have Antimicrobial Activity. *Zeitschriftfür Naturforschung C* **1999,** *54* (7-8), 605–609.

Saklani, A.; Upreti, D. K. Folk use of Lichens in Sikkim. *J. Ethnopharmacol.* **1992,** *3*, 229–233.

Shukla, P.; Upreti, D. K.; Tewari, L. M. Lichen Genus Usnea (Parmeliaceae, Ascomycota) in Uttarakhand. India. *CREAM* **2014,** *4* (2), 188–201.

Singh, K. P.; Pinokiyo, A. New Taxa of Foliicolous Lichens from Eastern India. *Lichenologist* **2008,** *40* (1), 23.

Singh, K. P.; Sinha, G. P. In *Indian Lichens: Annotated Checklist*; Botanical Survey of India: Kolkata, 2010.

Singh, P.; Singh, K. P. New Distribution Records of Lichens for the State of Meghalaya, North-East India. *Indian J. For.* **2016,** *39* (3), 245–254.

Sinha, G. P.; Nayaka, S.; Joseph, S. Additions to the Checklist of Indian Lichens After 2010. *Crypt. Biodivers. Assess.* **2018,** 197–206.

Singh, K. P.; Sigh, P.; Sinha, G. P. Lichen Diversity in the Eastern Himalaya Biodiversity Hotspot Region, India. *Crypt. Biodivers. Assess.* **2018,** 71–114.

Solhaug, K. A.; Gauslaa, Y.; Nybakken, L.; Bilger, W.; UV-Induction of Sun-Screening Pigments in Lichens. *New Phytol.* **2003,** *158* (1), 91–100.

Thangjam, N. M.; Kumar, A.; Kumar, K. S.; Shukla, A. C.; Upreti, D. K. Exploration of Lichens Diversity from Mizoram with Addition of *Pyrenula dissimulans* (Pyrenulaceae) a New Record to India. *Asian J. Conserv. Biol.* **2020,** *9* (1), 98–106.

Thangjam, N. M.; Kumar, A.; Shukla, A. C.; Upreti, D. K. Diversity and Distribution of Lichens in Murlen National Park of Mizoram, India. *Environ. Ecol.* **2019,** *37* (3), 664–672.

United States Department of Agriculture (USDA). *Why Lichens Matter, The Benefits of Lichens to Humans and Nature*; Forest Service, FS-1071, June 2016.

Upreti, D. K. Loss of Diversity in Indian Lichen Flora. *Envrion. Conserv.* **1995,** *22,* 362–363.

Upreti, D. K.; Nayaka, S. Need for Creation of Lichen Gardens and Sanctuaries in India. *Curr. Sci.* **2008,** *94* (8), 976–978.

Upreti, D. K.; Chatterjee, S. Significance of Lichens and Their Secondary Metabolites: A Review. In *Fungi Multifaceted Microbes*; Ganguli B. N., Deshmukh, S. K., Eds.; Anamaya Publishers: New Delhi, 2007; pp 169–188.

Upreti, D. K.; Debnath, R.; Uppadhyay, V.; Rout, J. Diversity and Distribution of Lichens in North and West Districts of Tripura. *Phytotaxonomy* **2015,** *14*, 122–129.

Upreti, D. K.; Divakar, P. K.; Nayaka, S. Commercial and Ethnic Use of Lichens in India. *Econ. Bot.* **2005,** *59* (3), 269.

Vartia, K. O. Antibiotics in Lichens. II. *Ann. Med. Exp. Biol. Fenniae.* **1950,** *28*, 7–19.

Vartia, K. O.; Antibiotics in Lichens. In *The Lichens*; Academic Press, 1973, pp 547–561.

Web1:https://envirotrips.in/wp-content/uploads/2016/03/Uses-of-Lichens.pdf. accessed on dated-11.04.2021

Web2:http://www.biologydiscussion.com/notes/lichens/lichens-eaningcharacteristics-and-classfication/46549.

Web3:http://wgbis.ces.iisc.ernet.in/biodiversity/sahyadri_enews/newsletter/issue16/identify.htm (accessed Apr 12, 2021).

White, P. A.; Oliveira, R.; Oliveira, A. P.; Serafini, M. R.; Araújo, A. A.; Gelain, D. P.; Moreira, J. C.; Almeida, J. R.; Quintans, J. S.; Quintans-Junior, L. J.; Santos, M. R. Antioxidant Activity and Mechanisms of Action of Natural Compounds Isolated from Lichens: A Systematic Review. *Molecules* **2014,** *19* (9), 14496–14527.

Yılmaz, M.; Türk, A. Ö.; Tay, T.; Kıvanç, M. The Antimicrobial Activity of Extracts of the Lichen *Cladonia foliacea* and its (-)-Usnic Acid, Atranorin, and Fumarprotocetraric Acid Constituents. *Zeitschriftfür Naturforschung C* **2004,** *59* (3-4), 249–254.

Index

A

Allele-specific primer extension, 157–159
Amplified fragment-length polymorphism (AFLP), 151–153
Anthelmintic, 37
Anti-aging activities, 83
Antiallergic activity, 11–12
Anticholinesterase activity, 10
Anti-diabetic effects, 84
Antigastric ulcer activity, 10
Anti-inflammation, 37
Anti-inflammatory activity, 7–8
Antimicrobial, 35–36
Antimicrobial activity, 125–126
Antimultidrug resistance activity, 12
Antineoplastic, 84
Anti-obesity activity, 15, 85–86
Antiosteoarthritis activity, 13–14
Antioxidant, 35
Antioxidant activity, 11
Antioxidant agent, 125
Antiplasmodial activity, 6–7
Antispasmodic activity, 10
Antitumor activity, 12
Antiviral activity, 7
Antixanthine oxidase activity, 10–11
Ascorbic acid (vitamin C), 31–32

B

Barak Valley, 205
Blood fluidity, 10
Broom grass *(Thysanolaena latifolia),* 230
 harvesting, 232–233
  marketing, 236–240
  processing, 234–235
 methodology, 231
 result and discussion
  cultivation, 231–232

C

Carotenoids (vitamin A), 30–31
Chenopodium album L., 142
 chloroplast genome, 159
 DNA-based molecular profiling
  amplified fragment-length polymorphism (AFLP), 151–153
  chloroplast RBCL, 155–156
  and estimation of evolutionary age, 155–156
  internal transcribed spacer region (ITS), 154
  MATK regions, 155–156
 Himalayan Chenopods, 143
 morphological characters, 143, 144, 146–149
 SNP-based species specific marker
  allele-specific primer extension, 157–159
 soluble seed proteins, 149–151
Consultative Group on International Agriculture (CGIAR), 140
Cytochrome P450 enzymes activity, 8

D

Diversified Agriculture Support Project (DASP), 185–186
DNA-based molecular profiling
 amplified fragment-length polymorphism (AFLP), 151–153
 chloroplast RBCL, 155–156
 and estimation of evolutionary age, 155–156
 internal transcribed spacer region (ITS), 154
 MATK regions, 155–156
Domestic resource cost (DRC), 192–193

E

Elaeagnus latifolia Linn.
 future prospects, 39–40
 material and methods, 24
 pharmaceutical properties
  anthelmintic, 37
  anti-inflammation, 37

antimicrobial, 35–36
antioxidant, 35
insecticidal, 36
nanoscience, 37–38
value addition, 38
research gaps, 39–40
results
ascorbic acid (vitamin C), 31–32
botanical description, 25
carotenoids (vitamin A), 30–31
economic importance, 28
ethnomedicinal uses, 25–27
folic acid (vitamin B9), 33
geographical distribution, 25
mineral composition, 33
niacin (B3), 32
nutritional composition, 29–30
pantothenic acid (vitamin B5), 32
physical properties, 29
phytochemical assessments, 34–35
propagation, 27
pyridoxine (vitamin B6), 33
riboflavin (vitamin B2), 32
species ecology, 27–28
taxonomy, 25
thiamine (vitamin B1), 32

F

Floral biology, 89
Assam, 91–92
karyotype study, 90–91
Folic acid (vitamin B9), 33

G

Garcinia L., 69
description, 71–72
distribution and taxonomy, 71–72
floral biology, 89
Assam, 91–92
karyotype study, 90–91
hydroxycitric acid (HCA)
functional food, 87–88
reaction mechanism, 86
therapeutic dosages, 87
weight loss process, 86
phyto-pharmacology, 72–77
proximate information
anti-aging activities, 83
anti-diabetic effects, 84
antineoplastic, 84
anti-obesity activity, 85–86
chemo-preventive effects, 84
free radical scavenging properties, 83
gastro-protective effects, 84
industrial use, 82
lipid-lowering effect, 84
phytochemistry, 82
traditional medicinal uses, 83
traditional management
in Africa, 80–81
in Bangladesh, 81–82
China, 77–79
distribution, studies, 79
ethnobotanical utilization, 79–80
phyto-harmacology, 79–80

H

Hepatoprotective activity, 8
Himalayan Chenopods, 143
Himalayan region
biological diversity, 258
humans, 256–257
litmus paper, 257
nutrient, 258
perfume, 257
pollution indicators, 258
sunscreen, 257
and water cycling, 258
wool and fabric, 256
Hydroxycitric acid (HCA), 69-70
functional food, 87–88
reaction mechanism, 86
therapeutic dosages, 87
weight loss process, 86

I

Indian organic basmati rice
export competitiveness, 184
India, 185
regional issues, 186–188
export performance
domestic resource cost (DRC), 192–193
export performance ratio, 193–194
growth trends, 190–191
nominal protection coefficient (NPC), 192

production, 185
suggestions, 199
enhancing export, 200–201
supply chain, 199–200
supply chain model, 194
certification, 198
direct marketing, 195
farmer's response, 197
government end, 198
strategies, 195–197
SWOT analysis, 195
trader's response, 197–198
Uttarakhand
aroma crisis, 189
Diversified Agriculture Support Project (DASP), 185–186
germplasm, loss, 189
increasing demand, 188
longer grain type, 188
marketability, 189–190
policies, 190
quality, 189
shrinkage in area, 188
Uttarakhand Organic Commodity Board (UOCB), 201
Internal transcribed spacer region (ITS), 154
International plant Genetic Resources Institute (IPGRI), 140
International Union for Conservation of Nature (IUCN), 55

K

Kaempferia parviflora, 1
adverse effects, 15
anti-obesity activity, 15
botanical description, 4
clinical studies, 15
geographical distribution, 5
pharmacological activity
antiallergic activity, 11–12
anticholinesterase activity, 10
antigastric ulcer activity, 10
anti-inflammatory activity, 7–8
antimultidrug resistance activity, 12
anti-obesity activity, 15
antiosteoarthritis activity, 13–14
antioxidant activity, 11
antiplasmodial activity, 6–7
antispasmodic activity, 10
antitumor activity, 12
antiviral activity, 7
antixanthine oxidase activity, 10–11
blood fluidity, 10
cardiogenic activity, 9
cytochrome P450 enzymes activity, 8
hepatoprotective activity, 8
neuroprotective activity, 12–13
sexual behavioral activity, 11
transdermal permeable activity, 14
vasorelaxation activity, 9
phytochemical constituents, 5
taxonomic study, 3
traditional uses, 6
Tsumura Suzuki obese diabetes (TSOD), 15

L

Lichens, 244
economical importance
spice ingredient, 253
traditional medicine, 252
traditional systems of medicine (TSM), 251
habit and habitat, 245
diversity and distribution, 246–247
ecological importance, 251
types, 246
Himalayan region
biological diversity, 258
humans, 256–257
litmus paper, 257
nutrient, 258
perfume, 257
pollution indicators, 258
sunscreen, 257
and water cycling, 258
wool and fabric, 256
pharmacological importance, 255–256
secondary metabolites, 253–255

N

Naga King Chili
clinical uses, 127–128
cultivation practices, 120–121
disease management studies, 128–130
diseases, 128
ethnomedicinal uses, 120

in Manipur
studies, 132–133
pharmacological effects
antimicrobial activity, 125–126
antioxidant agent, 125
cancer prevention, 123–124
cardiovascular disease treatment, 126
gastrointestinal benefits, 127
pain therapy, 122–123
weight management, 124–125
phytochemical constituents, 121–122
plant description, 119–120
scoville heat unit (SHU), 119
Niacin (B3), 32
Nominal protection coefficient (NPC), 192

P

Pantothenic acid (vitamin B5), 32
Parkia timoriana (DC.), 99
diseases of, 102–103
distribution, 100
economical importance, 102
edible uses, 101
morphology, 101
pharmacological activity, 104, 110–111
and biological activities, 112
reported, 105–110
phytochemical profile, 103–104
toxicology studies, 104
traditional uses, 101–102
vernacular names, 100–101
Pyridoxine (vitamin B6), 33

R

Riboflavin (vitamin B2), 32
Riparian areas, 205
materials and methods, 206–207
results and discussion, 207

S

Scoville heat unit (SHU), 119
Sea buckthorn (*Hippophae* L.)
applications
functional, 173–174
nutraceutical industries, 173–174
pharmacological, 173–174
area and production, 168–170
berries
processing, 176–177
uses, 172–173
biochemical and nutritional compositions, 171–172
climate and soil, 171
harvest and postharvest management, 175
India
status and potential, 177–178
traditional uses and recent developments, 174–175
plant, 170
Sexual behavioral activity, 11
Single Nucleotide Polymorphism (SNP), 139
Swertia chirayita, 53
CR medicinal plants, 54
cultivation
commercial products, 61
cost-benefit assessments, 60
cultural practices, 58–59
harvesting and yield, 59–60
insect, 60–61
International Union for Conservation of Nature (IUCN), 55
land preparation, 57–58
nursery preparation, 56–57
pests and diseases, 60–61
propagation, 56
soil type, 56
threats, 61
transplanting, 58
properties and uses, 54–55

T

Thiamine (vitamin B1), 32
Traditional systems of medicine (TSM), 251
Tsumura Suzuki obese diabetes (TSOD), 15

U

Uttarakhand
aroma crisis, 189
germplasm, loss, 189
increasing demand, 188
longer grain type, 188
marketability, 189–190
policies, 190
quality, 189
shrinkage in area, 188

Uttarakhand Organic Commodity Board
(UOCB), 183, 201
UttarakhandUttarakhand
Diversified Agriculture Support Project
(DASP), 185–186

V

Vasorelaxation activity, 9

Z

Zingiberaceae family, 2